Machine Learning Applications in Industrial Solid Ash

Woodhead Publishing Series in Civil and Structural Engineering

Machine Learning Applications in Industrial Solid Ash

Chongchong Qi

School of Resources and Safety Engineering, Central South University, Changsha, Hunan, China

Erol Yilmaz

Department of Civil Engineering, Recep Tayyip Erdoğan University, Rize, Türkiye

Qiusong Chen

School of Resources and Safety Engineering, Central South University, Changsha, Hunan, China

Woodhead Publishing is an imprint of Elsevier
50 Hampshire Street, 5th Floor, Cambridge, MA 02139, United States
The Boulevard, Langford Lane, Kidlington, OX5 1GB, United Kingdom

Copyright © 2024 Elsevier Inc. All rights reserved, including those for text and data mining, AI training, and similar technologies.

No part of this publication may be reproduced or transmitted in any form or by any means, electronic or mechanical, including photocopying, recording, or any information storage and retrieval system, without permission in writing from the publisher. Details on how to seek permission, further information about the Publisher's permissions policies and our arrangements with organizations such as the Copyright Clearance Center and the Copyright Licensing Agency, can be found at our website: www.elsevier.com/permissions.

This book and the individual contributions contained in it are protected under copyright by the Publisher (other than as may be noted herein).

Notices
Knowledge and best practice in this field are constantly changing. As new research and experience broaden our understanding, changes in research methods, professional practices, or medical treatment may become necessary.

Practitioners and researchers must always rely on their own experience and knowledge in evaluating and using any information, methods, compounds, or experiments described herein. In using such information or methods they should be mindful of their own safety and the safety of others, including parties for whom they have a professional responsibility.

To the fullest extent of the law, neither the Publisher nor the authors, contributors, or editors, assume any liability for any injury and/or damage to persons or property as a matter of products liability, negligence or otherwise, or from any use or operation of any methods, products, instructions, or ideas contained in the material herein.

ISBN: 978-0-443-15524-6 (print)
ISBN: 978-0-443-15525-3 (online)

For Information on all Woodhead Publishing publications
visit our website at https://www.elsevier.com/books-and-journals

Publisher: Glyn Jones
Editorial Project Manager: Namrata Aggarwal
Production Project Manager: Prem Kumar Kaliamoorthi
Cover Designer: Vicky Pearson Esser

Typeset by MPS Limited, Chennai, India

Dedication

Dedication of Chongchong Qi

For my wonderful wife Liu Yisha (刘艺莎), my confidante, my muse, my love.

For my great son Qi Yu (齐遇), who fills my heart with joy each and every day.

Dedication of Erol Yilmaz

Dedicated to my wife and daughter, for an unwavering of love and support

Elif Yilmaz and Ayse Elif Yilmaz

Dedication of Qiusong Chen

Dedicated to my wife and children, for an unwavering of love and support

Yuan Jingjing, Chen Nuoyan, Chen Nuochen

Contents

Preface

In the past several decades, the industry has rapidly developed and played a vital role in the global economy. Urbanization has been promoted to an unexpected rate, leading to the significant increase in living standards. To support the aforementioned rapid development, a large amount of energy is needed worldwide. Though the importance of renewable energy cannot be further emphasized, such as the Directive 2018/2001 and the Green Deal, around 38% of the world's electricity still comes from coal combustion. At present, coal is the world's most abundant, widely distributed, conventional, and inexpensive fossil energy source. Developing countries with rapid economy growth and urbanization, that is, China and India, rely more heavily on coal consumption for the supply of electricity.

Another environmental impact of urban expansion and economic development is the generation of the municipal solid wastes (MSWs) and biomass wastes. Taking MSW as an example, MSWs mainly come from residential life, commercial activities, and institutional activities. Globally, 3.5 million tons of MSW are generated in urban areas per day, which is expected to reach around 6.1 million tons per day by 2025. MSW, including but not limited to food waste, glass, textile, etc., cannot be naturally degraded and thus requires other disposal strategies to mitigate potential environmental problems.

Considering the waste disposal and energy requirement, incineration becomes a popular management option for MSW and biomass wastes. Incineration can recover energy from waste and reduce the mass and volume of waste by 70% and 90%, respectively. During the waste incineration, the hot gases produced from the burning process are used to produce electricity and/or heat. Due to its technical and economic advantages, waste incineration has been increasingly used worldwide. For example, 13 incinerators have been installed in the United Kingdom, which can process 2.9 million tons of waste per year.

However, either coal consumption in power plant or waste incineration in incinerators, two categories of solid residues are inevitably generated: bottom ash and air pollution control residues. The air pollution control residues are a mixture of fly ash, lime, and carbon. The wide utilization of coal and waste burning produced a large amount of bottom ash and fly ash, which are regarded as industrial solid ashes in this book, resulting in the urgent need for their recycling and disposal. However, the current recycling and management of solid ash rely heavily on lab experiments, hindering the optimization of the recycling strategies. The embracement of information technology, especially machine learning (ML) techniques, will be a promising option to speed up the efficient management of industrial solid ashes.

This book is meant to be a first introduction to industrial solid waste management with the help of ML, which will be a valuable source for students, scholars, and practitioners. It is directed primarily to those whose background is in solid waste recycling and management, materials science, machine learning, data science, or related disciplines. Emphasis is placed both on the detailed explanation of the basic concepts, that is, about industrial solid ash and machine learning, and on practical applications by providing straightforward examples. Thus we hope that this book will serve both as an introduction that assumes no previous specialist knowledge of solid ash and machine learning and as a guide to further research.

The book is organized as follows: Chapters 1–5 deal mainly with the introduction to industrial solid ashes, including its production, properties, current management strategies, emerging innovative techniques, and legal framework. Chapters 6–7 deal with the basics about machine learning and the ML modeling methodology for industrial solid ash management. Chapters 8–13 present seven application examples for the ML modeling in industrial solid ash management, covering a wide variety of ML modeling scenarios, such as regression, classification, and clustering. Chapter 14 summarizes the challenges and future perspectives, and sample codes to reproduce the application examples are provided in Appendix.

The literature of industrial solid ash is voluminous, and we cannot emphasize enough that there are many extremely good books on either industrial solid ash or machine learning. Of necessity, coverage in the present book has been selective, but it is hoped that the most important and relevant studies have been covered. Throughout this book, we will try to point you to other good references appropriately at the end of each chapter. Hope you enjoy the journey through machine learning in industrial solid ash.

Chongchong Qi
Erol Yilmaz

Acknowledgments

I am most grateful to my parents for their unfailing love and support. Raising a child in those days was no small feat and they made a lot of sacrifices to ensure I could receive a well education. I am forever grateful to my mentors, such as Liyuan Chai, Andy Fourie, and Guichen Li, who have been a guiding beacon throughout my academic career. Special acknowledgments are given to my students, especially Mengting Wu, who have given invaluable assistance during the preparation of the book chapters. I would like to express my profound gratitude to Qiusong Chen, Le Wang, KI-IL Song, Binh Thai Pham for all of their support and encouragement, and for that I am most grateful.

Chongchong Qi

As one of the authors of this book, I would like to sincerely thank two special people, Mr. Muhammet Sari and Mr. Tugrul Kasap, who are presently pursuing a PhD degree under my supervision at Recep Tayyip Erdogan University Institute of Graduate Studies (civil engineering) in Rize, Türkiye. They have made significant contributions to the execution of Chapters 1–5 with their invaluable suggestions, evaluations, and comments. These chapters would not have been so effective without the continuous support and contributions of these generous people.

Erol Yilmaz

The effective management of industrial solid ashes has been become a worldwide topic. I appreciate the opportunity to participate in the writing of this book and share useful research findings. I would like to take this opportunity to thank my family, my friends, my colleagues in the Central South University, cooperative enterprises, for their unwavering support and assistance.

The authors would like to express our appreciation, both to the Elsevier publisher, and to the Elsevier staff, for making this book come to fruition.

Qiusong Chen

Industrial solid ashes generation

1

Abstract

This chapter presents a brief introduction to industrial solid ashes. It begins with an overall introduction to solid waste management and energy consumption, which lead to the utilization of incineration, thus the generation of solid ashes. Then, various types of industrial solid ashes are introduced, together with their representative properties and characteristics. The chapter finishes by providing information on the potential production amounts of industrial solid ashes.

1.1 Introduction

In the past several decades, industry has rapidly developed and played a vital role in the global economy. Urbanization has been promoted to an unexpected rate, leading to the significant increase in living standards. To support the above-mentioned rapid development, a large amount of energy is needed worldwide. Although the worldwide energy consumption has increased approximately three times since 2000, it is known that it will remain raising steadily in years to come. Despite reductions in the energy demand triggered by COVID-19 virus in 2020, global energy demand increased by 5.8% in 2021 [1,2]. It is also estimated that energy consumption will reach 740 million terajoules (an additional growth of approximately 30%) by 2040 in the usual scenario [3]. While an increase of 8 EJ was observed in renewable energy between 2019 and 2021, the consumption of fossil fuels remained stable. Fossil fuels mostly dominate the worldwide energy supply [4]. Fossil fuels accounted for 82% of global energy use in 2020, 83% in 2019 and 85% in 2014. In addition, today, 80% of the total main energy supply is provided by crude oil and natural gas, primarily coal [5]. Fig. 1.1A shows the worldwide primary energy consumption in 2021. Fig. 1.1B also reflects the change of primary energy sources over the years.

Coal, being one of the most abundant organic-rich sedimentary rock, is an extremely important fossil fuel type used in different fields (e.g., manufacturing). So far, the main purpose of use of coal has been generated the electricity by creating steam by various combustion methods [6]. Since the first commercial coal-fired power plant was built in the United States (Pearl Street Station) in 1822, coal-based electricity generation has shown tremendous advances [7]. According to 2021 data, China, which alone undertakes 73.7% of the world coal production, is followed by India with 771 million tons (MT) of production and the Indonesia with a production of 545 MT. Though the importance of renewable energy cannot be further

Machine Learning Applications in Industrial Solid Ash. DOI: https://doi.org/10.1016/B978-0-443-15524-6.00012-1

© 2024 Elsevier Inc. All rights reserved.

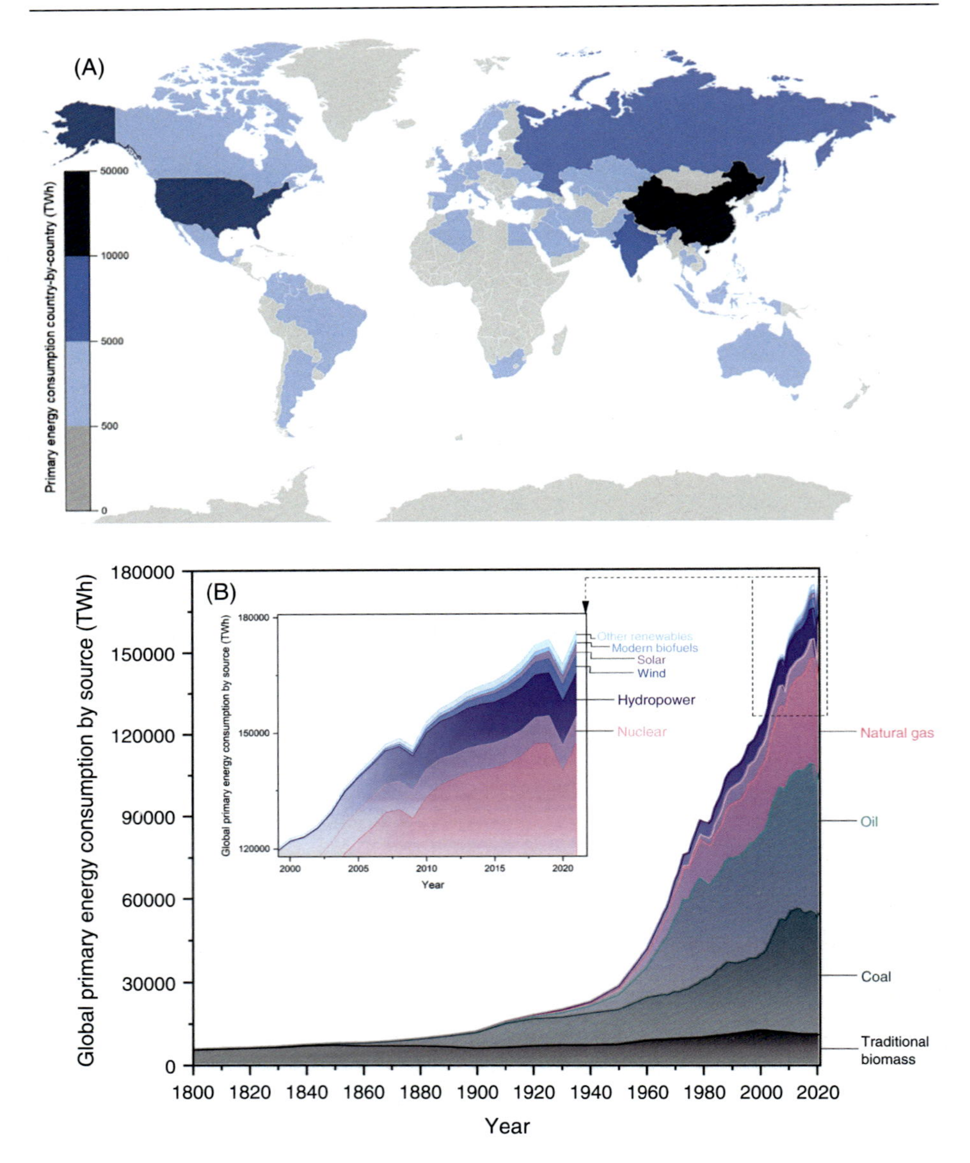

Figure 1.1 Global energy consumption by country (A) and source (B).

emphasized, such as the Directive 2018/2001 and the Green Deal [8], around 38% of the world's electricity still comes from coal combustion [9]. At present, coal is the world's broadly disseminated, conventional and inexpensive fossil energy basis [10]. Developing countries with rapid economy growth and urbanization, that is, China and India, rely more heavily on coal consumption for the supply of electricity.

Another environmental impact of urban expansion and commercial growth is the production of municipal solid wastes (MSW) and biomass wastes. Taking MSW as the example, MSW mainly come from housing lifecycle, viable actions, and institutional actions. Globally, 3.5 M tons of MSW is daily created in urban areas, getting a daily amount of ~6.1 M tons by 2025 [11]. MSW, including but not limited to food waste, glass, textile, etc., cannot be certainly ruined and thus entail other disposal strategies to mitigate potential environmental problems [12,13]. 685 TWh of electricity was annually produced from biomass in 2020. About 69% of this bioenergy was obtained from solid biomass, and 17% from domestic/industrial waste sources. In general, according to 2019 data, 39% of all bioenergy production worldwide takes place in Asia and 35% in Europe. As of 2020, it is estimated that 5.3% biomass is used for electricity generation in power plants all over the world [14]. Considering the waste disposal and energy requirement, incineration becomes a popular management option for MSW and biomass wastes. Incineration can recover energy from waste, and diminish the quantity and size of waste by 65% and 85%, respectively. Throughout the waste burning, the boiling airs created from burning process are utilized for cropping both electricity and heat. Due to its technical and economic advantages, waste incineration has been increasingly used worldwide. For example, 13 incinerators have been installed in the United Kingdom, which can process 2.9 MT of waste per year.

However, either coal consumption in power plant or waste burning in incinerators, two solid waste types are inevitably generated: air pollution control waste and bottom ash waste. The first is a blend of lime, fly ash, and carbon. The latter is a type of ash created in burning plant. The widely use of coal and waste burning produced a large amount of fly/bottom ash, which is regarded as industrial solid ashes in this book, resulting in the urgent need for their recycling and disposal.

1.2 Making and types of industrial solid ashes

As indicated before, various materials can be burnt within the electrical generating station in order to produce power. In the literature, these materials undergone the most investigation include coal, MSW, and biomass. To facilitate the burning process of different raw materials, the electrical generating station could be designed differently. Accordingly, the solid ashes generated by different raw materials and incineration process could be divided into different categories. In this section, a brief introduction to the generation of different solid ashes is provided.

1.2.1 Coal ashes

Coal bottom ash (CBA) and coal fly ash (CFA) are two important waste types formed through the burning of coal in power plants. Hence, the characteristics of coal ashes count on the sort of the technique employed in ash making [15]. The two boiler kinds are utilized for the creation and release of ashes before the drain of chimney fumes. These are traditional pulverized coal and fluidized bed boilers [16,17]. The most widely used method among the mentioned coal burning methods is the pulverized coal burning method. The pulverized coal method, also called entrained phase or suspension-fired combustion, is advantageous for large-scale power generation [18]. Although the power generation station varies for different coal burning methods, the general layout is similar. Pulverized coal combustion (PCC) is carried out by using powder burners to feed the coal at high temperatures (1450°C–1600°C) of the burning space where the fire core regions are located. To obtain the desired efficiency, the pulverized coal is rotated in the burning space [19]. While the heat within boiler is usually higher than the melting-point of most constituents, approximately 50% of the materials are released from melting [20]. Fig. 1.2 displays a typical chart of a coal thermic power facility. Before the coal is gusted with air into the boiling area the electrical generating station, also known as the furnace, it is initially powdered in grinding mills. In the combustion region, the flammable elements (i.e., C, H, O) of coal ignite. As a result, a large sum of heat is created, which is moved to pipelines covering great stressed liquid such as water. This high-pressure liquid is boiled to vapor and then the steam travels through a turbine to produce electricity.

Noncombustible minerals present within coal, such as SiO_2, $CaCO_3$, $CaSO_4.2H_2O$, FeS_2, and clay/feldspar minerals, flux in the furnace and create minor fluid drops. 85%–95% of them are delivered from the boiling region with chimney vapors and cool quickly, making small/spherical shiny grains when they get rid of the space. Mechanical/electrical precipitators are then employed to separate these from the chimney vapors, resulting in CFA's generation. About 5%–15% of the incombustible by-products are coarse and granular, which will fall to the water-filled hopper in the boiler's bottom and is termed as CBA. CBA is type of materials with dark gray, permeable, and sand characteristics. When bottom ash accumulates in the feed chamber, this accumulated mass,

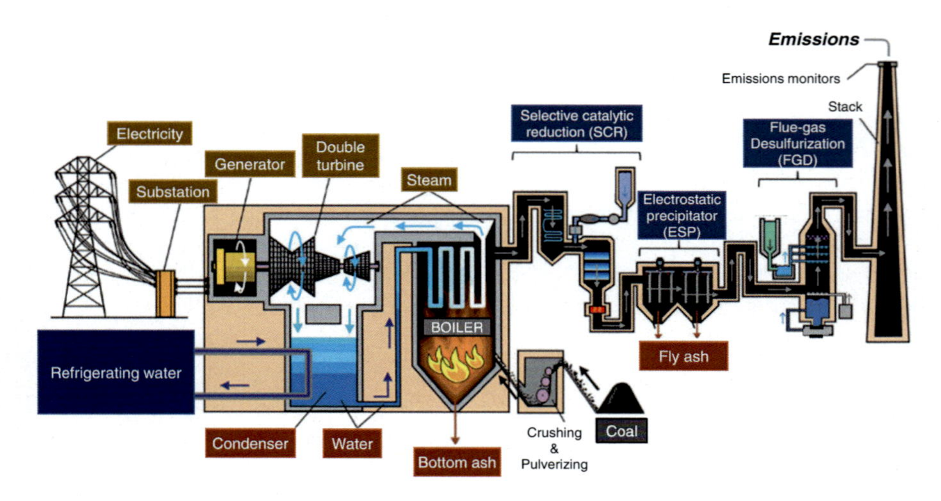

Figure 1.2 A typical schematic of a coal-fired power plant.

which adversely affects the flow, is removed using a water jet. CBA is then carried by the channel ways to a disposal pond or recycling facility [21].

Thanks to fluidized bed combustion (FBC), one of other coal combustion methods, low-grade fuels (e.g., coal and woody biomass) can be burned with high efficiency without the need for pulverization [22]. This method has a smaller combustion chamber compared to conventional methods such as pulverized coal (PCC). It also has various benefits in terms of cost and efficiency. FBC is basically divided into two main groups, atmospheric systems, and pressurized systems. Pressurized systems, on the other hand, are divided into two small subgroups, bubbled, and circulating fluidized bed boilers [23,24].

Combustion temperatures in FBC boilers range from 800°C to 900°C and consist of multi-step processes like PCC. The working principle of fluidized bed boilers is basically an advanced combustion method used to burn solid fuel particles suspended in a section [25]. To run the O_2 necessary for burning, other particulate materials such as sand, limestone, and other particles, by which air jets are blown, pass through bubbling fluidized bed. This supports rapid/complete mixing of gases/solids and rapid temperature transfer and chemical reactions in the bed [26]. After this stage, different processes are applied to purify the fuel particles from gas. First, fuels such as coal are heated to the softening temperature in the range of 350°C–480°C and the moisture in it is evaporated. Then, the coal enters the decomposition process, which will become plastic with the release of volatile substances in the boiler. In the last step, the coal is completely degassed, and the coke conversion happens [27]. Here, the condensation and polymerization of carbonaceous materials continues even though the removal of volatile substances from the coke is completed. Pure carbon in coke is oxidized and burned at temperatures of about 700°C. The creation of fly ash remaining in the chimney vapors discharged from the bed in the FBC system starts at this point [24]. Here, the combustion of small grains such as fly ash is not desired, and coke is eliminated in suspension with boiler ash. Burning efficiency is increased by keeping the temperature constant to achieve long-term mixing in the bed.

1.2.2 Municipal solid waste

MSW consists mainly of refuse collection, garden refuse, industrial refuse, hospital refuse, hazardous refuse, etc. Its production has recently increased significantly due to urbanization, improved lifestyle, and population growth. As stated by the World Bank data [28], total 3.5 MT of MSW are daily created. It is also estimated that this figure will more than double in 2030. This excessive production in MSW continues to increase exponentially every year and is projected to extent 2.5 billion tons by 2025. This corresponds to a rise of 1.45 kg per person worldwide, which indicates that effective disposal methods should be developed and employed urgently for sustainable development [29].

Fig. 1.3 shows the MSW generation amount of diverse countries, and one can see that the highest generation is in the United States. It should be emphasized that although the United States creates much more MSW than any other country globally, its recycling rate is significantly lower than in other countries. OECD Annual Report

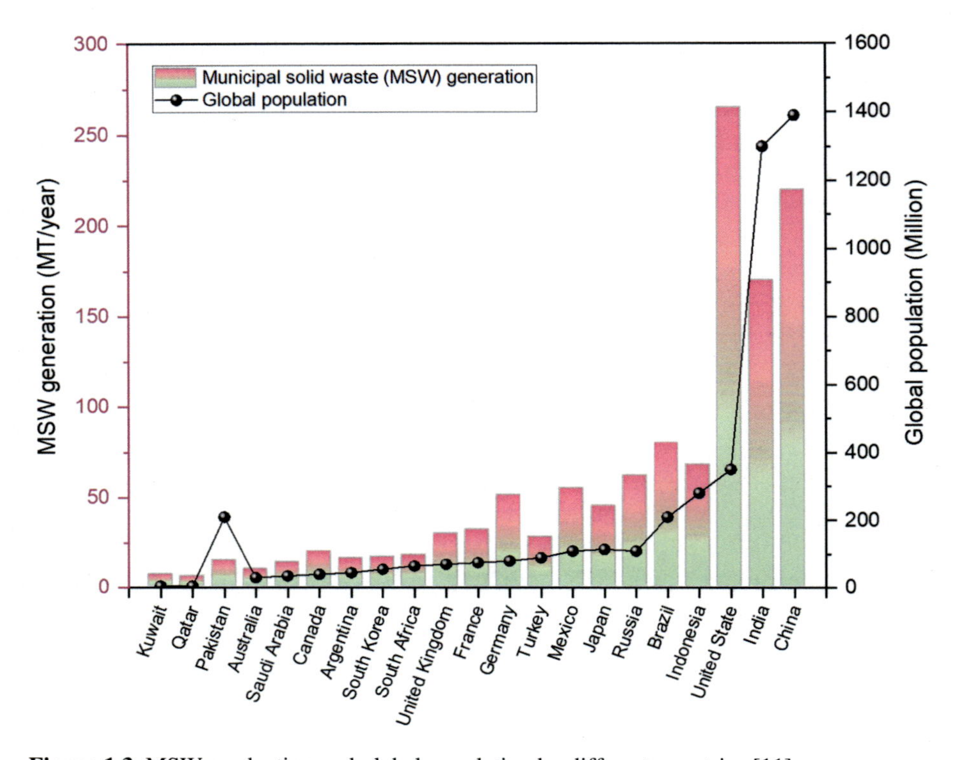

Figure 1.3 MSW production and global population by different countries [11].

[30]. Total MSW production in the United States has reached enormous amounts, increasing 250 MT each year. Only 35% of this waste is recycled, 12% is used to produce energy, and the remaining 53% continues to be stored in designated reserve areas at certain points [31]. The amount of MSW produced in China reached over 200 MT in 2018 and continues to increase regularly every year [32]. The current MSW production worldwide is more than 2 billion tons and it is estimated that this amount will more than double by 2050. Hence, it is not a secret that there are alternative ways of effective use for the sustainable development of countries [33].

In MSW management, incineration technology is adopted to reduce the gradual increase in waste generation volume (by about 60% and 90% by weight and volume, respectively). Incineration, one of the methods considered for the effective disposal of MSW, consists usually of three key fragments: burning, energy recovery and air trash control [34]. Three main techniques are used to effectively dispose of MSW: batch incineration, waste derived fuel and fluidized bed incineration. Over the past several decades, MSW incineration has evolved from a mere waste disposal strategy to a strategy with dual advantages, namely waste proposal, and energy recovery. Modern MSW fire facilities are presently equipped with energy retrieval units, enabling their fiscal benefits. Besides, MSW incineration has many environmental benefits, including but not limited to the saving of fossil fuels, lessened landfill disposal of MSW, stabilization of risky elements, and recycling of nutrients. In the MSW incineration plant, MSW can be either used as a principal fuel or supplementary fuel, depends on the properties of MSW and the purpose of MSW burning facility. The incineration of MSW is the major disposal method for most modern countries worldwide, such as Japan, Sweden, Denmark.

According to research, 2100 MSW burning plants (230 MT of combustion capacity) exist in the world, and more than 40% of them are located in the United States, the EU and other industrialized republics [35]. Although the sum of waste through incineration treatment has increased, the amount of waste from landfills has fallen below 50% by 2019. The number of incinerators in China rose from 109 in 2011 to 389 in 2019 [36]. The European Confederation of Waste-to-Energy Facilities declared that in 2011, about 85% of total MSW in Europe was disposed of at approximately 371 incinerators. It is also estimated that within the next 20 years, one hundred percent of their waste will be disposed of in a total of twice the facility (more than 800 incinerators) within the zero waste concept [37]. In addition, although a total of 86 MSW incinerators were run in 24 states across the United States in 2010 [38], the number of waste incinerators in countries across the EU remained at 75 in 2018 [39]. Consequently, despite the growing amount of waste, both incineration and recycling rates have increased steadily and the net zero target is expected to be reached by 2035. In other words, waste is thought to be a raw material used for energy and recycling rather than a problematic material [37].

Fig. 1.4 demonstrates the leading steps of both burning and waste generation for MSW. Even though there exist several different types of waste furnaces, the simple system of all MSW incinerators is generally similar. Incineration plants use a system that feeds MSW into the combustion chamber. The chimney vapor formed after the burning procedure must be adjusted to be exposed to a heat of no less than 850°C for the correct breakdown of the output products. Once cremating dangerous MSWs with halogenated organic matter content, the heat requests suddenly rise to 1100°C [41]. A low temperature pyrolysis treatment has also been found to be

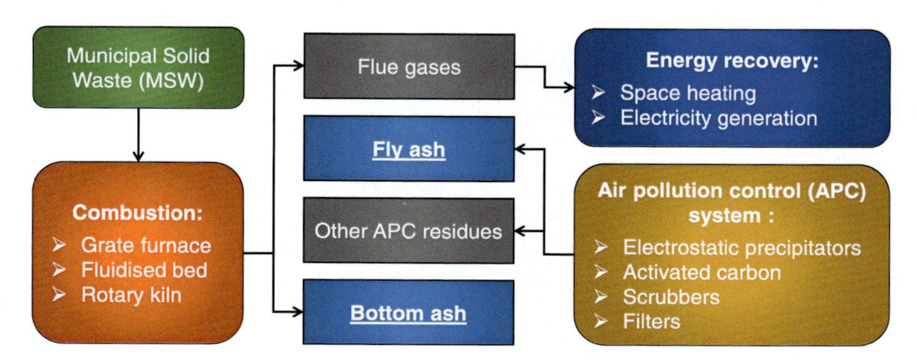

Figure 1.4 Main stages of MSW incineration [40].

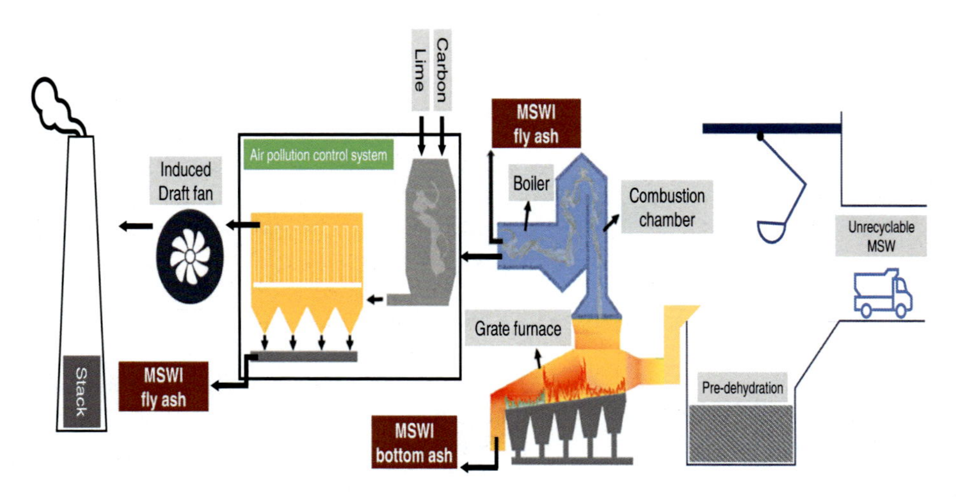

Figure 1.5 A typical MSW incineration plant [43].

operative in eliminating and curing in excess of 95% of the chlorine in the feed of MSW [42].

Fig. 1.5 also illustrates a typical MSW incineration plant. Briefly speaking, the MSW from the storage bunker is fed into the grate furnace, where the MSW incineration takes place. The MSW incineration time within the grate furnace usually ranges from 30 to 120 minutes, producing steam with temperatures of 850°C–1150°C and pressures of 35–75 bar. Like the coal powder plant, steam is generated by the high temperature to supply a steam turbine unit that generates electricity. Moreover, the steam is also used in the boiler water degassing process for heating purpose.

The fire-retardant quantities of the MSW cremated may result in a drop in the overall facility efficacy after their statement on temperature transfer exteriors, and may cause corrosion by salt deposits, sintered/molten ash. In most modern MSW incinerator, some separate ashes are created, including grate/boiler/scrubber/baghouse ashes. As a result of MSW incineration, ashes could be classified as fly/bottom ashes. Bottom ash can be described as a lumped combustion residue that does not rise with the gases because it is relatively heavy at the bottom. The bottom ash shaped throughout the burning of MSW is essentially the main solid fraction and constitutes nearly 75%–80% of the entire combined ash flow. MSW incineration bottom ash primarily contains grate ash (over 90%), which is porous, greyish, and encloses minor quantities of non-combustible organic materials and portions of metal.

Around 20%–25% of the ash stream are delivered to the heating reservoir and air trash system. As the burning vapor moves across the heating reservoir, the suspended particles cling to the heating reservoir conduits and wall (termed as boiler ash) are kept in the air trash control system (termed fly ash). The MSW burning fly ash is made up of too fine grains that rise with the vapors and composed from reactor and filter with a major fraction quantifying below 0.1 mm in size.

1.2.3 Biomass

Compared to fossil fuels, the burning of biomass in thermic power plants is one of the chief efficient and sustainable methods of generating electricity [44]. The increasing usage of biomass for electricity in power plants (an increase of 8% per year) has been adopted by most modern industrialized countries [45]. The global contribution of electricity produced using biomass is 130 GW, and the continents with the highest production are Asia, Africa, and America, respectively. Biomass incineration generates power or electricity using the same process as coal combustion, but instead of coal, biomass incineration plants burn biomass materials, such as plants, wood, and waste. In comparison with coal, biomass is however still a renewable energy source. Biomass raw materials have various properties and some feed stocks are forest/wood waste; agricultural waste (i.e., corn cobs and wheat straw); urban wood waste (i.e., packing crates and pallets) and food processing wastes [46]. Fig. 1.6 presents some biomass sources.

Including all plant-based waste, biomass is alienated into four key sets: herbaceous grasses, woody plants, aquatic plants, and fertilizers. Biomass materials, also called biomass feed stocks, come from living organisms, and therefore, they contain energy first derived from the sun. The biomass energy makes up an important fraction in the world energy supply. For example, the total biomass energy supply was

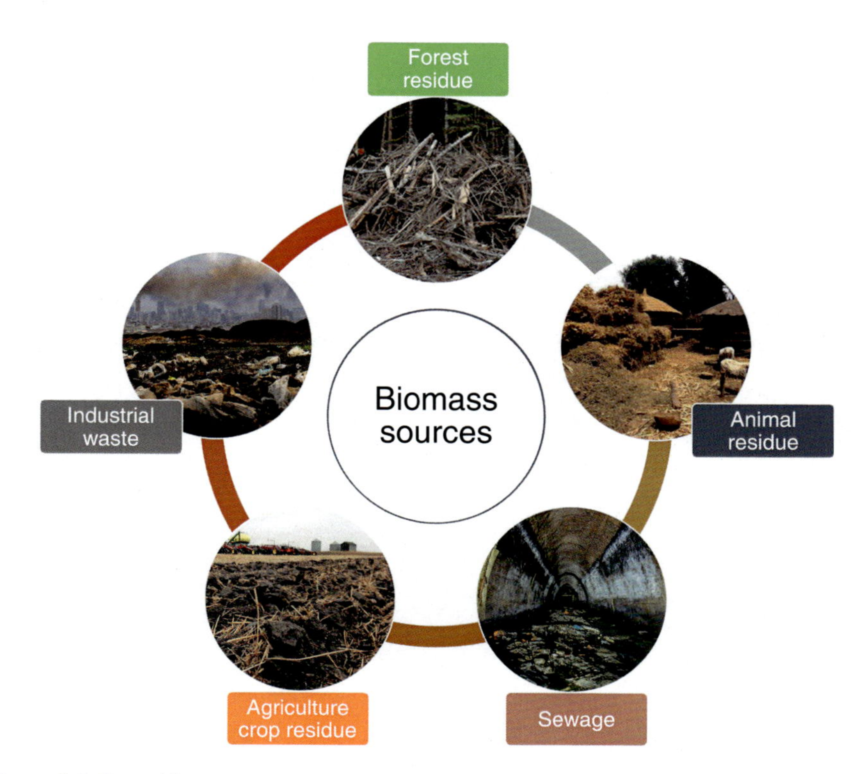

Figure 1.6 Some biomass sources.

over 50 EJ in 2010, corresponding to about 10% of the total world energy supply in that year [47]. The energy (bioenergy) obtained from biomass has an important place as it contributes to nearly 70% of the entire renewable energy sources.

Over the past decades, technologies to generate heat and power from biomass have been settled, covering direct biomass burning, co-firing with coal/biomass gasification. In 2017, 91% of biomass energy is obtained from solid biomass, 7% from liquid biofuel and 2% from biogas [45]. In 2012, the efficiency of biomass-based combined temperature and thermic power plants varies between 70% and 90% [48], which is continuously increased with further technology development. As stated by International Energy Agency IEA [49], the potential energy production from biomass is currently around 50 EJ per year, and this value is predicted to rise to 300 EJ per year by 2050, with a substantial rise in biomass predicted [49].

To improve biomass characteristics and make biomass handling, transport, and conversion more efficient and cost-effective, several pretreatment technologies can be employed. Commonly used pretreatment technologies include drying, pelletisation and briquetting, torrefaction, and pyrolysis. After pretreatment, the biomass feed stocks can be transported into the biomass-fired power plant for biomass combustion. Depending on the fuel-feeding system, the biomass-fired power plant is separated: fixed/FBC. Fig. 1.7 displays an example of the direct burning of biomass-fired power plant. The bubbling fluidized bed technology, which has some advantages compared to other combustion technologies, such as compact boiler design, possibility of burning variable fuel types, high combustion efficiency and reduced pollutant emission, well converts biomass into energy [51].

Though biomass combustion is measured to have key ecological benefits, it also has some weaknesses that it produces a huge amount of ash, which reduces the efficacy of

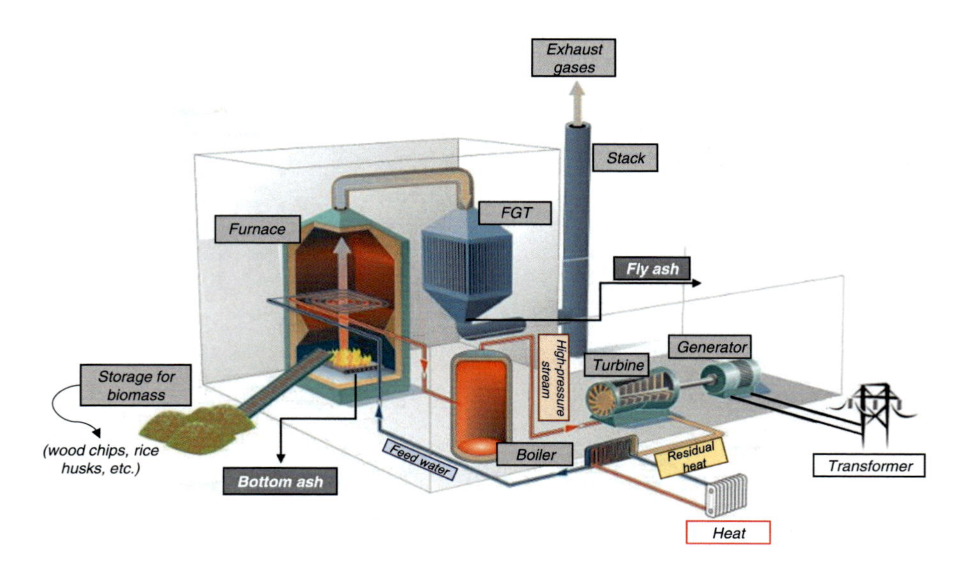

Figure 1.7 Direct combustion of biomass-fired power plant [50].

the methods used in the burning process and causes additional costs for boiler cleaning, and in addition to these, it prevents the further use of biomass materials in combustion [52]. Ash making can cause a natural angst if it is not well accomplished. Generally speaking, two ash types are created during the biomass burning [53]: biomass bottom/fly ashes. The first consists mainly of the coarse fraction of the total ash stream and is produced in the combustion chamber, which consists of sand particles (i.e., quartz) and other mineral impurities [54]. In contrast, the latter includes the ash particulates that are detached from the vapor stream outdoor the burning space. Specially designed equipment is utilized to retain these fine particulates to stop releasing their harmful vapors into the air [55]. Biomass fly ash consists of the finest fraction of ashes, and includes inert fraction and a slight plant portion.

Some ash retrieval strategies could be used to catch the bottom/fly ash produced by combustion. While bottom ash is tangibly parted in fluidized bed furnaces, it is composed by gravity, excluding these furnaces. Fly ash elements can also be composed by gravity in settling chambers. However, equipment such as filters must be employed to detach too fine fly ash from chimney vapor [56]. Since various biomass feedstocks could be used to biomass combustion, the amount and class of biomass ashes varied. The employed burning technique and working situations of burning also affect the ash generation [57]. For example, the burning of wood creates fewer quantities of biomass ashes than herbaceous biomass since the latter has a greater ash content [56].

1.3 Production amounts of industrial solid ashes

1.3.1 Coal ashes

CFA is an important source of the waste created in coal thermic power plants. Especially its abundance, low cost and significant increase in energy demand worldwide strengthen the place of coal in energy use. Coal burning represents nearly 40% of entire electricity production worldwide and it is thought that there are more than 13,000 coal-fired units in active and/or idle state [58]. Fig. 1.8 provides information on the locations, status (e.g., operating, canceled, and retired thermic plant) and quantities of coal-fired plants worldwide.

According to the BP [58] world energy statistical analysis report, the highest coal production/consumption takes place in the Asian region. Moreover, the coal consumption rate in a country's total primary energy in 2020 is 56.8%, 55%, 26% and 10% for China, India, Japan, and the United States, respectively [5]. Fig. 1.9 shows the coal production amounts in the world based on countries. This massive use of coal to meet global energy needs results in the production of enormous amounts of coal ash. In fact, coal ash making, which reached 500 MT in 2005, increased exponentially to 750 MT in 2015 [58,59]. Coal ash generally corresponds to 5%–20% of the coal used in production and is mainly produced in the coarse-sized bottom ash and fine-sized fly ash forms. In general, 85%–95% of the total coal ashes created consists of fine fly ash [60].

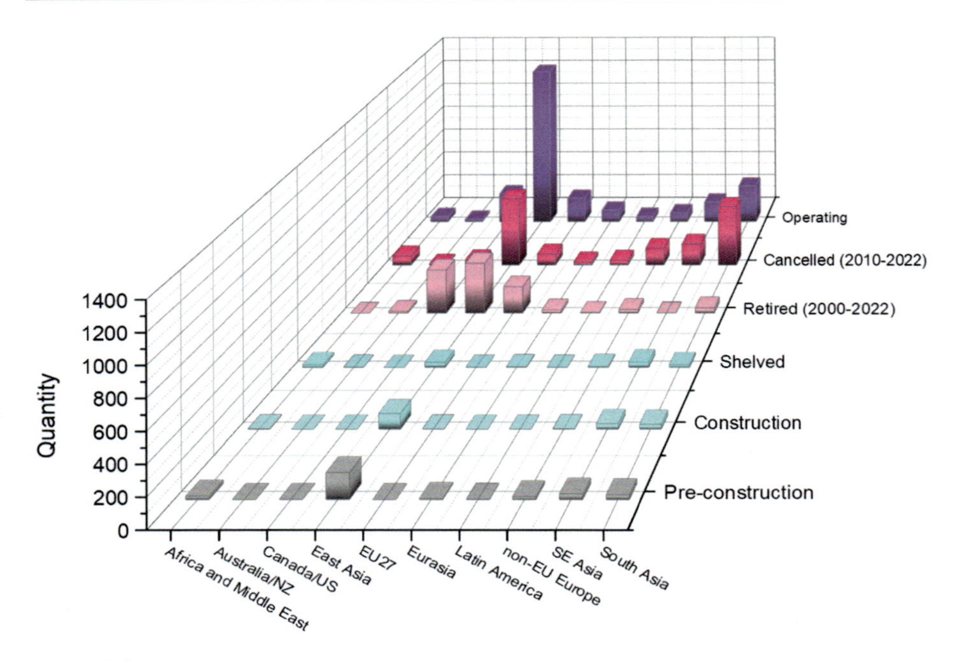

Figure 1.8 Information on quantities of coal-fired plants worldwide.

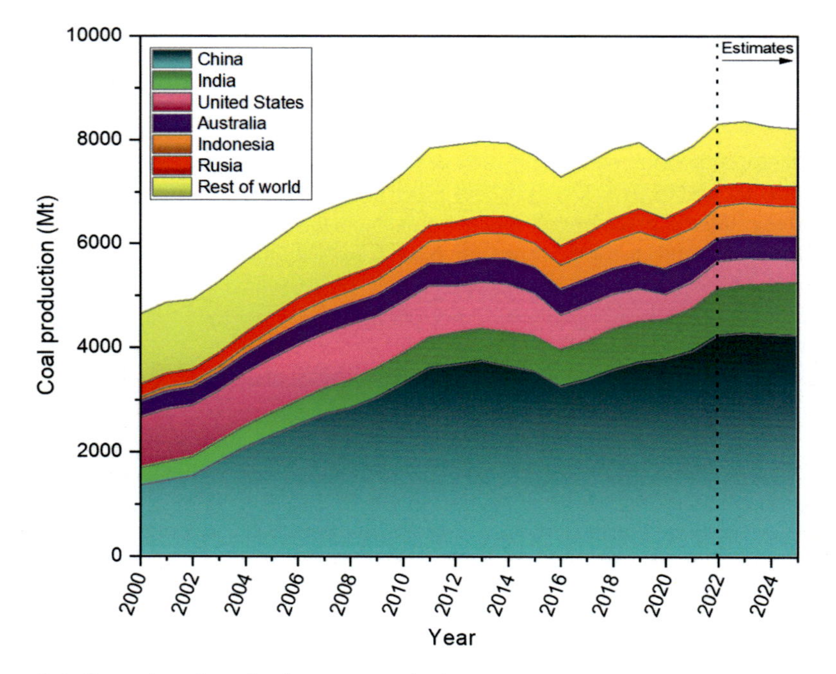

Figure 1.9 General coal production amounts in the world [5].

In the IEA [49] report, it is estimated that the coal demand (from coal power plants) will be around 21,000 MT in 2024 and the fly ash production will reach 720 MT. As stated by the American Coal Ash Association report, nearly 30 MT of CFA and 9 MT of CBA from coal thermic power facilities were generated in 2021 [61]. Moreover, coal making in India, which is one of the largest producer countries, was 40 MT between 1993 and 1994, while the CFA use was only 3% (1 MT). With the increase in the value given to coal ash, the CFA use reached the level of nearly 60% compared to 107.77 MT coal production in the 2015–16 period. Total coal production worldwide in 2019 increased by 1.5% (total 600 MT) compared to previous years. As a result of this situation, approximately 500 MT of CFA was produced [62]. In the light of the above information, the global use of coal ashes corresponds to about one-fourth of the total production amount. Fig. 1.10 shows the graph summarizing fly ash production and use data by country.

1.3.2 Municipal solid waste

Burning has become the most widely used way of protecting landfills and reducing the size of MSW [64]. Nevertheless, the disposal of the ashes resulting from the burning of waste is an important concern. Nearly 3%–5% of the waste products created after burning is moved to MSW incineration fly ash (MSWI-FA); this is equivalent to about 10–30 kg of MSWI-FA produced per ton of waste burning [65]. Volatile particles are transported from the burning space to air trash control, where they are detached from the chimney vapor and solid waste is produced. Only

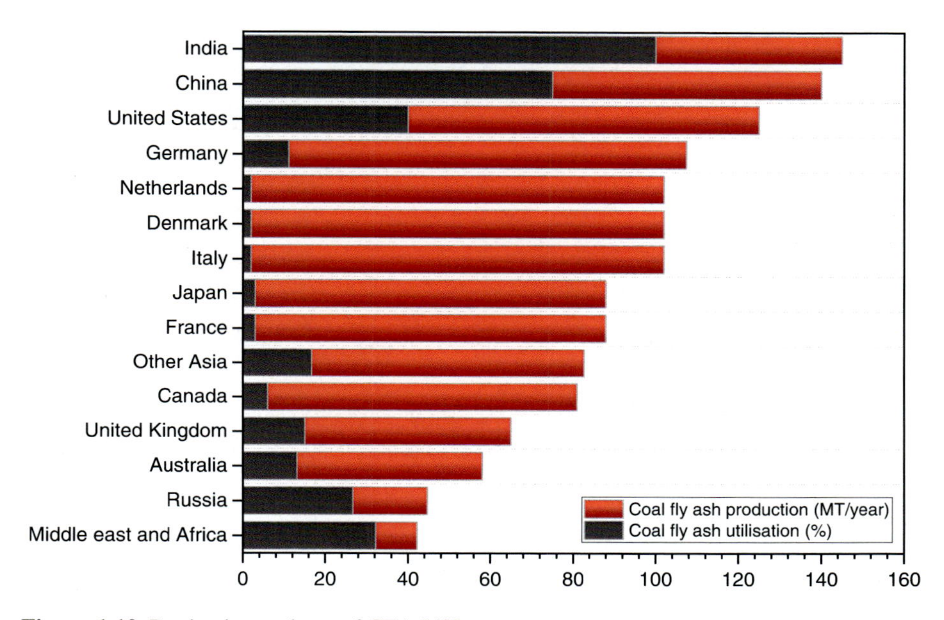

Figure 1.10 Production and use of CFA [63].

1.42–2.53 wt.% of MSW is composed as fly ash in burning plant, and bottom ash is 25–30 wt.% of total MSW cremated. Considering the mass of MSW in the initial state, the amount of MSWI-BA is on the point of 25% of this mass, and while the residual quantity of waste in the air trash control system may be less than 5%, 70% or more may be present in the chimney vapor. In general, the above-mentioned values vary according to the nature of MSW, combustion process and APC technology [66].

In short, MSWI-FA is produced in the burning method and constitutes 3–15 wt.% of the left-over material [43]. Let's say, in China, ~146 MT of MSW were cremated in 2020, and ~10 MT of MSWI-FA were generated [67]. In summary, some studies have concluded that burning 1 ton of MSW creates 250–300 kg of CBA and 25–30 kg of air trash control wastes and fly ash [68].

1.3.3 Biomass

Since many countries are in the shifting phase to tiny carbon emission and economy, current annual biomass ash making amounts are not issued regularly and there are changes in production [69]. This situation makes it difficult to estimate ash production amounts from sustainable fuels. Further complicating the situation, some reported production data include ash from partial burning of biomass with coal in addition to the amount of biomass ash [70]. Nonetheless, ash making can also be assessed in regions where biomass making data are reliably verified. As a result, it can be said that the yearly production of biomass ash worldwide is 500 MT [71]. The properties of the ash produced change according to the natural structure of

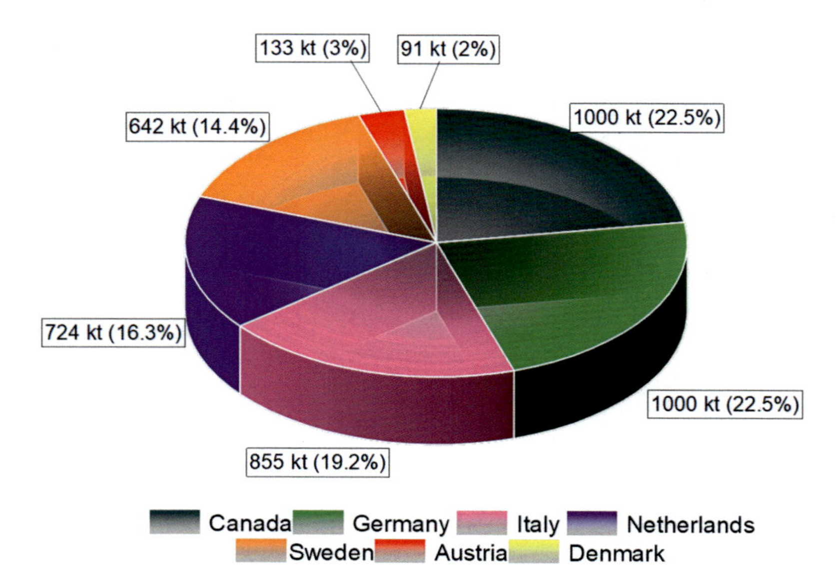

Figure 1.11 Biomass ash generation in diverse nations [70].

Table 1.1 Biomass ash types according to some countries [70].

Types of biomass	Liquid biomass	Manure	Paper sludge	Sewage sludge	Grown biomass	Straw	Demolition wood	Forestry wood
Sweden	✓		✓				✓	✓
Netherlands		✓	✓	✓			✓	✓
Italy							✓	✓
Germany	✓	✓	✓	✓	✓		✓	✓
Denmark						✓	✓	✓
Canada			✓				✓	✓
Austria					✓			✓

biomass as well as the combustion conditions. Thus the characteristics of the ash produced can be shaped according to the intended use. Grate/FBC is the most broadly employed way in ash making, while efficacy is low in grate burning and high in FBC. Some biomass ashes contain more toxic elements than others and therefore these ashes must be properly examined and disposed of in the most appropriate way. An estimate of biomass ash production and biomass ash types of some countries are given in Fig. 1.11 and Table 1.1, respectively.

References

[1] M.A. Aktar, M.M. Alam, A.Q. Al-Amin, Global economic crisis, energy use, CO2 emissions, and policy roadmap amid COVID-19, Sustainable Production and Consumption 26 (2021) 770–781.

[2] S. Saint Akadiri, A.A. Alola, G. Olasehinde-Williams, M.U. Etokakpan, The role of electricity consumption, globalization and economic growth in carbon dioxide emissions and its implications for environmental sustainability targets, Science of The Total Environment 708 (2020) 134653.

[3] S. Faisal, A. Zaky, Q. Wang, J. Huang, A. Abomohra, Integrated marine biogas: a promising approach towards sustainability, Fermentation 8 (10) (2022) 520.

[4] E. Loth, C. Qin, J.G. Simpson, K. Dykes, Why we must move beyond LCOE for renewable energy design, Advances in Applied Energy 8 (2022) 100112.

[5] British Petroleum BP. Statistical Review of World Energy, 2022.

[6] H. Nøhr-Hansen, G.K. Pedersen, P.C. Knutz, J.A. Bojesen-Koefoed, K.K. Śliwińska, J. Hovikoski, et al., The cretaceous succession of northeast Baffin Bay: stratigraphy, sedimentology and petroleum potential, Marine and Petroleum Geology 133 (2021) 105108.

[7] A.K. Singh, R.E. Masto, B. Hazra, J. Esterle, P.K. Singh, Ash from Coal and Biomass Combustion, Springer International Publishing, 2020.

[8] E. Monedero, J.J. Hernández, R. Collado, A. Pazo, M., Aineto, A. Acosta, Evaluation of ashes from agro-industrial biomass as a component for producing construction materials, Journal of Cleaner Production 318 (2021) 128517.

[9] Q. Wang, X., Song, Y. Liu, China's coal consumption in a globalizing world: insights from multi-regional input-output and structural decomposition analysis, Science of the Total Environment 711 (2020) 134790.

[10] M. Höök, W. Zittel, J., Schindler, K. Aleklett, Global coal production outlooks based on a logistic model, Fuel 89 (11) (2010) 3546–3558.

[11] M.A. Al-Ghouti, M. Khan, M.S. Nasser, K., Al-Saad, O.E. Heng, Recent advances and applications of municipal solid wastes bottom and fly ashes: insights into sustainable management and conservation of resources, Environmental Technology & Innovation 21 (2021) 101267.

[12] P.N. Ashani, M., Shafiei, K. Karimi, Biobutanol production from municipal solid waste: technical and economic analysis, Bioresource Technology 308 (2020) 123267.

[13] N. Wang, D. Chen, U., Arena, P. He, Hot char-catalytic reforming of volatiles from MSW pyrolysis, Applied Energy 191 (2017) 111–124.

[14] Y. Nie, J. Li, C. Wang, G. Huang, J. Fu, S. Chang, et al., A fine-resolution estimation of the biomass resource potential across China from 2020 to 2100, Resources, Conservation and Recycling 176 (2022) 105944.

[15] A. Brink, D. Lindberg, M. Hupa, M.E. de Tejada, M. Paneru, J. Maier, et al., A temperature-history based model for the sticking probability of impacting pulverized coal ash particles, Fuel Processing Technology 141 (2016) 210–215.

[16] M. Rafieizonooz, E. Khankhaje, S. Rezania, Assessment of environmental and chemical properties of coal ashes including fly ash and bottom ash, and coal ash concrete, Journal of Building Engineering 49 (2022) 104040.

[17] S. Zhu, J. Hui, Q. Lyu, Z. Ouyang, J. Liu, J. Zhu, et al., Experimental study on pulverized coal combustion preheated by a circulating fluidized bed: preheating characteristics for peak shaving, Fuel 324 (2022) 124684.

[18] T. De Riese, D. Eckert, L. Hakim, S. Fendt, H. Spliethoff, Modelling the capture of potassium by solid al-si particles at pulverised fuel conditions, Fuel 328 (2022) 125321.

[19] R. Cai, K. Luo, H. Watanabe, R. Kurose, J. Fan, Recent advances in high-fidelity simulations of pulverized coal combustion, Advanced Powder Technology 31 (7) (2020) 3062–3079.

[20] J. Watanabe, K. Yamamoto, Flamelet model for pulverized coal combustion, Proceedings of the Combustion Institute 35 (2) (2015) 2315–2322.

[21] Hecht, N., Duvall, D. Characterization and utilization of municipal and utility sludges and ashes. III. Utility coal ash. Environ Prot Technol Ser EPA US Environ Prot Agency, 1975.

[22] F. Scala, Particle agglomeration during fluidized bed combustion: mechanisms, early detection and possible countermeasures, Fuel Processing Technology 171 (2018) 31–38.

[23] A.M. Cormos, C.C. Cormos, Techno-economic evaluations of post-combustion CO2 capture from sub-and super-critical circulated fluidised bed combustion (CFBC) power plants, Applied Thermal Engineering 127 (2017) 106–115.

[24] L. Pang, Y. Shao, W. Zhong, H. Liu, Experimental investigation on the coal combustion in a pressurized fluidized bed, Energy 165 (2018) 1119–1128.

[25] H. Chi, M.A. Pans, C. Sun, H. Liu, An investigation of lime addition to fuel as a countermeasure to bed agglomeration for the combustion of non-woody biomass fuels in a 20kWth bubbling fluidised bed combustor, Fuel 240 (2019) 349–361.

[26] Z. Gong, Y. Shao, L. Pang, W. Zhong, C. Chen, Study on the emission characteristics of nitrogen oxides with coal combustion in pressurized fluidized bed, Chinese Journal of Chemical Engineering 27 (5) (2019) 1177–1183.

[27] K. Ohenoja, J. Pesonen, J. Yliniemi, M. Illikainen, Utilization of fly ashes from fluidized bed combustion: a review, Sustainability 12 (7) (2020) 2988.

[28] World Bank, Solid Waste Management, The World Bank Group, 2019.

[29] R.K.D. Obe, J. De Brito, C.J. Lynn, R.V. Silva, Sustainable Construction Materials: Municipal Incinerated Bottom Ash, Elsevier Science, 2017.
[30] OECD Annual Report. The organisation for economic co-operation and development (OECD), Paris, France, 2007.
[31] EPA. Municipal Solid Waste Generation, Recycling, and Disposal in the United States: Facts and Figures for 2010. U.S. Environmental Protection Agency (U.S. EPA): Washington, DC, USA, 2011.
[32] China NBS, China statistical yearbook of 2019, Chinese Statistical Bureau, China Statistical Press, Beijing, 2019.
[33] S. Kaza, L. Yao, P. Bhada-Tata, F. Van Woerden, What a Waste 2.0: A Global Snapshot of Solid Waste Management to 2050, World Bank Publications, 2018.
[34] R.P. Singh, V.V. Tyagi, T. Allen, M.H. Ibrahim, R. Kothari, An overview for exploring the possibilities of energy generation from municipal solid waste (MSW) in Indian scenario, Renewable and Sustainable Energy Reviews 15 (9) (2011) 4797–4808.
[35] Smart Core Industry Research Center. Market Research and Analysis on the Development of Global Waste Incineration Power Generation in 2020. Sweden: World Bioenergy Association, 2020.
[36] B. Huang, M. Gan, Z. Ji, X. Fan, D. Zhang, X. Chen, et al., Recent progress on the thermal treatment and resource utilization technologies of municipal waste incineration fly ash: a review, Process Safety and Environmental Protection 159 (2022) 547–565.
[37] N. Scarlat, F. Fahl, J.F. Dallemand, Status and opportunities for energy recovery from municipal solid waste in Europe, Waste and Biomass Valorization 10 (9) (2019) 2425–2444.
[38] Michaels, T. The 2010 ERC Directory of Waste-to-Energy Plants; Energy Recovery Council: Washington, DC, USA, 2010.
[39] Council, E.R. The 2018 ERC Directory of Waste-to-Energy Facilities; Energy Recovery Council: Washington, DC, USA, 2018.
[40] R.K. Dhir, J.D. Brito, C.J. Lynn, R.V. Silva, Municipal solid waste composition, incineration, processing and management of bottom ashes. Sustainable Construction, Materials (2018) 31–90.
[41] EU, Directive 2000/76/EC of the European parliament and of the council of 4 December 2000 on the incineration of waste, Official Journal of the European Communities (2000) 91–111.
[42] H. Jouhara, D. Ahmad, I. Van Den Boogaert, E. Katsou, S., Simons, N. Spencer, Pyrolysis of domestic based feedstock at temperatures up to 300C, Thermal Science and Engineering Progress 5 (2018) 117–143.
[43] Y. Zhang, L. Wang, L. Chen, B. Ma, Y. Zhang, W., Ni, et al., Treatment of municipal solid waste incineration fly ash: state-of-the-art technologies and future perspectives, Journal of Hazardous Materials 411 (2021) 125132.
[44] A. Briones-Hidrovo, J. Copa, L.A. Tarelho, C. Gonçalves, T.P., da Costa, A.C. Dias, Environmental and energy performance of residual forest biomass for electricity generation: gasification vs. combustion, Journal of Cleaner Production 289 (2021) 125680.
[45] World Bioenergy Association, Global bioenergy statistics 2020, World Bioenergy Association, Stockholm, Sweden, 2020.
[46] Irena. Renewable Energy Technologies: Cost Analysis Series. Wind Power. Volume 1: Power sector Issue 5/5. Biomass for Power Generation. International Renewable Energy Agency. IRENA, 2012.
[47] IRENA, W. Statistical Issues: Bioenergy and Distributed Renewable Energy. New York, USA, 2013.

[48] A. Eisentraut, A. Brown, Technology roadmap: bioenergy for heat and power, Technology Roadmaps 2 (2012) 1–41.
[49] International Energy Agency (IEA). Bioenergy for Heat and Power Technology Roadmap, 2012.
[50] S.B.A. Kashem, M.E. Chowdhury, A. Khandakar, M. Tabassum, A. Ashraf, J. Ahmed, A comprehensive investigation of suitable biomass raw materials and biomass conversion technology in Sarawak, Malaysia, International Journal of Technology 1 (2) (2020) 75–105.
[51] W. Żukowski, D. Jankowski, J., Wrona, G. Berkowicz-Płatek, Combustion behavior and pollutant emission characteristics of polymers and biomass in a bubbling fluidized bed reactor, Energy 263 (2023) 125953.
[52] L. Wang, J.E. Hustad, Ø. Skreiberg, G., Skjevrak, M. Grønli, A critical review on additives to reduce ash related operation problems in biomass combustion applications, Energy Procedia 20 (2012) 20–29.
[53] D. Picco, Technical assistance for the development and improvement of technologies, methodologies and tools for enhanced use of agricultural biomass residues, Energy Plant Report. Central European Initiative 53 (2010). Italy.
[54] R.C.E. Modolo, V.M. Ferreira, L.A. Tarelho, J.A. Labrincha, L., Senff, L. Silva, Mortar formulations with bottom ash from biomass combustion, Construction and Building Materials 45 (2013) 275–281.
[55] S. Maschio, G. Tonello, L., Piani, E. Furlani, Fly and bottom ashes from biomass combustion as cement replacing components in mortars production: rheological behaviour of the pastes and materials compression strength, Chemosphere 85 (4) (2011) 666–671.
[56] J. Koppejan, S. Van Loo, The Handbook of Biomass Combustion and Co-firing, Routledge, 2012.
[57] R. Rajamma, R.J. Ball, L.A.C. Tarelho, G.C. Allen, J.A., Labrincha, V.M. Ferreira, Characterisation and use of biomass fly ash in cement-based materials, Journal of Hazardous Materials 172 (2) (2009) 1049–1060.
[58] USEIA. Statistics on Global Coal Production, Consumption [WWW Document]. 2014. URL https://www.eia.gov/coal/data.php#production.
[59] Z.T. Yao, X.S. Ji, P.K. Sarker, J.H. Tang, L.Q. Ge, M.S. Xia, et al., A comprehensive review on the applications of coal fly ash, Earth-Science Reviews 141 (2015) 105–121.
[60] P. Kumar, N. Singh, Influence of recycled concrete aggregates and coal bottom ash on various properties of high volume fly ash-self compacting concrete, Journal of Building Engineering 32 (2020) 101491.
[61] ACAA. Coal Ash Recycling Rate Increases in 2021, 2021.
[62] CEA. Report on Fly Ash Generation at Coal/Lignite Based Thermal Power Stations and its Utilization in the Country for the Year 2020–21, 2021.
[63] A. Bhatt, S. Priyadarshini, A.A. Mohanakrishnan, A. Abri, M. Sattler, S. Techapaphawit, Physical, chemical, and geotechnical properties of coal fly ash: a global review, Case Studies in Construction Materials 11 (2019) e00263.
[64] J. Giro-Paloma, J., Formosa, J.M. Chimenos, Granular material development applied in an experimental section for civil engineering purposes, Applied Sciences 10 (19) (2020) 6782.
[65] L. Xu, Q. Sun, J. Zhang, J. Yan, K. Wu, F., Pan, et al., Status and development trend of harmless and resourceful disposal of municipal solid waste incineration fly ash, Advances in Environmental Protection 7 (2017) 414–422.

[66] P.H. Brunner, H. Rechberger, Waste to energy—key element for sustainable waste management, Waste Management 37 (2015) 3—12.
[67] National Bureau of Statistics of the People's Republic of China, China Statistical Yearbook, China Statistics Press, Beijing, 2012-2021.
[68] M.S. Ashraf, Z. Ghouleh, Y. Shao, Production of eco-cement exclusively from municipal solid waste incineration residues, Resources, Conservation and Recycling 149 (2019) 332—342.
[69] J. Zhai, I.T. Burke, D.I. Stewart, Beneficial management of biomass combustion ashes, Renewable and Sustainable Energy Reviews 151 (2021) 111555.
[70] IEA. Bioenergy. Options for Increased Use of Ash from Biomass Combustion and Cofiring, 2018.
[71] N.C. Cruz, F.C. Silva, L.A., Tarelho, S.M. Rodrigues, Critical review of key variables affecting potential recycling applications of ash produced at large-scale biomass combustion plants, Resources, Conservation and Recycling 150 (2019) 104427.

Properties of industrial solid ashes

2

Abstract

This section provides a brief overview of the potential hazards and characteristics of industrial solid ashes. It begins with a general introduction to industrial ash management and energy sources, leading to a good understanding of the production and properties of ash. Depending on the burning processes of industrial solid ashes in electrical and thermic power facilities, some physical, chemical, and mineralogical properties of these must be known in order to reduce or completely eliminate their detrimental effects on the environment. Moreover, knowing these basic properties well has become an important issue for their safe disposal and potentially their use in different sectors as a raw material resource. This chapter mainly draws attention to the properties of ash for the treatment, disposal, and environmental impact assessment of industrial solid ashes.

2.1 Introduction

In general, industrial ashes are pulverized coal, biomass, domestic solids, or a blend of these, which is burned in electrical and thermic power facilities. These ashes can be expressed as the principal wastes produced by the incinerating of fossil fuels and other energy bases [1]. Today, reducing the amount of landfills due to the making of industrial ashes has become an extremely important issue for environmental welfare. However, for the disposal and potential uses of industrial ashes, the properties of these ashes must be well understood. Overall, the characterization (e.g., physical, mineralogical, chemical) of industrial ashes from co-incineration should be determined comprehensively.

Coal ashes (e.g., CFA, CBA) formed during the coal's burning produce significant amounts of solid dust residues. Coal ash generally corresponds to 5%−20% of the coal used in production and is chiefly produced in coarse bottom ash and fine fly ash forms. In total, 85%−95% of the entire ash created consists generally of fine fly ash. CFA is fine particles of essentially glassy spherical shape, hollow, and/or solid [2]. In comparison, CBA has a structure like the grain size of coarser and fine aggregates compared with CFA. Although the chemical/mineralogical structures of CFA and CBA are diverse, they differ relying on the sort of the coal burned and the way/condition of burning [3,4]. In addition, the structure/content of metal oxides in CFA and CBA gains characteristics according to the region where the coal is located. Most of the time, all coal ash is rich in SiO_2 and Al_2O_3. Therefore the recovery of coal ash could be considered as an important waste treatment issue. CFA/CBA has potentially harmful effects on the environment due to its heavy metal content and is considered a risky waste [5].

Compared with coal, which is a fossil fuel, biomass is defined as a renewable energy basis. Biomass is herbal/animal-based material that is generally used in power plants to

Machine Learning Applications in Industrial Solid Ash. DOI: https://doi.org/10.1016/B978-0-443-15524-6.00001-7
© 2024 Elsevier Inc. All rights reserved.

generate electricity and/or heat [6]. There are many types of biomass raw materials with different properties. Forestry (wood and leaves), urban wood (pallets and packing crates), agricultural (wheat stalks and corn cobs), food processing wastes are among the most used biomass types [7]. In the course of the burning of biomass in power plants, products having different properties emerge, such as BFA and BBA. Overall, biomass ash's features rely on composition contamination, soil structure, type/time of harvest, pesticide usage, herbal age, and rising environments [8]. In addition, biomass ashes consist of multicomponent organic−inorganic materials and have a heterogeneous structure. Combustion of biomass produces ashes that contain less alumina (Al_2O_3) and more alkali (e.g., K and Na) than coal ash. Biomass ashes are known to be richer in trace elements and micro-nutrients compared with other ashes. The combustion process in power plants directly affects the sum of non-burned carbon in biomass ash. Primary concerns about biomass ash are processing/management, such as storage, use, transportation, and reduction of the amount of unburned carbon and environmental impacts [9].

Municipal solid waste (MSW) mainly includes garbage collection, garden waste, industrial waste, hospital waste, hazardous waste. Thanks to the cremating of these wastes in power facilities to generate electricity and/or heat, MSW combustion fly/bottom ash is created. Overall, 80%−90% CBA and 10%−20% CFA are produced after MSW is burned [10]. MSW ash generally consists of small grain size, large surface area gray powders. MSW burning bottom ash is employed as a raw product source in different sectors owing to its low heavy metal content or low polluting properties. On the contrary, MSW burning fly ash is considered as risky waste worldwide owing to its pollutants [11]. MSW burning fly ash contains many oxides and may contain high levels of heavy metals [12]. MSW ash must meet the requests for low-size addition rate and steady substance structure for the ultimate disposal. The stop of metals or deletion of dioxins is also an environmental essential issue for using MSW ash [13,14].

2.2 Characteristics of coal ashes

Characterization of coal ashes differs according to the nature of coal, burning process and working methods according to other factors. Many factors, including different types of coal and sources, have a substantial impact on making of coal ashes. In particular, type of coal that forms the source of its ashes and its variability; it depends on parameters such as degree of grinding (pulverization) of coal before burning, boiler type, the heat of burning, characteristics of ash collection-removal systems, additives added to coal, and features arising from the elements that can change over time. In addition, other compound level coals, mainly sub-bituminous and lignite used in coal thermic power plants, directly affect coal ash's sorts due to their variable carbon content. Anthracite coal (86%−97%) and bituminous coal (40%−80%) have the highest carbon content [15]. Although the properties of each coal and ash vary from country to country, the main components of coal are determined according to ASTM D388-18 and ASTM D4326-13 [16], ASTM-D388-18a [17]. Changes in coal types cause differences in physical/mineralogical features of coal ash, especially in chemical properties.

Coal ash, which is taken into account as a key waste product and raw material to be used in other sectors such as construction and civil engineering, is generally discarded into pools or stored in the surface built-dam structures. Although the structure and storage of coal ashes could create a severe risk to the surroundings, its harmful effects on human health should not be ignored. Consequently, a virtuous thought of the coal ash features is extremely imperative for ash reuse and disposal.

2.2.1 Coal fly ash

2.2.1.1 Physical features

Physical characteristics of coal fly ash (CFA) generally change as a role of coal sort, carbon quantity, burning temperature, and type/properties of boiler. The production of CFA particles begins because of the oxidation of coal minerals during combustion (1400°C–1700°C). CFA particles take a spherical shape as they pass into solid form in suspension with the flue gas. As a result, CFA exists as shiny fine grains of diverse diameters and characters outside of compound reactions [18–20]. They are usually dull or hard. Generally, color of these glassy grains can be black, gray, or beige as a function of the sum of unburned C in CFA. It is formed with light brown to gray colored CFA particles because of the combustion of lignite and/or bituminous coals [21]. Fig. 2.1 shows

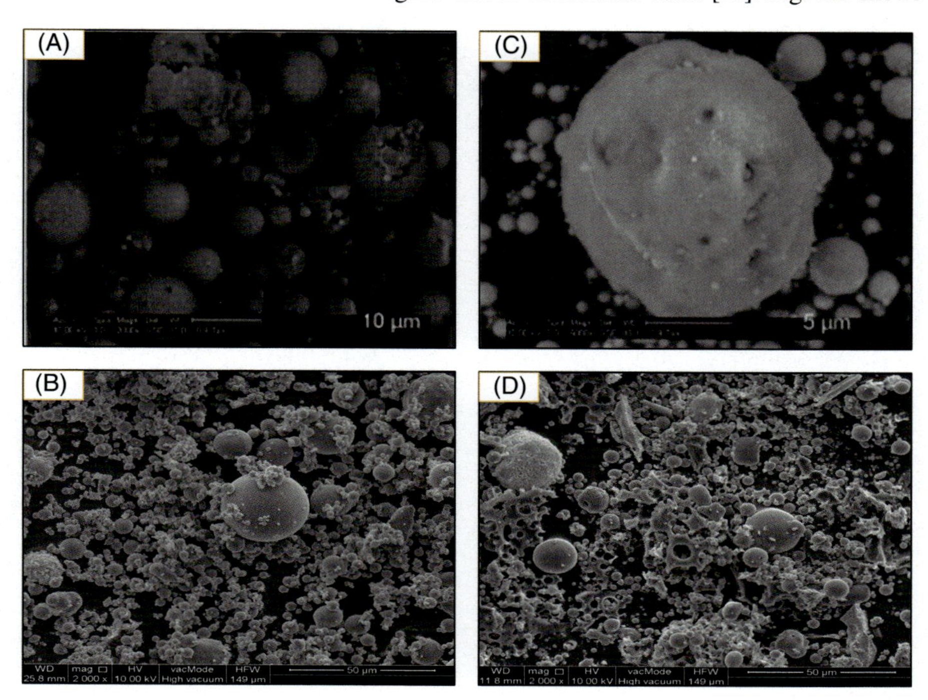

Figure 2.1 SEM views of coal fly ash: (A and B) type F CFA and (C and D) type C CFA [22,23].

SEM images of CFA samples containing different sizes of burnt-unburned coal grains and glassy spherical grains.

However, grain sizes/specific surface, which is vital physical pointer for evaluating CFAs, depends on the modes of formation of inorganic materials in coal and combustion ways (e.g., reservoir) and ash pool techniques. The dominant grain size distribution of CFAs mostly varies between 1 and 100 μm [9,24]. The coarse grains (>75 μm) are usually obtained from mechanical collectors during lignite/bituminous coal production. In addition, fine minerals (0.02–0.2 μm) in organic CFA grains support ash formation, while the disintegration of inorganic minerals creates grains of 0.2–10 μm size [25].

Fine CFA grains cause higher reactivity mostly thanks to the vast specific surface. Especially the fact that these particles are exposed to hydroxide attack and rapid cooling after the combustion space is the main reasons for this situation. Generally, the surface area of CFAs varies between 5 and 10 m^2/g, but the specific surface of CFAs obtained from lignite coal could reach 200 m^2/g [23,26]. In addition, it is predicted that the Ss values of CFAs obtained from electrostatic precipitators can reach up to 1200 m^2/g [27]. Densities of CFAs vary between 2.1 and 3 g/cm^3, and specific gravity values vary in the range of 1.90–2.55. The size of the S_s value and the point of particle surface development shrewdly govern the aptness of CFAs for use as sorbent and/or catalyst [28]. Table 2.1 provides detailed information about the general physical properties of CFAs.

2.2.1.2 Mineral features

CFAs are generally a complex material with multicomponent and heterogeneous composition. It contains organic (1%–9%) and inorganic (90%–99%) materials in its structure as phase composition. It also contains trace amounts of liquid

Table 2.1 Physical features of CFA [22,29].

Features	Coal fly ash
Color	Reddish-tan/black/greyish white
Specific gravity (no unit)	1.9–2.6
Specific surface, S_s, (m^2/g)	5–10
Apparent density (g/cm^3)	2.1–3.0
Max. dry density (g/cc)	0.9–1.6
Optimum water content (%)	38.0–18.0
Grain size (μm)	1–100
Porosity (%)	30–65
pH	1.2–12.5 (most are alkaline)
C_u (Coefficient of uniformity)	3.1–10.7
Permability (cm/s)	$8 \times 10^{-6} - 7 \times 10^{-4}$
Compression index (Cc)	0.05–0.4
C_v (Coefficient of consolidation, cm^2/s)	$1.75 \times 10^{-5} - 2.01 \times 10^{-3}$
Internal friction angle (j)	300°–400°
Cohesion (kN/m^2)/Plasticity	Negligible/non-plastic
Melting point (°C)	1200–1500
Thermal stability	0.42–17.32

components (<0.5%) in its structure [23]. Typically, unburned carbon forms its organic components. In addition, liquid components include gas–liquid additions linked with liquid, gas, and inorganic/organic mixtures. Structurally, inorganic components representing the majority are characterized by the amorphous phase and the crystalline phase (quartz, mullite, goethite, and calcite). CFA structural components are technogenic or of natural origin [9]. These components can be classified as follows

1. Sulfates, carbonates, oxides, silicates, and phosphates (coal minerals/phases)
2. Mineral phases formed during the combustion of coal
3. Minerals expected to be formed during the storage and transportation of CFA (e.g., calcite, gypsum, portlandite, and dolomite).

The CFA's key mineral element is amorphous aluminosilicate glass phases (spheres indicated in Fig. 2.2). There are also iron oxides (e.g., siderite, magnetite), unburned coal grains, and cryptocrystalline phases (e.g., lime, quartz, and mullite) in its structure. The amorphous phases in the CFA content directly affect the reactivity potential, geopolymer development, and toxic element release. Due to the formation of this amorphous phase, CFA also exhibits pozzolanic properties [28].

CFAs produced during production of bituminous coal generally contain minerals such as unburned carbon, iron oxides, quartz, pyrrhotite, goethite, spinel, and mullite. In addition, it is known that CFAs obtained from the combustion of lignite coal have a similar mineralogical structure compared with bituminous coal but have a more amorphous structure. In addition, sulfur minerals (e.g., anhydrite and calcite gypsum) are found in the CFAs formed during the burning of lignite coals where desulfurization is performed [30,31]. Fig. 2.2 displays CFA's XRD forms. As can be clearly seen here, the key constituents of CFAs are mullite ($Al_4Si_2O_{10}$) and quartz (SiO_2). the common mineral components of CFA are also given in Table 2.2.

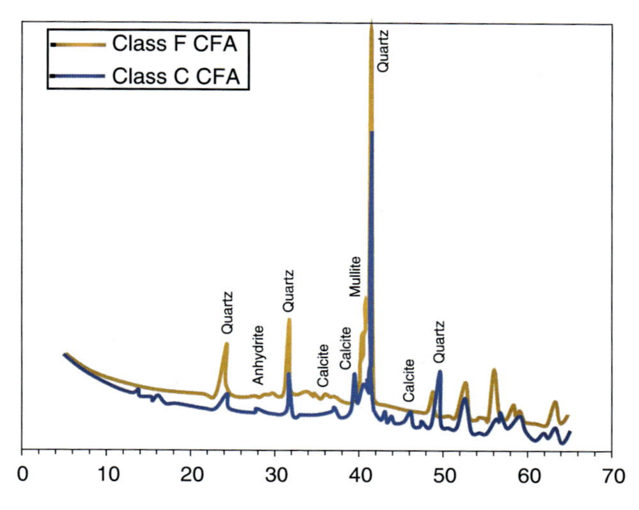

Figure 2.2 XRD profiles of coal fly ash [30].

Table 2.2 Widespread mineral stages found in CFA [29].

Oxides (wt.%) (2.9–28.1)	Mineral phases				
	SiO_2 (quartz)	Al_2O_3 (corundum)	Fe_2O_3 (hematite)	Fe_3O_4 (magnetite)	CaO (lime)
Aluminosilicates (3.3–37.4)	$3Al_2O_3 \cdot 2SiO_2$ (mullite)	$NaKAlSiO_4$ (nepheline)	$K/Na/Ca/AlSi_3O_8$ (feldspar)	$Ca_2Al_2SiO_7$ (gehlenite)	$Ca_2Al_2SiO_8$ (anorthite)
Silicates (<1)	Ca_3SiO_5, C_3S (tricalcium silicate)	Ca_2SiO_4, C_2S (wollastonite)	$Ca_3Mg(SiO_4)_2$ (merwinite)	$Ca/Mg/Al/Si_2O_7$ (akermanite)	–
Silicates (<1)	$CaCO_3$ (calcite)	$CaMg(CO_3)_2$ (dolomite)	$Ca/Fe/Mg, Mn/(CO_3)_2$ (ankerite)	–	–
Sulfates (<14.3)	$CaSO_4$, CS (anhydrite, gypsum)	–	–	–	–
Shapeless (19.9–94.5)	Alumino-silicate	Ca alumino-silicate	Fe	blended	coal

2.2.1.3 Chemical features

The structure/elemental groups of CFAs are like rocks in the Earth's high crust due to the inorganic raw materials found in coal (e.g., oxide, clay, quartz, and feldspar). Thus, it is often key to govern the CFA's chemical assembly, the source of coal deposits, and their geological effects [9].

The key components typically originated in CFA are silica, alumina, iron oxide, and CaO (calcium oxide), and their quantities vary according to fly ash's type. The quantity of C in CFA varies according to the coal type and the burning way and changes the amount of major components. The alkaline oxides (e.g., MgO and SO_3) are also present within CFA as minor components. CFA generally contains basic oxides at values of SiO_2 (25%–60%), Al_2O_3 (10%–30%), Fe_2O_3 (1%–15%), and CaO (1%–40%) [32–34]. These different values characterize the fly ash's sort. However, fly ash's chemical structure differs because coal is obtained from diverse locations. Fly ash's key constituents are typically defined as alumina as well as silica. CaO content of CFAs formed because of the combustion of bituminous coal is lower than that of lignite coal, but it is richer in Fe_2O_3 [35]. Table 2.3 displays the chemical content of CFAs produced according to diverse coal natures.

Elements in CFA content can be divided into major ($1\% <$), minor (0.5%–0.1%), and rare ($0.1\% >$). They could be also established in organic/inorganic materials in the CFA composition. In addition, these elements have dominant properties due to their proximity to the phases and minerals in CFA [37,38].

CFA's chemical structure includes Si, Al, Ca, and S as major elements. The minor elements in its structure are generally N, H, O, PBa, Cl, F, Sr, and Mn. In addition, the matrix of CFAs is mainly composed of alumina silicates and the rare earth elements Fe, Mg, Na, K, Ca, Ti and rare earths that can be found with them. However, it is known to contain many radionuclides (U, Th) [39]. Mo, As, Se, Cd, Pb, Ga, and Zn, which are volatile or form volatile oxides, do not tend to enter the matrix. These elements accumulate on fly ash's surfaces, whose concentration is in reverse related to particle diameter. Trace elements (e.g., Se, Hg, Cd, As, Pb) that

Table 2.3 Summary of chemical features of fly ash created by diverse coal forms [36].

Chemical component (wt.%)	**Coal type**			
	Bituminous	**Subbituminous**	**Lignite**	**Anthracite**
SiO_2	20–60	40–60	15–45	28–57
Fe_2O_3	10–40	4–10	4–15	3–16
Al_2O_3	5–35	20–30	10–25	18–36
CaO	1–12	5–30	15–40	1–27
MgO	0–5	1–6	3–10	1–4
SO_3	0–4	0–2	0–10	0–9
Na_2O	0–4	0–2	0–6	0–1
K_2O	0–3	0–4	0–4	0–4
LOI	0–15	0–3	0–5	1–8

have negative environmental effects are likely to be found within CFA. This situation has been seen in some countries worldwide as a risk that CFA causes groundwater, air pollution and soil acidification [29].

2.2.1.4 Classification

CFA is classified by different methods according to the burning technique and the kind of the coal cremated. Hence, the sorting established by Vassilev and Vassileva [40], primarily ASTM's standard operating procedure, is employed. Table 2.4 shows CFA's detailed classification. According to ASTM C-618, CFA is generally considered in two different categories as types F and C. The key variance between them is determined by variation in four critical components of CFA: calcium oxide (CaO 1%–35%), alumina (A_2O_3 10%–30%), silica (SiO_2 35%–60%), and iron (Fe_2O_3 4%–20%). In addition, although the components of CFA differ depending on the structure of the coal bed, it can contain many elements in trace concentrations [1,42].

As stated by ASTM C-618, type F (low calcium) is defined as CFA with a total alumina, silica, and iron content of over 70 wt.% and below 10 wt.% of lime, which are formed by burning bituminous and anthracite coal [43]. To obtain advanced pozzolanic features, cementitous materials such as quicklime and water must be added to type F fly ash [44]. In comparison, CFAs that are produced by the combustion of type C (high calcium) lignite/sub-bituminous low-quality coals, with more than 20% of the existing lime content and 50–70 wt.% of the total silica, iron, and alumina content, are called type C CFAs. The alkali and sulfate content of type C fly ash is greater than that of type F and shows lower pozzolanic properties [45].

2.2.2 Coal bottom ash

2.2.2.1 Physical features

It is a coarse-grained (>100 μm), porous and heterogeneous product that is deposited at the lowest part of coal boilers via airflow. CBA's physical features largely count on size and shape of its particles. The combustion temperature and pulverization rate during the formation of CBA also affect the physical properties [46]. Although CBA is coarser in size than CFA, it can be characterized as a light substance. Generally, the grain size of CBAs is like the particle diameter of fine aggregates having low silt and clay content. It has a low proportion of medium-coarse sand and a higher proportion of fine sand compared with fine aggregates. In addition, CBA has a lower density than CFA (1200–1620 kg/m^3) [47].

Although CBA's physical presence changes from gray to blackish, it has a spongy texture, angular and irregular structure (Fig. 2.3). The spongy and porous nature of CBA makes it daintier and more stiff than ordinary sands. It also has an almost similar particle distribution with fine aggregate passages with a grain size of 0.075–20 mm (Fig. 2.4) [49].

Table 2.5 shows the general physical properties (e.g., refinement modulus, specific gravity, specific surface, and water immersion) of CBA products produced in

Table 2.4 Detailed classification of CFA [41].

Oxides (wt.%)	Coal category			ASTM C618			ACAA (2003)		
	Bitu-minous	Subbituminous	Lignite	Type F	Type C	Type I	Type II	Type III	Type IV
SiO2	20–60	40–60	15–45	50–70	70 <	50 <	35–50	35 <	Very low
Al2O3	5–35	20–30	10–25			Medium	High	–	–
Fe2O3	10–40	4–10	4–15			Medium	Medium	–	–
CaO	1–12	5–30	15–40	20<	10<	7<	Type I <	Very high	Very high
MgO	0–5	1–6	3–10						
Na2O	0–4	0–2	0–6						
K_2O	0–3	0–40	0–4						
SO_3	0–4	0–2	0–10						
LOI	0–15	0–3	0–5	6<	12<				

Figure 2.3 Raw (A) and ground (B) form of CBA [48].

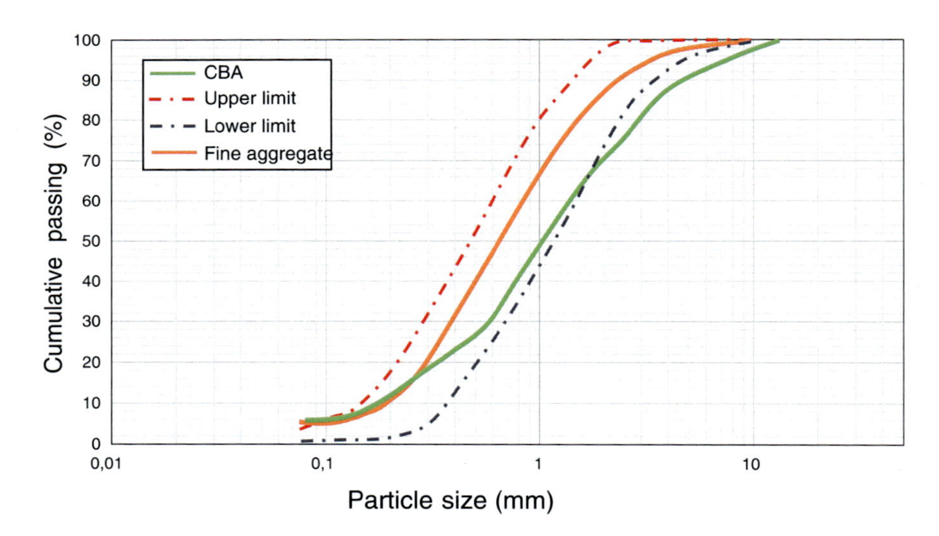

Figure 2.4 Grain size curves of CBA/aggregate.

Table 2.5 Physical properties of CBA by several countries [49].

Properties	**Source of coal**				
	Turkey	**China**	**India**	**Malaysia**	**South Korea**
Specific gravity	2.2	2.2	1.9	2.6	1.9
Specific surface (m^2/kg)	93	–	600	316	–
Fineness modulus	–	1.8	2.4	2.9	5.6
Water content, ω (%)	–	28.9	8.1	1.0	4.1

different countries. CBA's specific gravity changes from 1.39 to 2.60. This relatively low value in the specific gravity is mainly as a result of CBA's porous structure and the type of combustion of coal. In addition, high porosity causes higher water absorption (11.61%–32.23%) compared with natural fine aggregates. CBA's fineness modulus varies between 1.97 and 3.55. These fineness modulus values are in a well agreement for orthodox fine aggregates [50].

Although the size of the surface area of CBA grains varies from country to country, it can be developed by crushing-milling. It improves the response between CBA's grains and chemicals, when high specific surface values will increase the creation of hydration materials such as CSH and CASH. According to previous scientific studies, the largest known surface area value for CBA is 7799 m^2/kg, while the surface area for a typical OPC is approximately 4050 m^2/kg [51]. Fig. 2.5 shows SEM micrographs of CBA particles. From here, it is clearly seen that the fresh CBA's grains are bony, rough, and in the size of sand-gravel. This irregularity mostly varies according to the amount of unburned carbon, anhydrate, and calcite contained in CBA. It is also thought that the methods used in CBA production (e.g., rinsing, sintering, alkali fusion) may cause some degree of microstructural change.

2.2.2.2 Chemical/mineralogical features

CBA's chemical content is directly affected by coal's source, type, and combustion conditions. Especially because of the lignite's burning and sub-bituminous coal, CBA formation rich in calcium and alkali elements occurs [15]. The presence of alkaline elements causes CBAs to display an alkaline (basic) property. The alkalinity of CBA is evaluated in three different categories: slightly alkaline (6.5–7.5), medium basic (7.5–8.5), and high basic (8.5 <) [49].

The CBAs obtained from anthracite coal have low calcium and high iron content. In general, the CBA composition consists of silica (45%–75%) and alumina

Figure 2.5 SEM images under different magnification [52].

(10%–35%). This composition allows CBA to be considered as a pozzolanic product as stated by the ASTM C618 standards. One can also say that CBA is considered as type F according to ASTM C618-15 due to its reactive lime content (~7%) and LOI value (<8%) [48,53]. Silicon oxide (SiO_2) is one of the main oxides that make up CBAs. It can be also found significantly in Fe/Al/Mg/Ca oxides. Among these oxides, SiO_2 and Al_2O_3 are the most important oxides for pozzolanic reactivity and are found in similar proportions to CFA. In addition, the particles of CBAs containing high amounts of SiO_2 generally tend to show hydrophilic properties [52].

When examined from the view of mineralogical features, the key constituents of CBAs are C, Si, Ca, and Al. Among the trace elements, the elements with the highest concentration are Co, Pb, Hg, and Se. In addition, As, Hg, Sb, and Se are defined as volatile trace elements because they are concentrated on the surface of CFA grains rather than CBA [47,54]. Major environmental problems with CFA are usually as a result of occurrence of metals (e.g., Co, Ni, Cu, Zn, and Pb). Thanks to the burning process within coal thermic power plants, phyllosilicates, sulphides, and carbonates, especially quartz, are formed in the CBA. Among these, quartz, silicon phosphate ($SiPO_4$), and mullite ($Al_6Si_2O_3$) are the predominant crystal structures in CBA. In addition, iron (Fe) in CBAs is mostly found in magnetite (Fe_3O_4) and hematite (Fe_2O_3) forms [55,56].

2.3 Characteristics of MSW bottom/fly ash

As stated thoroughly in Chapter 1, two main ash types are created by the incineration of MSW wastes: bottom/fly ashes. MSW bottom ash makes up 25% of the mass of MSW incinerated, and its properties vary depending on the MSW's own characteristics, the type of furnace in which the incineration is carried out, and the combustion efficiency. In addition, the material composed of the chimney vapor is typically named fly ash, and its residues collected in the filters form APC. As the properties of even the ashes obtained from the same incineration unit may differ among themselves, it is extremely important to determine the characteristics of these ashes depending on variable parameters and to interpret them according to works. Physical/chemical/elemental/mineralogical features of MSW ashes are presented as follows.

2.3.1 MSW bottom ash

2.3.1.1 Physical properties

MSW burning bottom ash (MSW-BA), which establishes a great part (80 wt.%) of the total solid waste created from MSW, has a heterogeneous structure with a particle size ranging from fine sizes such as silt and clay to gravel. This type of ash is a combination of different wastes (such as slag, metal, stone, glass, and ceramic and unburnt organic matter) and can be expressed as solid granular. Approximately

60%–90% of the bottom ash fraction consists of sand and gravel, 5%–15% is silt and clay, and the remainder consists of fine particles larger than 10 mm in size [57]. MSW-BA can be up to 100 mm in particle size, but in this case, size reduction and sieving can be applied to meet the standard. Moreover, particle sizes can be subjected to applications such as sieving and grinding according to the different areas where the ash will be evaluated (e.g., as a fine aggregate or cement component) as shown in Fig. 2.6.

MSW-BA involves principally of glass/ceramics (15–30 wt.%), ferrous/non-ferrous metals (2–5 wt.%), minerals (50–70 wt.%) and others [59]. The bottom ash consists primarily of alumina, amorphous silica, and CaO. When coal-BA and MSW-BA are compared, one can say that both ashes morphologically and physically have dark, angular, irregular, and grainy texture, but these properties can differ by counting on the source of ash and the burning's operational situations.

MSW-BA stands out as a lighter material compared with natural sand and aggregates thanks to its bulk specific gravity value. While the bulk specific gravity values are nearly 2.6–2.8 for orthodox aggregates, 1.5–2.2 for fine aggregate/sand-sized materials, and 1.9–2.4 for coarse aggregates, for MSW-BA these values range from 1.8 to 2.8 with an average value of 2.3 [60]. In addition, [61] stated that density values of MSW-BA varied between 2.58 and 2.84 g/cm^3, and therefore these values were less than those of cement. The vapor foams shaped by the effect of warming/cooling throughout the burning of MSW wastes greatly affect the morphology of the bottom ashes and may cause the formation of irregular, coarse-textured pieces. This also changes the porosity of the bottom ash, so MSW-BA has a porous structure, and its water absorption capacity increases with increasing

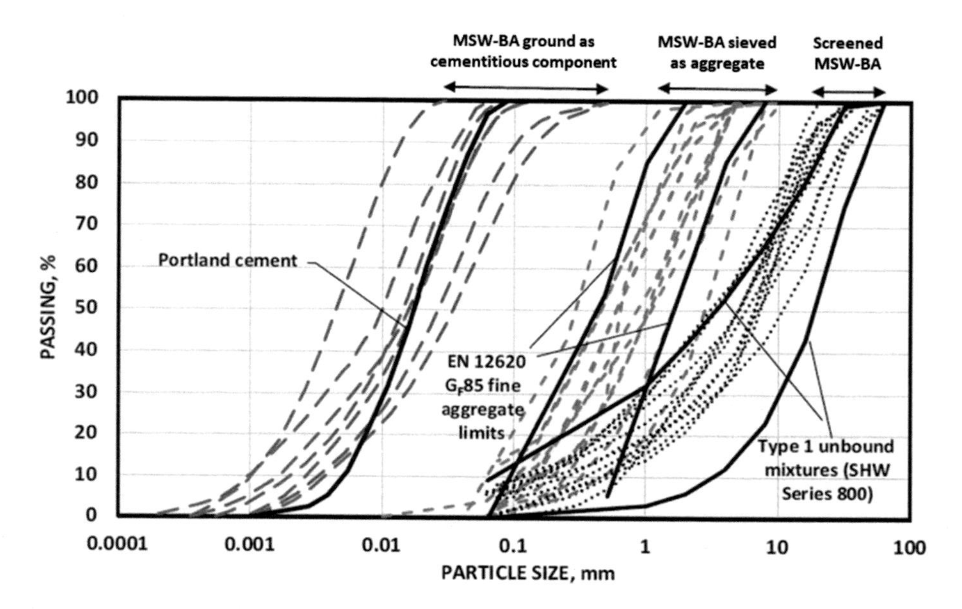

Figure 2.6 Grain size profiles of MSW-BA cremated bottom ash [58].

fineness (due to increase in surface area). The water engagement feature of MSW-BA is greater than that of ordinary aggregates, and it has been found to be between 9.3% and 9.7% for most applications [58,62].

2.3.1.2 Chemical features

Usually, it is very difficult to determine the chemical properties of MSW-BA exactly due to various factors (such as different chemical compositions of the wastes, type, and characteristics of the furnace in which the incineration process is carried out). Therefore, due to these difficulties, some differences may occur between studies on the MSW-BA's chemical features.

Being one of the prevalent alkalines in nature, bottom ash consists of highly toxic organic materials [63]. The MSW-BA's key chemical element is SiO_2, and CaO, Al_2O_3, and Fe_2O_3 are other important components. Determination of bottom ash's oxidized components, especially after grinding process, is extremely important in terms of their use in the making of diverse materials (e.g., cement/ceramics). Table 2.6 presents some chemical results obtained from various studies existing in the literature.

According to Dhir et al. [58], in light of the data obtained from a large number of MSW-BA specimens (total 187), the average contents of the plentiful oxides in the bottom ash are as follows: SiO_2 (silicon oxide, 37.4), CaO (calcium oxide, 22.2%), Al_2O_3 (aluminum oxide, 10.2%) and Fe_2O_3 (iron III oxide, 8.3%), Na_2O (sodium oxide, 2.8%), SO_3 (sulfur trioxide, 2.7%), P_2O_5 (phosphorus pentoxide, 2.3%), MgO (2.0%), and K_2O (potassium oxide, 1.4%).

Amounts of the most abundant (primary) elements (Si, Ca, Fe, and Al) in the bottom ash are closely interrelated to grain size and vary as stated by incinerators. Although the main components that make up the MSW-BA are Al, Ca, Fe, K, Mg, Na, and S, some trace basics (e.g., Ba, Co, Cr, Cu, Mn, Ni, Se, Sr, and V) are also present in small amounts. Apart from these, volatile elements (e.g., Cd, Zn, Pb, As, and Sb) can be uninvolved from bottom ash because of evaporation (from MSW ash). Although bottom ash is fewer toxic than fly ash and air trash control, it covers some harmful metals (e.g., Zn, Cr, Pb, Cu, Ni, Cd etc.). With the changing environmental conditions, these metals could be separated from bottom ash and cause a dangerous environment. This situation, which is defined as leaching, is closely

Table 2.6 Chemical features of MSW-BA (wt.%).

Composite	SiO_2	Al_2O_3	Fe_2O_3	CaO	MgO	SO_3	Na_2O	K_2O
Gao et al. [64]	19.12	12.04	9.31	43.12	2.12	2.40	2.36	0.85
Kuo and Gao [65]	24.84	8.42	30.63	27.80	1.50	–	–	–
Tang et al. [61]	35.98	9.00	11.54	19.34	1.81	4.95	1.35	1.15
Zhen et al. [66]	22.75	4.16	5.91	24.40	2.32	2.96	1.71	1.26

related to the physical property (e.g., grain size) of MSW-BA and the water content in the environment.

2.3.1.3 Mineralogical features

MSW bottom ash's mineralogical properties are of major consequences considering the impact of this ash type on environment, as well as the tendency of the ground forms to enter diverse reactions and the dynamism of the components it contains. XRD analysis is a frequently employed method for mineral determination of MSW-BA. Key crystalline phases of MSW-BA are SiO_2 (most abundant), $CaCO_3$, Fe_2O_3, Fe_3O_4, and gehlenite ($Ca_2Al[AlSiO_7]$). There also exist anhydrite, calcium chloride hydroxide, and salts linked with Cl form other main crystals [61]. In addition, there are more than 30 types ofminerals (e.g., aluminates, silicates, sulfates, oxides, phosphaottom ash. MSW-FA tes, and diverse salts) that are defined as less common [60,67].

As a result of their XRD analysis, they observed various minerals, such as quartz and calcite. Due to the decline in grain size, the peak diffraction density of calcite increased although that of quartz diminished [68]. In addition, it has been interpreted that some minerals (Fe_2O_3 and Al_2O_3) contained in MSW-BA are formed by oxidation of wastes during the combustion process. In addition, it can be said that there is a minor quantity of unburned C in bottom ash.

MSW-BA may exhibit diverse mineralogical features under different environmental conditions. Bottom ash stored outdoors may suffer carbonation or biodegradation because of its reaction with CO_2 in the air. Thanks to the changing mineralogy of bottom ash, pH properties of ash may change and movement of different components in the ash may be restricted. This situation, which is especially effective on leaching properties, determines the environmental impact of bottom ash, which has a large amount of heavy metal content, although it is less polluting than fly ash.

2.3.2 MSW fly ash

2.3.2.1 Physical properties

MSW fly ash (MSW-FA) constitutes approximately 3% of MSW's combustion by-products (small-scale burned or partially burned organic particles) and approximately 5%–20% of the total waste mass volume burned [69,70]. The amount of FA produced is highly affected by the sort of furnace, such as bottom ash. MSW-FA has a gray or dark gray powdery appearance with minor grain diameters (1–500 μm) and outsized specific surface/strong hydrophilicity. In addition, MSW-FA has low water content, unequal form, and irregular grain size, with a mean grain size changing between 10 and 150 μm, often <250 μm ([71,72]). As stated by Hwang and Ro [73], FA particle size ranges from 0.3 to 2.5 μm. The fact that FA grains are small can be shown as a reason that facilitates the dispersion of these ashes into air and affects human health [74].

Fly ash is mostly spherical and may have different crystal formations depending on burning conditions. Fig. 2.7 shows SEM images of some MSW-FA samples found in the literature. By looking at SEM images, one can interpret that the creation of MSW-FA contains loosely packed grains, and there may be large openings between one another. This morphology can trigger metals to slide through the ash particles and leak into the environment. The MSW-FA's density varies between 1.7 and 2.4 t/m^3, which is less than that of standard sand (2.65 t/m^3). As stated by López-Zaldívar et al. [77], MSW-FA can engage water by 30% of its weight. The MSW-FA's high specific surface enables it to be easily adsorbed on surfaces of some volatile metals such as Pb, Cd [78]. In addition, MSW-FA has a permeable construction consisting of minor slums, which facilitates easy penetration of metals in MSW-FA into the atmosphere.

2.3.2.2 Chemical features

MSW-FA's chief elements are Ca, Si, A1, C1, Na, K, S, Fe, Mg, and P [70]. In addition to these elements, the high amount of solvable constituents in MSW-FA causes this ash type to not exhibit chemically inert behavior. Unlike bottom ash, MSW-FA contains higher concentrations of metals (e.g., Pb, Cd, Hg, Mo, Ni, and Se), salts, and others (e.g., dioxins, furans, sulfate, and chloride) due to evaporation/condensation of different elements during combustion. and acids [79,80].

MSW-FA cannot be used as a raw product thanks to its high values of harmful soluble components and is seen as a risky waste in numerous republics. Consequently, either storage or disposal of MSW-FA may raise some concerns for the environment and human health. Because any error that may occur during the storage of FA will cause harmful heavy metal pollutants contained in FA to harm the environment. Countless works have been undertaken on determination of fly ash's chemical constituents. Table 2.7 displays some of the results present in the literature.

MSW-FA's key oxide constituents are SiO_2, CaO, and Al_2O_3, and the type of incinerator used is one of the most vital issues influencing the structure of MSW-

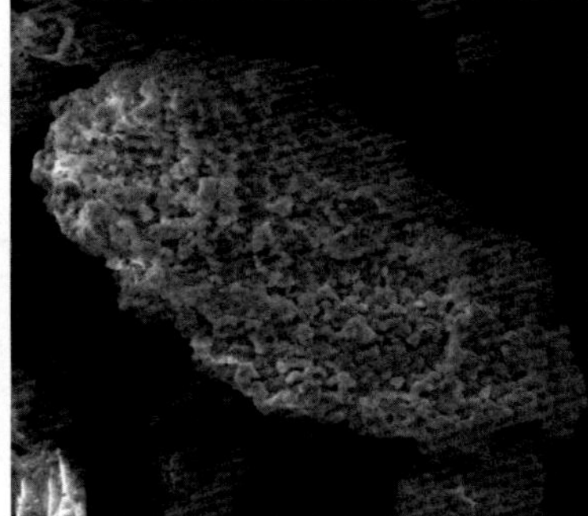

Figure 2.7 MSW-FA's SEM view [75,76].

Table 2.7 Chemical compositions of MSW-FA (wt.%).

Oxide analysis	SiO_2	Al_2O_3	Fe_2O_3	CaO	MgO	SO_3	Na_2O	K_2O
Fan et al. [71]	2.10	0.40	0.80	45.30	1.20	8.60	9.90	8.20
Chen et al. [81]	3.10	1.20	–	41.80	1.70	4.20	7.30	–
Xinghua et al. [82]	23.00	7.30	2.10	36.90	3.30	14.80	–	–
Cristelo et al. [83]	3.25	2.31	0.39	38.70	1.67	4.59	11.57	8.35

FA [84]. When the data in the literature are examined, it can be said that the high CaO content of FA is caused by the excessive lime water used to reduce (neutralize) the harmful (acidic) gases released after the combustion process, and the chlorine content is caused by the high amount of plastic and kitchen waste.

2.3.2.3 Mineralogical features

MSW-FA's mineralogical structure usually reflects a blend of crystalline/amorphous phases. FA's minerals are formed because of various events (e.g., evaporation, melting, crystallization, compression, and precipitation) that happen through the fire of MSW. Key difficulties in determining the mineralogical structure of FA are thanks to the complexity/heterogeneity of ash mineralogy and MSW's diversity.

There are many studies examining the mineralogical structure of MSW-FA and indicate that the ash is mainly composed of amorphous chlorine salts and sulphates [71,85]. XRD results of fly ash from two different sites [86] showed that it entails shapeless phases rich in calcareous raw materials ($Ca(OH)_2$, CaClOH, $CaCO_3$, $CaSO_4$) and chloride salts (NaCl, KCl). Wang et al. [87] stated that CaClOH is a transitional stage that occurs through the engagement of HCl and that $CaCO_3$ may be formed due to the partial, early carbonization of the ash. In addition, Cristelo et al. [83] characterized MSW-FA with their XRD analysis, and as a result, fly ash was composed of silvite (KCl), halite (NaCl) and calcite ($CaCO_3$), and traces of $Ca(OH)_2$, anhydrite ($CaSO_4$) and CaO components have reached the conclusion. It can also be said that MSW-FA is rich in Ca and Cl, and these elements form the basis of the mineral phases [88].

2.4 Characteristics of biomass bottom/fly ash

Biomass ashes (BFA and BBA) produced in biomass thermic power facilities refer to the residues formed after the burning of biomass. They involve various mineral materials in oxidized form and unburned organic materials. It is possible to separate these organic substances (up to 20%) into four key constituents: carbohydrates, lignin, lipids and proteins, and most biomass wastes can be said to be lignocellulosic-based [8,89]. The amount, yield, and properties of the ash produced vary depending on various factors (such as biomass type, burning chamber type, incineration temperature, and operating conditions) [90,91]. BBA deposited under the combustion chamber grate is generally the heaviest part, such as dark sand and forming by-products, while BFA is the thinnest part suspended in flue gases because of combustion and gasification.

Physical features of biomass ashes (for example, grain size, density, water absorption, and moisture content) depend on the properties of the material to be burnt and the sort of reservoir in which biofuel/biomass is incinerated [92]. Biomass ash, unlike coal burning ash, consists of irregularly shaped particles. The reason why biomass ashes do not form globally could be explained by the low heats of biomass burning in comparison with coal combustion chambers. Particle size as well as grain shapes differs between BBA and BFA: the maximum nominal size of BFA is about 200 μm (like coal FA), while the maximum nominal dimensions of BBA can range from 2 to 9 mm [93]. [94] found the average grain size of BFA of wood/forest and eucalyptus bark obtained using different combustion technologies as 52.92 μm and 16.04 μm, respectively. The bulk density, which is another physical property, was respectively designated to be 2.59 g/cm^3 and 2.54 g/cm^3. Eliche-Quesada et al. [95] stated the average grain size of BBA shaped thanks to burning a blend of pine and olive pruning as 258 μm. Maschio et al. [96] and Modolo et al. [97] also found the densities of BBA of wood and eucalyptus bark to be 2.60 g/cm^3 and 2.65 g/cm^3, respectively.

The moisture content of biomass wastes is another important physical property, which is usually the limit value of 30%, and residues below this value can be expressed as dry and above this value as wet [98]. Humidity values for rice straw, pineapple peel, and mixed food waste can be given as examples, and these values are 11.7%, 79.7%, and 76.5%, respectively [99–101]. Chemical features of BFA/BBA from biomass burning rely on biomass properties/source, physical structure of feedstock and furnace, combustion conditions, and applied technique [93,102]. Waste fragments are largely rich in Si, Fe, Al, K, Ca, P, Mg, and Na. The compositions of these elements cannot be considered independent of inorganic compounds found in biomass waste [103]. Several works have been conducted examining the chemical compositions of BBA and BFA, some of which are presented in Table 2.8. One can interpret that the ash's properties vary widely and therefore require special studies according to the type of biomass waste.

It is possible to divide BFA into classes (C and F) similar to coal-FA. BFA, which contains components such as SiO_2, CaO, Al_2O_3, MgO, and Fe_2O_3, is classified according to the percentage of CaO it contains. For example, if the CaO percentage is about 8%, it is called C, and if it is lower than this value, it is called F class [93].

The ashes shaped due to the burning of biomass contain poisonous elements (Pb, Cd, Cr, Cu, As, Ni, Zn), which are soluble in water, and elements existing in plants. Although these elements are especially effective in the use of BFA in different

Table 2.8 Biomass ash's chemical features [104–106].

Compounds	BBA (wt.%)		BFA (wt.%)	
	Olive plant	Forest biomass	Rice husk	Corn cobs
SiO_2	6.84	72.20	94.38	27.65
Al_2O_3	2.73	3.32	0.21	2.49
Fe_2O_3	1.39	0.78	0.22	1.55
CaO	31.41	17.16	0.97	13.19
MgO	2.45	1.97	0.19	2.05
SO_3	–	–	0.92	7.14
K_2O	12.31	0.75	–	–

applications, they are vital as they can cause health problems. The mineralogical structure of biomass ash consists of diverse mineral phases (such as silicate, sulphate, phosphate, carbonate, chloride) [106]. The oxide and hydroxide groups contained in BBA/BFA generally occur due to the decomposition/oxidation of carbonates, sulphates, phosphates, nitrates, and chlorides. The mineralogical structure of biomass can also be expressed by the existence of sulphates and phosphates, which are designed by the interaction of the oxides of organic substances it has, and the gases released during the combustion stage. For example, Fe_2O_3 (hematite) is formed because of the oxidation of organic iron.

Major minerals in various works: arcanite and calcite for BA of olive [107]; CaO, SiO_2 and $CaCO_3$ for BA of pine-olive pruning [95]; quartz for BA of waste paper [108]; quartz + albite for FA of cornstalks + peanut shells + wood chips [109]; quartz + portlandite for FA of wood [110] was found. Considering these cases, one can close that mineralogical structure of BBA and BFA is shaped and varies relying on biomass type/burning processes.

References

[1] X. Li, C. Bai, Y. Qiao, X. Wang, K. Yang, P. Colombo, Preparation, properties and applications of fly ash-based porous geopolymers: a review, Journal of Cleaner Production 359 (2022) 132043.

[2] T. Kasap, E., Yilmaz, M. Sari, Effects of mineral additives and age on microstructure evolution and durability properties of sand-reinforced cementitious mine backfills, Construction and Building Materials 352 (2022) 129079.

[3] K.K. Le Ping, C.B. Cheah, J.J. Liew, R. Siddique, W. Tangchirapat, M.A.B.M. Johari, Coal bottom ash as constituent binder and aggregate replacement in cementitious and geopolymer composites: a review, Journal of Building Engineering 52 (2022) 104369.

[4] M. Rafieizonooz, E. Khankhaje, S. Rezania, Assessment of environmental and chemical properties of coal ashes including fly ash and bottom ash, and coal ash concrete, Journal of Building Engineering 49 (2022) 104040.

[5] M. Rathnayake, P. Julnipitawong, S. Tangtermsirikul, P. Toochinda, Utilization of coal fly ash and bottom ash as solid sorbents for sulfur dioxide reduction from coal fired power plant: life cycle assessment and applications, Journal of Cleaner Production 202 (2018) 934–945.
[6] F.C. Silva, N.C. Cruz, L.A. Tarelho, S.M. Rodrigues, Use of biomass ash-based materials as soil fertilisers: critical review of the existing regulatory framework, Journal of Cleaner Production 214 (2019) 112–124.
[7] M.A. Munawar, A.H. Khoja, M. Hassan, R. Liaquat, S.R. Naqvi, M.T. Mehran, et al., Biomass ash characterization, fusion analysis and its application in catalytic decomposition of methane, Fuel 285 (2021) 119107.
[8] K. Wang, J.W. Tester, Sustainable management of unavoidable biomass wastes, Green Energy and Resources 1 (1) (2023) 100005.
[9] C. Belviso, State-of-the-art applications of fly ash from coal and biomass: a focus on zeolite synthesis processes and issues, Progress in Energy and Combustion Science 65 (2018) 109–135.
[10] P. Tang, H.J.H. Brouwers, Integral recycling of municipal solid waste incineration (MSWI) bottom ash fines (0–2 mm) and industrial powder wastes by cold-bonding pelletization, Waste Management 62 (2017) 125–138.
[11] X. Fan, R. Yuan, M. Gan, Z. Ji, Z. Sun, Subcritical hydrothermal treatment of municipal solid waste incineration fly ash: a review, Science of the Total Environment 865 (2022) 160745.
[12] Y. Zhang, L. Wang, L. Chen, B. Ma, Y. Zhang, W. Ni, et al., Treatment of municipal solid waste incineration fly ash: state-of-the-art technologies and future perspectives, Journal of Hazardous Materials 411 (2021) 125132.
[13] T. Lan, Y. Meng, T. Ju, Z. Chen, Y. Du, Y. Deng, et al., Synthesis and application of geopolymers from municipal waste incineration fly ash (MSWI FA) as raw ingredient-a review, Resources, Conservation and Recycling 182 (2022) 106308.
[14] L. Zheng, W. Wang, X. Gao, Solidification and immobilization of MSWI fly ash through aluminate geopolymerization: based on partial charge model analysis, Waste Management 58 (2016) 270–279.
[15] M.I. Al Biajawi, R. Embong, K. Muthusamy, N. Ismail, I.I. Obianyo, Recycled coal bottom ash as sustainable materials for cement replacement in cementitious composites: a review, Construction and Building Materials 338 (2022) 127624.
[16] ASTM D4326-13, Standard Test Method for Major and Minor Elements in Coal and Coke Ash by X-Ray Fluorescence, (2013). https://www.astm.org/Standards/D4326.
[17] ASTM-D388-18a, Standard Classification of Coals by Rank, (2018). https://www.astm.org/DATABASE.CART/HISTORICAL/D388-18A.
[18] I. Cavusoglu, E., Yilmaz, A.O. Yilmaz, Additivity effect on properties of cemented coal fly ash backfill containing water-reducing admixtures, Construction and Building Materials 267 (2021) 121021.
[19] T. Ju, Y. Meng, S. Han, L. Lin, J. Jiang, On the state of the art of crystalline structure reconstruction of coal fly ash: a focus on zeolites, Chemosphere 283 (2021) 131010.
[20] Z. Li, G. Xu, X. Shi, Reactivity of coal fly ash used in cementitious binder systems: a state-of-the-art overview, Fuel 301 (2021) 121031.
[21] Z.T. Yao, X.S. Ji, P.K. Sarker, J.H. Tang, L.Q. Ge, M.S. Xia, et al., A comprehensive review on the applications of coal fly ash, Earth-Science Reviews 141 (2015) 105–121.
[22] M. Amran, R. Fediuk, G. Murali, S. Avudaiappan, T. Ozbakkaloglu, N. Vatin, et al., Fly ash-based eco-efficient concretes: a comprehensive review of the short-term properties, Materials 14 (15) (2021) 4264.

[23] E. Grabias-Blicharz, W. Franus, A critical review on mechanochemical processing of fly ash and fly ash-derived materials, Science of the Total Environment 860 (2022) 160529.
[24] M. Sari, E. Yilmaz, T., Kasap, N.U. Guner, Strength and microstructure evolution in cemented mine backfill with low and high pH pyritic tailings: effect of mineral admixtures, Construction and Building Materials 328 (2022) 127109.
[25] R.S. Blissett, N.A. Rowson, A review of the multi-component utilisation of coal fly ash, Fuel 97 (2012) 1−23.
[26] A. Dindi, D.V. Quang, L.F. Vega, E. Nashef, M.R. Abu-Zahra, Applications of fly ash for CO2 capture, utilization, and storage, Journal of CO2 Utilization 29 (2019) 82−102.
[27] G. Xu, X. Shi, Characteristics and applications of fly ash as a sustainable construction material: a state-of-the-art review, Resources, Conservation and Recycling 136 (2018) 95−109.
[28] A.R. Gollakota, V. Volli, C.M. Shu, Progressive utilisation prospects of coal fly ash: a review, Science of the Total Environment 672 (2019) 951−989.
[29] K. Gao, M.C. Iliuta, Trends and advances in the development of coal fly ash-based materials for application in hydrogen-rich gas production: a review, Journal of Energy Chemistry 73 (2022) 485−512.
[30] E.M. Kumar, P. Perumal, K. Ramamurthy, Alkali-activated aerated blends: interaction effect of slag with low and high calcium fly ash, Journal of Material Cycles and Waste Management 24 (4) (2022) 1378−1395.
[31] S. Gjyli, A. Korpa, V. Teneqja, D. Siliqi, C. Belviso, Siliceous fly ash utilization conditions for zeolite synthesis, Environmental Sciences Proceedings 6 (1) (2021) 24.
[32] I. Cavusoglu, E., Yilmaz, A.O. Yilmaz, Sodium silicate effect on setting properties, strength behavior and microstructure of cemented coal fly ash backfill, Powder Technology 384 (2021) 17−28.
[33] K.M. Klima, K. Schollbach, H.J.H. Brouwers, Q. Yu, Thermal and fire resistance of class F fly ash based geopolymers−a review, Construction and Building Materials 323 (2022) 126529.
[34] D.K. Nayak, P.P. Abhilash, R. Singh, R. Kumar, V. Kumar, Fly ash for sustainable construction: a review of fly ash concrete and its beneficial use case studies, Cleaner Materials 6 (2022) 100143.
[35] W. Franus, M.M. Wiatros-Motyka, M. Wdowin, Coal fly ash as a resource for rare earth elements, Environmental Science and Pollution Research 22 (2015) 9464−9474.
[36] D.J. Arent, A. Wise, R. Gelman, The status and prospects of renewable energy for combating global warming, Energy Economics 33 (4) (2011) 584−593.
[37] S. Dandin, M. Kulkarni, M. Wagale, S. Sathe, A review on the geotechnical response of fly ash-colliery spoil blend and stability of coal mine dump, Cleaner Waste Systems 3 (2022) 100040.
[38] L. Wen, C. Yan, X. Yang, L. Li, Effect of fly ash addition on slurry dewatering by electroosmosis combined with mechanical pressure, Drying Technology 40 (5) (2022) 827−834.
[39] S.S. Alterary, N.H. Marei, Fly ash properties, characterization, and applications: a review, Journal of King Saud University-Science 33 (6) (2021) 101536.
[40] S.V. Vassilev, C.G. Vassileva, A new approach for the classification of coal fly ashes based on their origin, composition, properties, and behaviour, Fuel 86 (10-11) (2007) 1490−1512.
[41] Z. Zimar, D. Robert, A. Zhou, F. Giustozzi, S. Setunge, J. Kodikara, Application of coal fly ash in pavement subgrade stabilisation: a review, Journal of Environmental Management 312 (2022) 114926.
[42] J. Xie, O. Kayali, Effect of superplasticiser on workability enhancement of class F and class C fly ash-based geopolymers, Construction and Building Materials 122 (2016) 36−42.

[43] N.H.M. Nasir, F. Usman, A. Saggaf, Development of composite material from recycled polyethylene terephthalate and fly ash: four decades progress review, Current Research in Green and Sustainable Chemistry, 5, 2022, p. 100280.
[44] B. Makgabutlane, M.S. Maubane-Nkadimeng, N.J. Coville, S.D. Mhlanga, Plastic-fly ash waste composites reinforced with carbon nanotubes for sustainable building and construction applications: a review, Results in Chemistry 4 (2022) 100405.
[45] T. Hemalatha, A. Ramaswamy, A review on fly ash characteristics—towards promoting high volume utilization in developing sustainable concrete, Journal of Cleaner Production 147 (2017) 546—559.
[46] S.A. Mangi, M.H.W. Ibrahim, N. Jamaluddin, M.F. Arshad, R.P. Jaya, Short-term effects of sulphate and chloride on the concrete containing coal bottom ash as supplementary cementitious material, Engineering Science and Technology, An International Journal 22 (2) (2019) 515—522.
[47] N. Singh, A. Bhardwaj, Reviewing the role of coal bottom ash as an alternative of cement, Construction and Building Materials 233 (2020) 117276.
[48] N. Ankur, N. Singh, Performance of cement mortars and concretes containing coal bottom ash: a comprehensive review, Renewable and Sustainable Energy Reviews 149 (2021) 111361.
[49] H. Zhou, R. Bhattarai, Y. Li, B. Si, X. Dong, T. Wang, et al., Towards sustainable coal industry: turning coal bottom ash into wealth, Science of the Total Environment 804 (2022) 149985.
[50] K. Muthusamy, M.H. Rasid, G.A. Jokhio, A.M.A. Budiea, M.W. Hussin, J. Mirza, Coal bottom ash as sand replacement in concrete: a review, Construction and Building Materials 236 (2020) 117507.
[51] S. Gooi, A.A. Mousa, D. Kong, A critical review and gap analysis on the use of coal bottom ash as a substitute constituent in concrete, Journal of Cleaner Production 268 (2020) 121752.
[52] M. Rafieizonooz, J. Mirza, M.R. Salim, M.W. Hussin, E. Khankhaje, Investigation of coal bottom ash and fly ash in concrete as replacement for sand and cement, Construction and Building Materials 116 (2016) 15—24.
[53] B. Zhang, C.S. Poon, Use of furnace bottom ash for producing lightweight aggregate concrete with thermal insulation properties, Journal of Cleaner Production 99 (2015) 94—100.
[54] S. Abbas, U. Arshad, W. Abbass, M.L. Nehdi, A. Ahmed, Recycling untreated coal bottom ash with added value for mitigating alkali—silica reaction in concrete: a sustainable approach, Sustainability 12 (24) (2020) 10631.
[55] T.C. Esteves, R. Rajamma, D. Soares, A.S. Silva, V.M. Ferreira, J.A. Labrincha, Use of biomass fly ash for mitigation of alkali-silica reaction of cement mortars, Construction and Building Materials 26 (1) (2012) 687—693.
[56] H.K. Kim, Utilization of sieved and ground coal bottom ash powders as a coarse binder in high-strength mortar to improve workability, Construction and Building Materials 91 (2015) 57—64.
[57] X. Dou, F. Ren, M.Q. Nguyen, A. Ahamed, K. Yin, W.P. Chan, et al., Review of MSWI bottom ash utilization from perspectives of collective characterization, treatment and existing application, Renewable and Sustainable Energy Reviews 79 (2017) 24—38.
[58] R.K. Dhir, J.D. Brito, C.J., Lynn, R.V. Silva, Municipal solid waste composition, incineration, processing and management of bottom ashes, Sustainable Construction Materials (2018) 31—90.

[59] A. Nithiya, A., Saffarzadeh, T. Shimaoka, Hydrogen gas generation from metal aluminum-water interaction in municipal solid waste incineration (MSWI) bottom ash, Waste Management 73 (2018) 342–350.
[60] C.J. Lynn, G.S., Ghataora, R.K.D. Obe, Municipal incinerated bottom ash (MIBA) characteristics and potential for use in road pavements, International Journal of Pavement Research and Technology 10 (2) (2017) 185–201.
[61] P. Tang, M.V.A. Florea, P., Spiesz, H.J.H. Brouwers, Application of thermally activated municipal solid waste incineration (MSWI) bottom ash fines as binder substitute, Cement and Concrete Composites 70 (2016) 194–205.
[62] C.J. Lynn, R.K.D., Obe, G.S. Ghataora, Municipal incinerated bottom ash characteristics and potential for use as aggregate in concrete, Construction and Building Materials 127 (2016) 504–517.
[63] Y. Li, L., Hao, X. Chen, Analysis of MSWI bottom ash reused as alternative material for cement production. Procedia, Environmental Sciences 31 (2016) 549–553.
[64] X. Gao, B. Yuan, Q.L., Yu, H.J.H. Brouwers, Characterization and application of municipal solid waste incineration (MSWI) bottom ash and waste granite powder in alkali activated slag, Journal of Cleaner Production 164 (2017) 410–419.
[65] W.T. Kuo, Z.C. Gao, Engineering properties of controlled low-strength materials containing bottom ash of municipal solid waste incinerator and water filter silt, Applied Sciences 8 (8) (2018) 1377.
[66] G. Zhen, X. Lu, Y. Zhao, J. Niu, X. Chai, L. Su, et al., Characterization of controlled low-strength material obtained from dewatered sludge and refuse incineration bottom ash: mechanical and microstructural perspectives, Journal of Environmental Management 129 (2013) 183–189.
[67] T. Astrup, A. Muntoni, A. Polettini, R. Pomi, T., Van Gerven, A. Van Zomeren, Treatment and Reuse of Incineration Bottom Ash, Environmental Materials and Waste Academic Press, 2016, pp. 607–645.
[68] Y. Xia, P. He, L., Shao, H. Zhang, Metal distribution characteristic of MSWI bottom ash in view of metal recovery, Journal of Environmental Sciences 52 (2017) 178–189.
[69] H.M. Alhassan, A.M. Tanko, Characterization of solid waste incinerator bottom ash and the potential for its use, International Journal of Engineering Research and Applications 2 (4) (2012) 516–522.
[70] B. Huang, M. Gan, Z. Ji, X. Fan, D. Zhang, X. Chen, et al., Recent progress on the thermal treatment and resource utilization technologies of municipal waste incineration fly ash: a review, Process Safety and Environmental Protection 159 (2022) 547–565.
[71] C. Fan, B. Wang, H. Ai, Y., Qi, Z. Liu, A comparative study on solidification/stabilization characteristics of coal fly ash-based geopolymer and Portland cement on heavy metals in MSWI fly ash, Journal of Cleaner Production 319 (2021) 128790.
[72] D. Bernasconi, C. Caviglia, E. Destefanis, A. Agostino, R. Boero, N. Marinoni, et al., Influence of speciation distribution and particle size on heavy metal leaching from MSWI fly ash, Waste Management 138 (2022) 318–327.
[73] H. Hwang, C.U. Ro, Single-particle characterization of municipal solid waste (MSW) ash particles using low-Z particle electron probe X-ray microanalysis, Atmospheric Environment 40 (16) (2006) 2873–2881.
[74] M.H. Wu, C.L. Lin, W.C., Huang, J.W. Chen, Characteristics of pervious concrete using incineration bottom ash in place of sandstone graded material, Construction and Building Materials 111 (2016) 618–624.

[75] M. Čarnogurská, M. Lázár, M. Puškár, M. Lengyelová, J., Václav, Ľ. Širillová, Measurement and evaluation of properties of MSW fly ash treated by plasma, Measurement 62 (2015) 155–161.
[76] S. Rémond, P., Pimienta, D.P. Bentz, Effects of the incorporation of municipal solid waste incineration fly ash in cement pastes and mortars: i. experimental study, Cement and Concrete Research 32 (2) (2002) 303–311.
[77] O. López-Zaldívar, R.V. Lozano-Díez, A., Verdú-Vázquez, N. Llauradó-Pérez, Effects of the addition of inertized MSW fly ash on calcium aluminate cement mortars, Construction and Building Materials 157 (2017) 1106–1116.
[78] G.M. Kirkelund, L., Skevi, L.M. Ottosen, Electrodialytically treated MSWI fly ash use in clay bricks, Construction and Building Materials 254 (2020) 119286.
[79] C.H. Lam, A.W. Ip, J.P., Barford, G. McKay, Use of incineration MSW ash: a review, Sustainability 2 (7) (2010) 1943–1968.
[80] J. Yu, Y. Qiao, L. Jin, C. Ma, N., Paterson, L. Sun, Removal of toxic and alkali/alkaline earth metals during co-thermal treatment of two types of MSWI fly ashes in China, Waste Management 46 (2015) 287–297.
[81] D. Chen, Y. Zhang, Y. Xu, Q. Nie, Z. Yang, W., Sheng, et al., Municipal solid waste incineration residues recycled for typical construction materials—a review, RSC Advances 12 (10) (2022) 6279–6291.
[82] H. Xinghua, Z., Shujing, J.Y. Hwang, Physical and chemical properties of MSWI fly ash, Characterization of Minerals, Metals, and Materials 2016 (2016) 451–459.
[83] N. Cristelo, L. Segadães, J. Coelho, B. Chaves, N.R., Sousa, M. de Lurdes Lopes, Recycling municipal solid waste incineration slag and fly ash as precursors in low-range alkaline cements, Waste Management 104 (2020) 60–73.
[84] Z. Chen, S. Lu, M. Tang, J. Ding, A. Buekens, J. Yang, et al., Mechanical activation of fly ash from MSWI for utilization in cementitious materials, Waste Management 88 (2019) 182–190.
[85] J. Tan, J. De Vlieger, P. Desomer, J. Cai, J. Li, Co-disposal of construction and demolition waste (CDW) and municipal solid waste incineration fly ash (MSWI FA) through geopolymer technology, Journal of Cleaner Production 362 (2022) 132502.
[86] Z. Liu, J. Li, L. Hu, X. Zhang, S., Ding, H. Li, Strength and environmental behaviours of municipal solid waste incineration fly ash for cement-stabilised soil, Sustainability 15 (1) (2023) 364.
[87] L. Wang, Y., Jin, Y. Nie, Investigation of accelerated and natural carbonation of MSWI fly ash with a high content of Ca, Journal of Hazardous Materials 174 (1-3) (2010) 334–343.
[88] M. Nikravan, A.A., Ramezanianpour, R. Maknoon, Study on physiochemical properties and leaching behavior of residual ash fractions from a municipal solid waste incinerator (MSWI) plant, Journal of Environmental Management 260 (2020) 110042.
[89] R. Melotti, E. Santagata, M. Bassani, M., Salvo, S. Rizzo, A preliminary investigation into the physical and chemical properties of biomass ashes used as aggregate fillers for bituminous mixtures, Waste Management 33 (9) (2013) 1906–1917.
[90] M.A. Munawar, A.H. Khoja, S.R. Naqvi, M.T. Mehran, M. Hassan, R. Liaquat, et al., Challenges and opportunities in biomass ash management and its utilization in novel applications, Renewable and Sustainable Energy Reviews 150 (2021) 111451.
[91] X. Yao, K., Xu, Y. Liang, Comparative analysis of the physical and chemical properties of different biomass ashes produced from various combustion conditions, BioResources 12 (2) (2017) 3222–3235.
[92] J.I. Odzijewicz, E. Wołejko, U. Wydro, M., Wasil, A. Jabłońska-Trypuć, Utilization of ashes from biomass combustion, Energies 15 (24) (2022) 9653.

[93] F. Agrela, M. Cabrera, M.M. Morales, M., Zamorano, M. Alshaaer, Biomass fly ash and biomass bottom ash, New Trends in Eco-Efficient and Recycled Concrete, Woodhead Publishing, 2019, pp. 23–58.
[94] R. Rajamma, L. Senff, M.J. Ribeiro, J.A. Labrincha, R.J. Ball, G.C. Allen, et al., Biomass fly ash effect on fresh and hardened state properties of cement based materials, Composites Part B: Engineering 77 (2015) 1–9.
[95] D. Eliche-Quesada, M.A., Felipe-Sesé, M.J. Fuentes-Sánchez, Biomass bottom ash waste and by-products of the acetylene industry as raw materials for unfired bricks, Journal of Building Engineering 38 (2021) 102191.
[96] S. Maschio, G. Tonello, L., Piani, E. Furlani, Fly and bottom ashes from biomass combustion as cement replacing components in mortars production: rheological behaviour of the pastes and materials compression strength, Chemosphere 85 (4) (2011) 666–671.
[97] R.C.E. Modolo, V.M. Ferreira, L.A. Tarelho, J.A. Labrincha, L., Senff, L. Silva, Mortar formulations with bottom ash from biomass combustion, Construction and Building Materials 45 (2013) 275–281.
[98] G. Ravindiran, P. Saravanan, R.M., Jeyaraju, J. Josephraj, Water—conventional and novel treatment methods, Solar-Driven Water Treatment, Academic Press, 2022, pp. 37–66.
[99] B. Biswas, N. Pandey, Y. Bisht, R. Singh, J., Kumar, T. Bhaskar, Pyrolysis of agricultural biomass residues: comparative study of corn cob, wheat straw, rice straw and rice husk, Bioresource Technology 237 (2017) 57–63.
[100] W.H. Chen, Y.Y. Lin, H.C., Liu, S. Baroutian, Optimization of food waste hydrothermal liquefaction by a two-step process in association with a double analysis, Energy 199 (2020) 117438.
[101] N.U. Saqib, S., Baroutian, A.K. Sarmah, Physicochemical, structural and combustion characterization of food waste hydrochar obtained by hydrothermal carbonization, Bioresource Technology 266 (2018) 357–363.
[102] N.C. Cruz, F.C. Silva, L.A., Tarelho, S.M. Rodrigues, Critical review of key variables affecting potential recycling applications of ash produced at large-scale biomass combustion plants, Resources, Conservation and Recycling 150 (2019) 104427.
[103] T.P. da Costa, P. Quinteiro, L.A. Tarelho, L., Arroja, A.C. Dias, Life cycle assessment of woody biomass ash for soil amelioration, Waste Management 101 (2020) 126–140.
[104] R.E. Modolo, T. Silva, L. Senff, L.A.C. Tarelho, J.A. Labrincha, V.M. Ferreira, et al., Bottom ash from biomass combustion in BFB and its use in adhesive-mortars, Fuel Processing Technology 129 (2015) 192–202.
[105] V. Sklivaniti, P.E. Tsakiridis, N.S. Katsiotis, D. Velissariou, N. Pistofidis, D. Papageorgiou, et al., Valorisation of woody biomass bottom ash in Portland cement: a characterization and hydration study, Journal of Environmental Chemical Engineering 5 (1) (2017) 205–213.
[106] S.V. Vassilev, D. Baxter, L.K., Andersen, C.G. Vassileva, An overview of the composition and application of biomass ash. part 1. phase–mineral and chemical composition and classification, Fuel 105 (2013) 40–76.
[107] M.G. Beltrán, A. Barbudo, F. Agrela, J.R., Jiménez, J. de Brito, Mechanical performance of bedding mortars made with olive biomass bottom ash, Construction and Building Materials 112 (2016) 699–707.
[108] M. Cabrera, J.L. Díaz-López, F., Agrela, J. Rosales, Eco-efficient cement-based materials using biomass bottom ash: a review, Applied Sciences 10 (22) (2020) 8026.

[109] J. Jin, G. Zhang, Z. Qin, T. Liu, J., Shi, S. Zuo, Viscosity enhancement of self-consolidating cement-tailings grout by biomass fly ash vs. chemical admixtures, Construction and Building Materials 340 (2022) 127802.

[110] J. Fořt, J. Šál, R. Ševčík, M. Doleželová, M. Keppert, M. Jerman, et al., Biomass fly ash as an alternative to coal fly ash in blended cements: functional aspects, Construction and Building Materials 271 (2021) 121544.

Ash management, recycling, and sustainability

3

Abstract

This chapter covers in detail the issues focused on the management, recycling, and sustainability of solid ashes. It begins with an introduction, describing the increase in the production amount of industrial ashes depending on the energy needs, which causes an austere maltreatment to both atmosphere and community. Subsequently, conventional recycling/reuse methods of numerous solid ashes are explained comprehensively. The chapter ends by drawing attention to the management and environmental characteristics of industrial solid ashes.

3.1 Introduction

Depending on the rapid population growth and urbanization in countries, energy demand is rapidly growing progressively. Accordingly, demand for fuel has increased the interest in coal/oil/gas fuels, especially after industrial revolution. Today, approximately 80% of the electricity produced in the energy sector is based on fossil fuels [1]. But, to reduce sustainable growth and environmental problems, the tendency from traditional energy sources to renewable resources is increasing. Especially biomass and municipal waste are widely used in electric power plants around the world, and their scope is increasing day by day. Although these traditional and renewable fuels are found cheap and easy to find, the demand for each one leaves behind, and their prices are increasing. In addition to the plants in the current power plants, the increasing energy demand causes the establishment of many conventional power plants working with these fuels [2,3]. Hence, depending on the fuels used, industrial waste/ash (coal ashes, biomass ashes, and municipal solid waste ashes) is produced in energy power plants each year. Considering the partial scope of ecological guidelines and rules for these industrial ashes, appropriate management strategies should be determined. Otherwise, they will order an immense menace due to damages caused by the atmosphere and human health.

General concerns about energy power plants operating with traditional/renewable fuels include the disposal, storage, use, and transport of industrial ashes. It also includes the processing and management options owing to the amount of burned carbon contained in ashes and the minimum reduction of environmental effects. Industrial ashes are usually accumulated in regular storage areas, ash ponds, dumps, or precipitation pools. However, these disposal paths cause serious environmental problems due to toxicity, radioactivity, and infiltration of industrial ashes. In addition, emptying these ashes into a waste storage area occupies efficient and large agricultural land by reducing recyclable resources. It is thought that the increases in ash capacities due to time will

Machine Learning Applications in Industrial Solid Ash. DOI: https://doi.org/10.1016/B978-0-443-15524-6.00014-5
© 2024 Elsevier Inc. All rights reserved.

cause a decrease in storage areas of ashes and their costs to increase significantly [4–6]. Therefore industrial ashes must be eliminated without injuring the environment, for example, converting them into a product containing added value.

Today, like coal ashes, municipal solid waste ashes are mainly used as a raw product in cement/concrete production, but it is actively converted and evaluated in many sectors such as road building materials, embankments, ground stabilization, ceramics, and glass ([7,8]). Biomass ashes are generally employed actively in the construction sector, agriculture, and recycling fuel. In addition, it is used in various applications from the environment to fuel production [9]. The recycling-utilization process and applicability of industrial ash are directly related to chemical and physical properties. Although the key motive for employing ashes in different sectors is economic and environmental, in addition to these, resource supply crisis is another issue that should be given prominence to both recycling and reuse of ashes. Along with the making of great value-added crops from these problematic and initially considered noneconomic wastes in parallel with technological developments, their usage methods should be sustainable and environmentally harmless.

3.2 Management of CFA and CBA

3.2.1 Hazards of CFA and CBA

Today, there are differences of opinion on whether coal ash can be considered as a risky unwanted product. EPA [10] characterizes these wastes as harmless, as coal ash is beneficially used in many industries [10]. In addition, coal ash is not categorized as risky waste in numerous republics, especially in China. However, the fact that coal ash contains high levels of toxic elements (e.g., nickel and lead) indicates that it could be qualified as a perilous waste in keeping with many standards. Release of many toxic elements with the burning of coal in power facility indicates that CFA and CBA have potentially harmful effects on the environment (e.g., groundwater, soil, and air) [11]. Improper management of coal ash often causes accumulations in large areas, causing soil damage and degradation. Moreover, fine particles of CFA and CBA powders cause serious atmospheric pollution due to air circulation and may cause irritation in the throat/respiratory tract of living things. Leachates of CFA and CBA, conversely, cause the release of metal elements, such as Cu, Ni, As, Zn, and Pb, which are extremely noteworthy from the ecological/social/economic considerations for many nations.

Undeniably, these wastes may cause the release of toxic elements in contact with water, depending on grain diameter of CFA and CBA, concentrations of trace elements, and pH values of extraction solution and their chemical speciation [8]. In addition, it is evaluated that the radiations that can be caused by the radioactive elements (e.g., 222Rn and 40 K) contained in CFA and CBA may have dangerous effects. Although the radionuclides contained in CFA and CBA can be kept under control within a certain range, the damages that their radioactivity may cause to the environment should be monitored [12]. Today, CFA is deposited together with CBA in landfills, ash ponds, landfills, or settling ponds. These disposal methods are far from environmental due to leaching from the radioactivity and toxicity of coal ash. Because it creates significant risks for the health of living things by causing pollution of the soil and natural water resources. In addition, landfills and ponds occupy large areas, occupying the settlement of fertile agricultural lands [13]. Considering the gradually stringent disposal laws/restrictions, storage and other disposal costs, and reduced access to landfills, more environmental and economic methods are needed for the management of CFA and CBA. Maximizing sustainable reuse areas instead of storing and/or disposing of coal ash by various methods is one of the most important ways to reduce ecological concerns and create alternative economic opportunities. In fact, only 25% of the CFA produced worldwide is considered as alternative products (e.g., Portland cement, zeolite, fiber, ceramics, and filling material) in different sectors. However, the current usage rate is gradually growing compared with accumulative production of CFA and CBA [14,15]. For this reason, there is a need for the widespread use of ecologically problematic ashes to transform them into valuable assets with sustainable methods and to find alternative solutions.

3.2.2 Application and utilization of CFA and CBA

With the utilization/recycling of coal ashes in industrial areas, potential ecological and economic problems arising from the disposal of these wastes can be eliminated. Considering the physical/chemical features of CFA and CBA, it is possible to produce many different value-added products cheaply in industrial areas.

Fig. 3.1 shows the most common industrial applications of CFA and CBA, which are frequently used today. Furthermore, some of its common applications in the world are pointed out in the subsequent subdivisions.

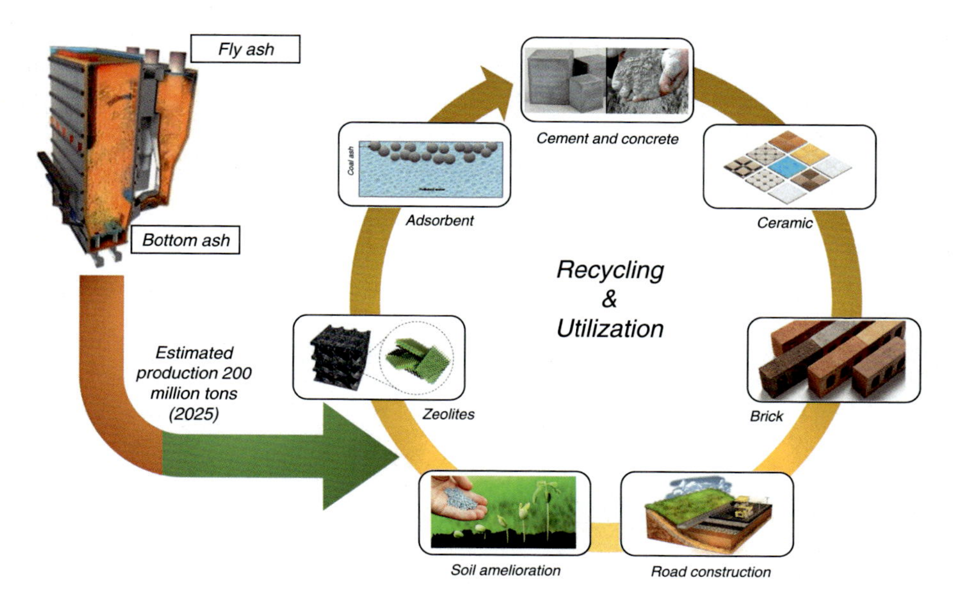

Figure 3.1 Common application areas of CFA and CBA.

3.2.2.1 Cement and concrete

CFA is widely recycled in construction sector because it has pozzolanic/cementitious features relying on the mass of CaO it contains (ASTM C618, types C and F). Especially thanks to CFA's chemical/mineralogical features, it is an indispensable product for the cement/concrete industries. Pozzolanic and structural features of CFA allow it to be used as a substitute for clinker, the main part of OPC. Additionally, reactions between the silica and calcium hydroxide contained in CFA cause the production of CSH, which is primarily accountable for strength of cementitious materials [16]. Due to the binding property of CFA, it increases workability and drainage in fresh concrete. When F class CFA is used as a cement substitute, it reduces the production costs by reducing the hydration temperature of the concrete mixture, eliminating the risk of cracking in the early stages. The use of CFA provides high durability and ultimate strength values in concrete [17]. Improving concrete performance, it also has environmental benefits. It guarantees fewer usage of cement residue and thus fewer toxic emissions, fewer energy depletion. Besides, it allows less contamination of the environment as the usage of thermic power facility waste. Overall, using CFA within concrete is up to 15%–35%. In addition, CFA-based geopolymers have become recently the most important extra for cement in building material field [18,19]. One can see that concretes containing CFA-based geopolymer exhibit similar durability and strength compared with concrete prepared with conventional cement. The suitability of CFA for various structural applications is proven day by day.

Like CFA, CBA is a material that has been employed in the building sector for a long period. Recently, employing CBA as cement/fine aggregate within concrete production has become gradually widespread. With the evaluation of CBA as fine faction in concrete, one can interpret that the concrete density decreases, compressive strength and pozzolanic reactions improve [20]. Due to the large grain diameter of CBA, its reactivity is quite low. Therefore, to be used as a cement substitute, its surface area and reactivity must be grinded to reach appropriate proportions. As the fineness of the CBA increases, the performance and quality of cement mortars potentially improve. As well, it is possible to lessen harmful impacts of alkali–silica reactions in concrete with the grinding process. However, the grinding process inevitably increases the costs of concrete production [21,22]. With the safe treatment of CBA, its ecological impacts in building sector, especially within cement/concrete, are decreasing. Latest studies reveal that cementitious mixtures created with CBA have analogous and/or improved ecological/mechanical behavior in comparison with CFA. It is thought that employing CBA as an extra in concrete compared with natural aggregate or other alternative products will cause less CO_2 releases [23]. Today, although benefits of CBA within cement/construction industries are known, its environmental impacts and cost benefits have not been clearly determined.

3.2.2.2 Ceramic industry

Ceramic tiles are thin plates that are mostly used for decorative purposes for floors, ceilings, and walls. It usually consists of a combination of 50% clay, 35% feldspar, and 15% quartz. At this point, clay and other components come to the fore with their binding properties. It also increases the plasticity of the composition to help give the desired shape for ceramic tiles. In addition, feldspar melts at low temperatures, filling the pores of the composition and supporting the formation of a dense material. Quartz, on the other hand, helps to prevent shrinkage and deterioration with its stable structure at high temperatures. CFA and CBA, low cost and rich in oxides such as Fe_2O_3, CaO, Al_2O_3, and SiO_2, make it an indispensable material for ceramic-tile making [24,25]. These oxides are used for thin ceramic paste applications without pretreatment. However, chemical additives and natural minerals are frequently used to surge crystallization and strength features of ceramic tiles for different productions. Sintered materials produced with CFA and CBA are also widely used in glass-ceramic/glass making. Ceramic tiles or glass is produced owing to the thermic initiation of raw CFA and CBA with heat change and auxiliaries. The structure of both CFA and CBA directly affects the properties (e.g., porosity, strength, density, and absorbency) of ceramic composites [26]. Especially with the addition of CFA and CBA, the decrease in the density of ceramic mixtures creates positive effects on the thermal insulation potential.

3.2.2.3 Brick production

Lime-silicate bricks were started to be produced in 1898 with the use of lime and sand together. Steam curing process is applied during the production phase. At first, fine sand was utilized in the production process, and then it was seen that coal ash could be employed as an extra for fine fraction. As coal ash shows pozzolanic properties, it combines with lime to support the formation of high-quality and -performance bricks. Today, clay bricks are widely used. Especially in India, in excess of 100 billion bricks are produced yearly, and it fills an important gap in the construction sector. Bricks produced with clay are weak in terms of strength and insulation (sound/heat). In addition, it lags far behind other alternatives owing to its quality, appearance, and high mortar requirement [27].

Bricks prepared with coal ash are formed by mixing and compacting fly ash/bottom ash, fine aggregate, cement/lime, gypsum in appropriate proportions. Water or steam cured water is added during mixing. Types F and C containing CFAs are used primarily in the production of ash bricks. The high pozzolanic properties of C-class CFAs increase the quality and performance in brick production. Cement or lime is added to the bricks produced, depending on the situation. Bricks prepared with coal ash have several important advantages over other bricks (e.g., conventional bricks) [28]. These bricks cause less mortar and plaster usage during application. In addition, it can be produced with the help of hydraulic press even in the field environment with less energy and machine equipment compared with traditional bricks. Generally, these bricks are produced by mixing ash, cement/lime, sand, and water in appropriate proportions and placing them in hydraulic press molds. After the pressing process, the ash bricks are cured in an open environment for 2 days and in a wet environment for 14 days.

The behavior and properties of bricks made with CFA and CBA vary depending on structure/type of ash, the amount of raw product, and the production steps (e.g., mixing, pressing, and curing). In addition, it differs according to the amount of lime, especially cement, used in the mixture. Although many technologies are currently available for brick production from CFA and CBA, not all of them are viable and economical [29,30]. For example, vapor cure of bricks produced with ash/sand/lime results in great operating costs and is therefore not a frequently used method. Instead, a more applicable method emerges because of using clay and coal ash together. Because the molding and firing process in the production process of these bricks is easily adopted by the manufacturers due to its low cost compared with steam curing. In addition, the weight of the bricks prepared with clay and coal ash is 15%–20% lower than the traditional bricks and the thermal insulation progress increases its suitability. High strength, machinability, and low fuel costs during making are other advantages. Fig. 3.2 shows the construction technologies of bricks produced with coal ash.

3.2.2.4 Road construction

CFA and CBA are qualified as a waste material suitable for recycling in road and embankment structures. Usually, CFA is used as concrete, filling material, subfloor, and road base aggregates in road construction, while CBA is used as filling material in road concrete, embankments, stabilized bases, and asphalt (cold mix) aggregate. The use of natural fills (surface soil) in structures such as roads and embankments causes high costs and environmental destruction. Hence, using high amounts of coal ash is supported in studies. Many features of CFA/CBA, such as lightness and high stability, highlight these materials for use as fillers [6]. Concrete properties improve significantly with CFA/CBA use in road concrete. Coal ash pointedly rises the strength, durability, and workability of road concrete. It also reduces bleeding, permeability, and costs in concrete. In stabilized road foundations, coal ash is mixed with lime and various aggregates instead of cement. In this way, a high-strength and durable foundation is formed, and energy-efficient productions are realized [32,33].

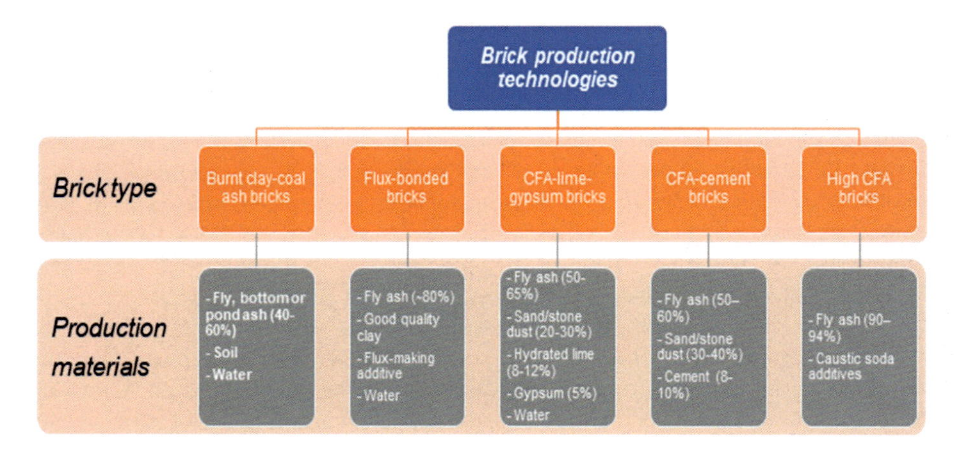

Figure 3.2 Different methods used in the production of bricks with coal ashes [31].

Fillings prepared using coal ashes on roads generally consist of mixtures containing ash, cement, and liquid. In some cases, materials such as sand and quarry dust can be added to the prepared mixtures. Compared with traditional filling methods (e.g., sand, gravel, and soil), it is one of its most important features to be applied without the need for compaction and vibration. In addition, it is easily applied in areas that cannot be reached by traditional filling methods, increases soil carrying capacity, prevents settlements, and has less heterogeneity/workmanship. It is employed as a key product for dam construction in CFA and CBA, such as fill. These structures serve to support road shoulders, major embankments, and structures around the road. It is low in cost, can be applied effortlessly, prevents plowing of natural fill, and could be easily applied on fills having poor bearing strength. CFA/CBA is frequently used to stabilize soils. It allows durable construction by improving the density, strength, water content, and plasticity of the stabilized soil. Among its important advantages is shortening the construction period by improving poor quality soil [34,35]. It helps to reduce the thickness and the need for natural aggregates on the floors covered for pavements. CFA is used as filling material in hot mix asphalts, and CBA is used as aggregate in cold mix asphalts. With the addition of coal ash to asphalt, strength of blending matrix, the rutting resistance, and the peeling tendency (hydrophobic nature of the ash) increase significantly.

3.2.2.5 Soil amelioration

Today, dolomite and lime-based materials are frequently used for the improvement of soils. However, these materials are not environmentally friendly and require significant stage to increase the structure/features of soils. Using CFA and CBA instead of materials such as lime can cause significant reductions in CO_2 emissions from lime production, slowing the effects of global warming, improving the structure of the soil, buffering its pH, increasing its water holding capacity, acting as a substrate for plants, and helping the proliferation of nutrients and microorganisms. In addition to improving pH environment of the soils, it has shown an important performance in many areas such as mining wastes, infertile and barren lands. CFA and CBA contain many minerals and beneficial nutrients (e.g., Ca, Mg, Mn, P, Cu, Zn) necessary for the growth, development, and crop yield of plants. The fact that coal ash is generally alkaline plays an important role in maintaining the pH value of the soil [18,36]. The fact that ash is generally basic has a key point in preserving pH value of soil. This alkaline structure differs relying on coal source as well as operational systems of thermic power plants. Differences in pH values and the effect on the soil are related to the neutralization ability of ash and soil's shielding ability.

Especially type C containing CFAs (CaO $\geq 15\%$) are frequently used because they tend to raise soil's pH. Type F containing CFAs (CaO $< 15\%$) show poor performance in improving the alkaline environment of the soil. Compared with CFA, the higher porosity and permeability of CBA make it a better product for soil improvement and nutrients. In addition to the properties mentioned earlier, CFA and CBA also have benefits such as infiltration, reduction of soil density, water holding ability/aeration. The presence of silica (corrosive) acting as an insecticide in coal ash also limits the availability and movement of metals [37,38]. However, along with its benefits, the sum of CFA and CBA added for soil improvement should be well determined. Owing to its basic features, adding coal ash can cause a quick rise in pH values of soils. Toxic side impacts and gentle deprivation of coal ash in the soil can be an environmental problem. Thus it is crucial to use CFA and CBA in a proper and environmentally friendly way.

3.2.2.6 Zeolites

Coal ash, which is one of the leading natural zeolites, and materials with volcanic properties constitute an effective area of use for the synthesis of zeolites due to their similar compositions. Zeolites are broadly employed in several diverse fields such as gas adsorption, farming, animal feed, green chemistry, oil, and water treatment. Natural zeolites are of critical commercial importance as they are easily contaminated by sulfate, ferrous, other zeolites, and amorphous glass. It is known that the central market shares of zeolites in 2020 are approximately US$ 5.2 billion. Until today, studies have been carried out on the fusion of zeolites with different properties from ash, especially CFA. Coal ash's structural features allow making of zeolites, which are commercially valuable as catalysts or adsorbents [39,40]. Especially due to the synthesis of coal ash and zeolite, enormous amounts of less costly adsorbent can be produced.

Zeolites are aluminosilicates with microporous crystals, generally characterized by tetrahedral SiO_4 and AlO_4 networks. Coal ash and zeolites have the same chemical composition. To produce zeolite from CFA and CBA, a hydrothermal (low temperature) conversion is usually done with low energy costs. This process consists of dissolution, condensation/gelling, and crystallization stages, respectively. Then, in the fusion process, coal ash is cured by robust basic solutions (e.g., NaOH/KOH) and heated between 20 and 200°C to produce zeolitic [21]. But, quartz/mullite, which is found in coal ash and has an inert structure, is not easy to dissolve in solution. Thus when the hand-made products are zeolites with unique crystalline phases like quartz/mullite, the creation of zeolites results in failure. Often zeolite clumps form in an egg white consistency casing coal ash's grains, which weakens the yield and transformation process. Zeolite synthesis can be carried out in a two-step manner, in which Si is taken out of coal ash, respectively, and silicon/aluminum rate is attuned by an outer aluminum [18,41]. By this way, it could be raised to obtain vastly pure/crystalline zeolites. It is also available in cases where it is not necessary to use water for the synthesis of zeolites. Microwave warming could be employed for zeolite fusion with measured crystallization heat and time. In general, as current zeolite-making process is slow and involves large amounts of reagents, it is then of great importance to simplify and advance the synthesis methods. Fig. 3.3 shows zeolites with CFA and CBA formed by different production methods (e.g., melting with NaOH and hydrothermal treatments).

3.2.2.7 Environmental protection (absorbent)

CFA and CBA perform well for the rehabilitation of inorganic and organic pollutants due to their porous structure, water retention potential, surface area, and metal

Figure 3.3 SEM views of zeolites made with CFA (A and B) and CBA (C and D) [42,43].

oxide content. Generally, coal ash shows an alkaline property, and its particles are covered with high-pH negative ions. This situation causes the removal of ions from precipitation or electrostatic adsorption. It is possible to use CFA and CBA as an absorbent that removes harmful metals (e.g., Cu) from industrial wastewater [44]. The large surface area and porosity indicate that CFA and CBA will be an important substrate for biofilm. This may subsidize the bioremediation of wastewater and reduce environmental damage. Coal ash's unburned C content also contributes to its high adsorption capacity. Studies show that the harmful boron element is removed in high amounts ($>90\%$) from desalinated sea water from coal ash. It is observed that the deletion of Cu/Zn/Pb/Cd metals from solutions made with coal ash decreases by nearly 95% [17,45]. It is possible to capture harmful fumes (CO_2, SO_2, NOX) from flue gases together with the rehabilitation of waste water. Especially cheap and efficient CO_2-holding capacity of amine-based CFAs can be evaluated. Its use and management as an absorbent for CFA and CBA constitute an important way to protect the environment [46]. But the contaminant reduction ability differs relying on the features of coal ash. Although mostly coal ash has a relatively low ability to prevent pollutants, better results can be obtained by improving it with the help of various methods.

3.3 MSW bottom ash and fly ash

MSW production is growing steadily due to growth in people, urbanization, rise in supply/demand for good/product, and socioeconomic development. The wastes produced in urban areas are likely to spread nearly 6.1 million tons/day in 2025 [47].

Thus it is vital to implement a naturally friendly and sustainable waste management policy. Incineration is an effective waste management technique widely used worldwide to treat/dispose of MSW (due to removal of organic harmful materials, volume reduction, and recovery of energy/metals) [48,49]. However, bottom/fly ashes and air trash emissions released because of incineration cannot eliminate environmental concerns [50]. This situation reduces the efficiency of the incineration method and requires effective management of the ashes formed. Therefore most countries encourage the appropriate disposal of MSW-BAs and MSW-FAs through strategic management plans and some legislation [51].

Although the commonly employed system for the removal/disposal of MSW ashes made in densely populated areas is landfill, these landfills cannot prevent the occurrence of various adverse events [52,53]. For example, heavy metals and water-soluble ions contained in BA/FA can damage the ecosystem by decomposing these ashes in storage areas. This requires viable/sustainable management of MSW ash disposed of by landfill [54]. Bottom/fly ashes contain harmful substances such as organic contaminants/heavy metals, and therefore they need safe disposal methods, which are suitable treatment and recycling after storage [55]. In general, treatment methods can be summarized in three categories, as shown in Fig. 3.4. These purification methods are separation and extraction solidification/stabilization and thermal management. The aim of separation is to enhance the value of postcombustion MSW residues, as well as to increase their use potential by separating some harmful components. The solidification/stabilization's key aim is to lessen the solubility/leachability/toxicity of pollutants. The heat management mostly contains vitrification/melting/sintering steps [56].

MSW ash is highly recyclable as it has many applicable properties [57]. Recycling and reuse of MSW ash are extremely important in reducing the volume of the areas where the residues are stored, thus extending their useful life. MSW-BA is broadly employed as a tributary raw product in cement/construction industry

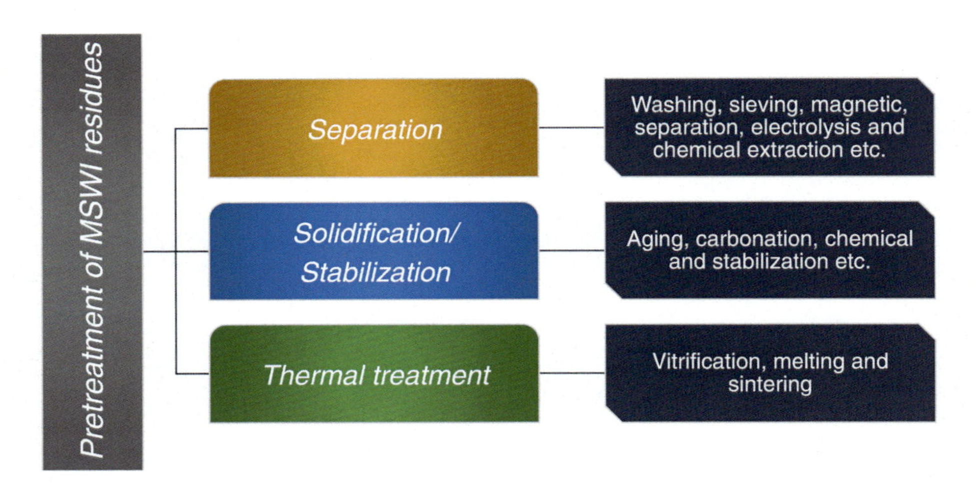

Figure 3.4 Treatment methods of MSW ashes.

thanks to its small pollutant concentrations [58]. MSW-BA/MSW-FA wastes could be employed in cement/concrete making, way pavements, glass/ceramic making, aggregates in embankments and road foundations, and as fillers for coal/salt mines. Table 3.1 shows the usage areas of MSW ash in different countries, and the use of MSW ashes in some different areas is briefly summarized as follows.

3.3.1 Cement production

Cement is a basic structure product that is extensively employed in various fields (e.g., plaster and concrete). The most common type of this thin material, consisting of lime and clay, which hardens because of contact with water, is Portland cement. Although cement is often a preferred product especially in construction industry, the cement sector is in an environmental bottleneck regarding the reduced CO_2 emissions (about 0.85 tons of CO_2 is released for 1-ton cement production) [60]. This negative situation can be eliminated by employing MSW ashes as a cement substitute/cementitious product [61].

MSW ashes (bottom/fly ash) could be employed as a raw product in cement making thanks to the chemical components; they contain such as CaO, SiO_2, Al_2O_3, and Fe_2O_3 [62,63]. For instance, one can urge that MSW-FA is utilized in low-energy special cements (Ca sulfo-aluminate one) with fast-hardening properties [64]. The inclusion of ash in cement production will adversely affect product quality, especially due to heavy metals and high chloride content of MSW-FA. As mentioned before, these harmful substances will also cause an environmental concern, as they leak into groundwater and damage the ecosystem. Therefore environmental safety of MSW ashes should be ensured by various pretreatment processes to minimize heavy metals and chloride in cement [65]. Several works [66,67] have designated that trace metal elements can be stabilized by mineralogical, physical, or a combination of these.

3.3.2 Concrete

It is likely to define concrete as a broadly utilized building product type involved in cement, water, and coarse/fine natural fractions and sometimes chemical/mineral

Table 3.1 Applications of MSW ashes in different countries [59].

Country	Application of MSW ashes
China	Road construction
Czech Republic	Soil surface
Denmark	Subbase layer
Netherlands	Building product
Poland	Way constructing
Spain	Fills/cements/concretes
United Kingdom	Structural platforms
Japan	Cement clinker

additives. The fact that it is used in a wide variety of fields and has different contents necessitates the production of concrete with sustainable approaches. The most attractive method used in production is the usage of recycled materials (e.g., fly ash/slag/silica fume) instead of the products that make up the concrete [68]. MSW ash (MSW-BA and MSW-FA) could be employed as an aggregate in cementitious product, either individually or as a mix of ashes, in addition to its use as an added part for cement in concrete making. For example, in the United States and the Netherlands, MSW-BA ashes are combined with other concrete components as an aggregate replacement and are frequently used in the construction of concrete blocks [63].

As stated by Zhang and Zhao [69], using MSW-BA as a lightweight aggregate in construction sector may affect concrete's strength even though it complies with standards due to its low heavy metal concentration. However, MSW-FA can be more useful in concrete making about its features such as density, permeability, and chemistry [32]. MSW-FA can be treated as pellets and certainly employed to be a light aggregate. However, the correct determination of the sum of fly ash is a key issue in terms of product quality and should be considered. In addition, Chuang et al. [70] examined using MSW ashes in the making of lightweight aggregates and emphasized the importance of the kiln type in this regard. One can summarize that MSW ash could be employed as an extra for fine/coarse fraction in concrete, and the resultant concrete always has low density and low compressive strength performance.

3.3.3 Road pavement

MSW burning ash can be reused in road construction thanks to its material properties. While MSW-BA ash has high silicon and calcium content and is more suitable for road construction in terms of its physical properties, MSW-FA is usually a second alternative due to the heavy metals it contains [71]. It has also been stated that using them in way constructing is advantageous due to both reducing the cost and contributing to the environment [72].

Bottom ash or mixed ash with suitable particle sizes is frequently used in road construction as natural aggregate. For example, the United States used the mixture ash in asphalt pavement trials in different cities between 1970 and 1980 and achieved successful results. In addition, with these studies, the values that the ash content can reach according to the places of use (e.g., roadbed and surface layer) have been clearly determined [63]. The irregular shapes of the ashes may allow these ashes to be used in road pavements (especially in the surface layer) by increasing compliance. Many countries, especially France, Spain, and the Netherlands, employ BA in way pavement and have some legal regulations for this application [73]. Owing to different studies, in which MSW-BA was used in asphalt mixtures instead of natural aggregates in different rates, MSW-BA increased elastic modulus and tensile strength of asphalt mix [74]. It has been observed that it produces low rutting resistance [75] and improves permanent deformation resistance [76].

MSW-FA has relatively smaller grain diameter, huge specific surface, and high basic properties. MSW-FA can increase the great heat solidity of subfloor product

[77]. Metals contained within MSW-FA employed in asphalt mix can be stabilized thanks to the impermeability and good adhesion of the asphalt [78]. In addition, fly ash reduces the penetration and creep rate of bitumen, while increasing the rutting factor, creep stiffness, and tensile strength. Yan et al., [79]. Thus MSW burning residues (BA/FA) could be employed as a replacement of aggregate in road pavements or asphalt pavements after proper treatment suitable for their chemical features has been conducted.

3.3.4 Embankment

Structures made to hold water using soil or stone materials are called embankments. Undesirable geotechnical properties of soils can commonly be removed by stabilizing them with cement/lime. Thanks to its pozzolanic features, MSW-FA can easily replace lime or cement and can be used in embankments after necessary purification processes [80]. The fact that fly ash's density is not more than other fills employed in embankment's construction allows it with a lower compacted density than conventional earth fills. Additionally, MSW-FA's granular material structure and free-draining feature increase its usability for this application [64]. Besides fly ash, MSW-BA can also be used for bunds and storage. The biggest challenge in the use of MSW ash for weirs is the possibility of damaging the harmful chemical content of these ashes to soil and groundwater. Therefore, MSW residues should be treated before dam construction.

3.3.5 Soil stabilization

MSW residues (FA and BA) are frequently used not only in the construction industry but also in soil improvement, thanks to their various chemical contents [81]. In particular, the fact that MSW-FA contains elements (such as nitrogen, potassium, and phosphorus) necessary for the growth of plants and supplied by fertilizer strengthens its usage to be fertilizer. Amount of the ash used in soil improvement should be adjusted very well as it will affect the toxicity of soil and therefore plants. Otherwise, the elements that are effective in the development of plants will cause different problems (e.g., phytotoxic). If the amount of ash usage is not adjusted well, this will cause some elements in the plant tissues to go beyond the acceptable limits and may even affect the natural life of these animals [82].

3.3.6 Ceramic and glass

Ceramic is a product formed owing to the hardening of nonmetallic inorganic materials by firing processes. The intensive use of silicate-based raw materials in ceramic production makes MSW ashes a potential source in this material industry [64]. The high silicate content of MSW burning ash makes it easy and very reasonable to use these ashes in ceramic production. MSW-FA can be melted at high temperatures into a glass-like product through a process called vitrification. Owing to the vitrification process, in addition to the glassy structure of the ash, the organic pollutants and heavy metals contained in the ash are separated in various ways (e.g., evaporation). The resulting ashes could be employed in various implementations like ceramic tiles. Glass ceramics is a product with a high market volume due to its valuable and unique usage areas. Glass ceramics are generally products with high wear and corrosion resistance, as well as good thermal stability and shock resistance. It is hence broadly employed in cement, machinery, and construction sectors [83]. Glass ceramics, the making and usage of which have amplified recently, can be obtained from wastes such as MSW-FA. The usage of fly ash in glass ceramic market contributes not only to the reduction of the ash volume but also to the steadiness of metals it contains.

3.4 Biomass bottom/fly ashes

Using biomass in power plant has augmented quickly recently, and it is assumed that ~50% of the world's energy production will be met by biomass by 2050 [84]. However, the amount of ash formed as a result of the burning of biomass has also increased. Relying on the features of the burnt biomass and the boiling condition, this excess amount of ash with diverse chemical contents is largely deposited in landfills. Metals and soluble salts contained in biomass ash cause extremely serious problems for landfills. Hence, transportation and storage of these ashes should be well managed by countries and necessary legal arrangements should be made. Biomass ashes, which complicate the management of storage areas and cause extra costs, can be used in different sectors (i.e., construction) and fields, ensuring their sustainability, and being beneficially evaluated through recycling and reuse. While biomass bottom ash generally does not have a direct use (without pretreatment) as it contains unburned carbon, biomass fly ash has a high use potential, especially in construction sector. For example, biomass ash could be employed in diverse applications such as concrete, cement, ceramics, brick, and fertilizer. As the features of ash are vastly reliant on biomass burnt, the material characteristics should be well investigated before using these ashes in any application.

There are different studies in the literature where biomass ashes are used in cement production due to their pozzolanic properties [3]. The pozzolanic feature of the biomass used may vary according to the characteristics of the waste, in other words, the raw material. The use of biomass ash waste, which takes up space in landfills and causes environmental pollution, in the cement industry reduces the cost of cement production and CO_2 emissions. These ashes, which help to eliminate the FA deficiency in cement production, also contribute to concrete production. For example, rice husk ash with appropriate grain diameter increases the concrete's strength remarkably while reducing water adsorption [85]. Oppositely, palm oil ash reduces the concrete's permeability by reducing the concrete's pores and increases sulfate resistance [86]. There are also studies presenting the contribution of ground coconut shells to concrete's strength properties [87].

While BBA poses some problems in terms of its use as it contains components of different sizes and unburned products, some types with suitable material properties can be considered as aggregates. Woody biomass ash could be used as aggregate or binder in cement mortar making [88]. Relying on its physical/chemical features, BFA is utilized as a binder within mortar, while BBA can be used to replace sand in blend [89]. Recent works [90,91] have reported that palm oil shells are used to replace the fine fraction in concrete. By using biomass ash as a fertilizing product in forestry, it can meet the decreasing basic nutritional needs of the soil [92]. Owing to the burning of biomass wastes at high temperatures, the ash mineralizes, and the cations it contains turn into oxides that are hydrated and then carbonated under outside settings. Ca, Mg, and K minerals, in carbonates form, are transferred to ground. Owing to its positive basic pH, biomass ash could be considered as a soil basic product [93]. Use of biomass ash as a fertilizer has become popular recently, and studies have focused on the aim of possible nutrient value within ashes and the impact of the ash effect on the growth of plants.

One of the sustainable forms of biomass ash management is to use it in brick production. Biomass ash, which has the properties (physical/chemical) of brick building materials, can be partly used to replace natural resources. The features of the bricks used as partitions and load-bearing walls in buildings are directly dependent on the structure of the raw materials and the way of production. Hence, the structure of brick varies in line with the characteristics of the biomass ashes used to lessen the feasting of clay beds. One can state that using rice husk ash in brick making increases the strength, while sugar cane bagasse ash reduces the density and enables the bricks' making [94]. In addition, Carrasco et al. [95] observed a decline in the conductivity of bricks prepared with wood biomass bottom ash, and similar reductions were seen in clay bricks based on rice husk ash and on sugarcane bagasse ash. Ceramics formed with biomass base ashes also show low density, light material, and low thermal conductivity, similar to bricks. In addition, the obtained ceramics offer a wider choice of colors compared with traditional ceramics. This application meets the raw product requirement in the making of ceramics, serves to reduce the ecological pollution by keeping ash in ceramic matrix.

Another issue is the durability of asphalt pavement used in road engineering under different conditions. Asphalt binder is a mix of various chemical constituents with well viscoelastic features. Biomass ash can be shown as an example of these materials, and it is possible to come across examples in the literature. Tahami et al. [96] used rice husk ash and palm kernel ash in different proportions of hot asphalt mix as fill and concluded that asphalt mixtures showed higher modulus of stability and stiffness, and that these ashes improved the adhesion force between aggregate and asphalt. Xue et al. [97] observed that wood sawdust ash, which they used in the modification of asphalt binder, increased the rutting factor and binder viscosity at high temperatures.

Biomass residues with high carbon content are low-cost sources that are often used as raw materials to produce activated carbon and biochar. Recently, while biomass-based adsorbents are common in postcombustion CO_2 capture, agricultural-based adsorbents are effective in water purification [98]. Many studies are directed toward sustainable adsorbent production from biomass wastes such as rice husk, corn straw, and hazelnuts. Ruiz et al. [99] stated that ash created due to forest biomass burning can be initiated by

Table 3.2 Examples of some application fields of biomass ash.

Reference	Biomass ash type	Application field
Tosti et al. [101]	Cacao husk	Cement
Ketov et al. [102]	Rice husk	Brick
Padhi et al. [103]	Rice husk	Concrete
Cabrera et al. [104]	Olive	Road
Velay-Lizancos et al. [105]	Eucalyptus wood	Concrete
Akarsh et al. [106]	Sugarcane bagasse	Asphalt
Maeda et al. [107]	Wood	Fertilizer
Wu and Lu [108]	Corn stalk	Brick
Carević et al. [109]	Wood	Cement
Gunawan et al. [110]	Coconut shell	Adsorbent

KOH at a heat of 750°C to obtain adsorbent material. Santhosh et al. [100] investigated the utilization of palm oil fuel ash in diverse applications and found that this ash was effective in removing some heavy metals when used as an adsorbent at different rates. Biomass ash exhibits different properties due to combustion conditions and diverse biomass, and this needs detailed analysis in the use of these ashes in diverse areas. Table 3.2 indicates some case studies using biomass ashes in diverse implementations.

References

[1] F. Johnsson, J. Kjärstad, J. Rootzén, The threat to climate change mitigation posed by the abundance of fossil fuels, Climate Policy 19 (2) (2019) 258–274.

[2] A. Karmakar, T. Daftari, K. Sivagami, M.R. Chandan, A.H. Shaik, B. Kiran, et al., A comprehensive insight into waste to energy conversion strategies in India and its associated air pollution hazard, Environmental Technology & Innovation 29 (2023) 103017.

[3] G. Sua-Iam, N. Makul, Utilization of coal-and biomass-fired ash in the production of self-consolidating concrete: a literature review, Journal of Cleaner Production 100 (2015) 59–76.

[4] X. Fan, R. Yuan, M. Gan, Z. Ji, Z. Sun, Subcritical hydrothermal treatment of municipal solid waste incineration fly ash: a review, Science of The Total Environment 865 (2022) 160745.

[5] S. Voshell, M. Mäkelä, O. Dahl, A review of biomass ash properties towards treatment and recycling, Renewable and Sustainable Energy Reviews 96 (2018) 479–486.

[6] N. Wang, X. Sun, Q. Zhao, Y. Yang, P. Wang, Leachability and adverse effects of coal fly ash: a review, Journal of Hazardous Materials 396 (2020) 122725.

[7] R. Siddique, Utilization of municipal solid waste (MSW) ash in cement and mortar, Resources, Conservation and Recycling 54 (12) (2010) 1037–1047.

[8] K. Gao, M.C. Iliuta, Trends and advances in the development of coal fly ash-based materials for application in hydrogen-rich gas production: a review, Journal of Energy Chemistry 73 (2022) 485–512.

[9] Y. Niu, H. Tan, Ash-related issues during biomass combustion: alkali-induced slagging, silicate melt-induced slagging (ash fusion), agglomeration, corrosion, ash utilization, and related countermeasures, Progress in Energy and Combustion Science 52 (2016) 1–61.

[10] U.S. Environmental Protection Agency (USEPA), Hazardous and Solid Waste Management System; Disposal of Coal Combustion Residuals From Electric Utilities, 2014.
[11] B.O. Oboirien, V. Thulari, B.C. North, Enrichment of trace elements in bottom ash from coal oxy-combustion: effect of coal types, Applied Energy 177 (2016) 81–86.
[12] J. Temuujin, E. Surenjav, C.H. Ruescher, J. Vahlbruch, Processing and uses of fly ash addressing radioactivity (critical review), Chemosphere 216 (2019) 866–882.
[13] Z. Chen, S. Lu, M. Tang, X. Lin, Q. Qiu, H. He, et al., Mechanochemical stabilization of heavy metals in fly ash with additives, Science of the Total Environment 694 (2019) 133813.
[14] A.R. Gollakota, V. Volli, C.M. Shu, Progressive utilisation prospects of coal fly ash: a review, Science of the Total Environment 672 (2019) 951–989.
[15] N.B. Singh, A. Agarwal, A. De, P. Singh, Coal fly ash: an emerging material for water remediation, International Journal of Coal Science 9 (2022) 44.
[16] H. Jiang, M. Fall, E. Yilmaz, Y., Li, L. Yang, Effect of mineral admixtures on flow properties of fresh cemented paste backfill: assessment of time dependency and thixotropy, Powder Technology 372 (2020) 258–266.
[17] C. Belviso, State-of-the-art applications of fly ash from coal and biomass: a focus on zeolite synthesis processes and issues, Progress in Energy and Combustion Science 65 (2018) 109–135.
[18] Z.T. Yao, X.S. Ji, P.K. Sarker, J.H. Tang, L.Q. Ge, M.S. Xia, et al., A comprehensive review on the applications of coal fly ash, Earth-Science Reviews 141 (2015) 105–121.
[19] B. Makgabutlane, M.S. Maubane-Nkadimeng, N.J. Coville, S.D. Mhlanga, Plastic-fly ash waste composites reinforced with carbon nanotubes for sustainable building and construction applications: a review, Results in Chemistry 4 (2022) 100405.
[20] K. Muthusamy, M.H. Rasid, G.A. Jokhio, A.M.A. Budiea, M.W. Hussin, J. Mirza, Coal bottom ash as sand replacement in concrete: a review, Construction and Building Materials 236 (2020) 117507.
[21] H. Zhou, R. Bhattarai, Y. Li, B. Si, X. Dong, T. Wang, et al., Towards sustainable coal industry: turning coal bottom ash into wealth, Science of the Total Environment 804 (2022) 149985.
[22] C. Argiz, A. Moragues, E. Menéndez, Use of ground coal bottom ash as cement constituent in concretes exposed to chloride environments, Journal of Cleaner Production 170 (2018) 25–33.
[23] E. Menéndez, A.M. Álvaro, M.T. Hernández, J.L. Parra, New methodology for assessing the environmental burden of cement mortars with partial replacement of coal bottom ash and fly ash, Journal of Environmental Management 133 (2014) 275–283.
[24] T.A. Otitoju, P.U. Okoye, G. Chen, Y. Li, M.O. Okoye, S. Li, Advanced ceramic components: materials, fabrication, and applications, Journal of Industrial and Engineering Chemistry 85 (2020) 34–65.
[25] C. Wang, G. Xu, X. Gu, Y. Gao, P. Zhao, High value-added applications of coal fly ash in the form of porous materials: a review, Ceramics International 47 (16) (2021) 22302–22315.
[26] Z. Zhang, J. Wang, L. Liu, J. Ma, B. Shen, Preparation of additive-free glass-ceramics from MSW incineration bottom ash and coal fly ash, Construction and Building Materials 254 (2020) 119345.
[27] C. Chen, Q. Li, L. Shen, J. Zhai, Feasibility of manufacturing geopolymer bricks using circulating fluidized bed combustion bottom ash, Environmental Technology 33 (11) (2012) 1313–1321.

[28] W. Zhang, S. Wang, J. Ran, H. Lin, W. Kang, J. Zhu, Research progress on the performance of circulating fluidized bed combustion ash and its utilization in China, Journal of Building Engineering 52 (2022) 104350.
[29] Z. Zhang, J. Qian, C. You, C. Hu, Use of circulating fluidized bed combustion fly ash and slag in autoclaved brick, Construction and Building Materials 35 (2012) 109–116.
[30] V. Gupta, H.K. Chai, Y. Lu, S. Chaudhary, A state of the art review to enhance the industrial scale waste utilization in sustainable unfired bricks, Construction and Building Materials 254 (2020) 119220.
[31] Singh, A.K., Masto, R.E., Hazra, B., Esterle, J., Singh, P.K. Ash from Coal and Biomass Combustion, Springer International Publishing, 2020.
[32] M. Ahmaruzzaman, A review on the utilization of fly ash, Progress in Energy and Combustion Science 36 (3) (2010) 327–363.
[33] E.K. Anastasiou, A. Liapis, I. Papayianni, Comparative life cycle assessment of concrete road pavements using industrial by-products as alternative materials, Resources, Conservation and Recycling 101 (2015) 1–8.
[34] ACAA (2003) Fly ash facts for highway engineers. American Coal Ash Association. US Department of Transportation, Federal Highway Administration.
[35] R. Chowdhury, D. Apul, T. Fry, A life cycle based environmental impacts assessment of construction materials used in road construction, Resources, Conservation and Recycling 54 (4) (2010) 250–255.
[36] G.W. Kim, M.I. Khan, P.J. Kim, H.S. Gwon, Unexpectedly higher soil organic carbon accumulation in the evapotranspiration cover of a coal bottom ash mixed landfill, Journal of Environmental Management 268 (2020) 110659.
[37] R.E. Masto, T. Sengupta, J. George, L.C. Ram, K.K. Sunar, V.A. Selvi, et al., The impact of fly ash amendment on soil carbon, Energy Sources, Part A: Recovery, Utilization, and Environmental Effects 36 (5) (2014) 554–562.
[38] S.M. Shaheen, P.S. Hooda, C.D. Tsadilas, Opportunities and challenges in the use of coal fly ash for soil improvements—a review, Journal of Environmental Management 145 (2014) 249–267.
[39] M. Yoldi, E.G. Fuentes-Ordoñez, S.A. Korili, A. Gil, Zeolite synthesis from industrial wastes, Microporous and Mesoporous materials 287 (2019) 183–191.
[40] F. Collins, A. Rozhkovskaya, J.G. Outram, G.J. Millar, A critical review of waste resources, synthesis, and applications for zeolite LTA, Microporous and Mesoporous Materials 291 (2020) 109667.
[41] J.H. Wee, A review on carbon dioxide capture and storage technology using coal fly ash, Applied Energy 106 (2013) 143–151.
[42] A.R. Gollakota, V.S. Munagapati, V. Volli, S. Gautam, J.C. Wen, C.M. Shu, Coal bottom ash derived zeolite (SSZ-13) for the sorption of synthetic anion alizarin red s (ARS) dye, Journal of Hazardous Materials 416 (2021) 125925.
[43] X. He, B. Yao, Y. Xia, H. Huang, Y. Gan, W. Zhang, Coal fly ash derived zeolite for highly efficient removal of Ni2 + inwaste water, Powder technology 367 (2020) 40–46.
[44] M. Ahmaruzzaman, Industrial wastes as low-cost potential adsorbents for the treatment of wastewater laden with heavy metals, Advances in Colloid and Interface Science 166 (1–2) (2011) 36–59.
[45] A.D. Papandreou, C.J. Stournaras, D. Panias, I. Paspaliaris, Adsorption of Pb (II), Zn (II) and Cr (III) on coal fly ash porous pellets, Minerals Engineering 24 (13) (2011) 1495–1501.
[46] M. Sarmah, B.P. Baruah, P. Khare, A comparison between CO2 capturing capacities of fly ash based composites of MEA/DMA and DEA/DMA, Fuel Processing Technology 106 (2013) 490–497.

[47] World Bank, Solid Waste Management, The World Bank Group, 2019. Available from: https://www.worldbank.org/en/topic/urbandevelopment/brief/solid-wastemanagement.
[48] N. Neehaul, P., Jeetah, P. Deenapanray, Energy recovery from municipal solid waste in mauritius: opportunities and challenges, Environmental Development 33 (2020) 100489.
[49] G. Weibel, U. Eggenberger, S., Schlumberger, U.K. Mäder, Chemical associations and mobilization of heavy metals in fly ash from municipal solid waste incineration, Waste Management 62 (2017) 147–159.
[50] D. Wang, X. Zhou, Y., Meng, Z. Chen, Durability of concrete containing fly ash and silica fume against combined freezing-thawing and sulfate attack, Construction and Building Materials 147 (2017) 398–406.
[51] F. Hasselriis, Waste-to-energy ash management in the United States, Encyclopedia of Sustainability Science and Technology (2012) 11737–11758.
[52] C.H. Tsai, Y.H., Shen, W.T. Tsai, Analysis of current status and regulatory promotion for incineration bottom ash recycling in Taiwan, Resources 9 (10) (2020) 117.
[53] Y. Wei, T. Shimaoka, A., Saffarzadeh, F. Takahashi, Mineralogical characterization of municipal solid waste incineration bottom ash with an emphasis on heavy metal-bearing phases, Journal of Hazardous Materials 187 (1–3) (2011) 534–543.
[54] X. Gao, B. Yuan, Q.L., Yu, H.J.H. Brouwers, Characterization and application of municipal solid waste incineration (MSWI) bottom ash and waste granite powder in alkali activated slag, Journal of Cleaner Production 164 (2017) 410–419.
[55] F. Huber, D., Laner, J. Fellner, Comparative life cycle assessment of MSWI fly ash treatment and disposal, Waste Management 73 (2018) 392–403.
[56] D. Lindberg, C., Molin, M. Hupa, Thermal treatment of solid residues from WtE units: a review, Waste Management 37 (2015) 82–94.
[57] R. Forteza, M. Far, C., Segui, V. Cerdá, Characterization of bottom ash in municipal solid waste incinerators for its use in road base, Waste Management 24 (9) (2004) 899–909.
[58] K. Yin, A., Ahamed, G. Lisak, Environmental perspectives of recycling various combustion ashes in cement production–a review, Waste Management 78 (2018) 401–416.
[59] D. Chen, Y. Zhang, Y. Xu, Q. Nie, Z. Yang, W. Sheng, et al., Municipal solid waste incineration residues recycled for typical construction materials—a review, RSC Advances 12 (10) (2022) 6279–6291.
[60] S. Zhang, Z., Ghouleh, Y. Shao, Use of eco-admixture made from municipal solid waste incineration residues in concrete, Cement and Concrete Composites 113 (2020) 103725.
[61] Z. Yang, S. Tian, L. Liu, X., Wang, Z. Zhang, Recycling ground MSWI bottom ash in cement composites: long-term environmental impacts, Waste Management 78 (2018) 841–848.
[62] Q. Qiu, X. Jiang, G. Lü, Z. Chen, S. Lu, M. Ni, et al., Degradation of PCDD/Fs in MSWI fly ash using a microwave-assisted hydrothermal process, Chinese Journal of Chemical Engineering 27 (7) (2019) 1708–1715.
[63] X. Sun, J. Li, X. Zhao, B., Zhu, G. Zhang, A review on the management of municipal solid waste fly ash in American, Procedia Environmental Sciences 31 (2016) 535–540.
[64] R. Siddique, Waste Materials and By-Products in Concrete, Springer Science & Business Media, 2008.
[65] Y. Li, L., Hao, X. Chen, Analysis of MSWI bottom ash reused as alternative material for cement production, Procedia Environmental Sciences 31 (2016) 549–553.
[66] Y. Mao, H. Wu, W. Wang, M., Jia, X. Che, Pretreatment of municipal solid waste incineration fly ash and preparation of solid waste source sulphoaluminate cementitious material, Journal of Hazardous Materials 385 (2020) 121580.

[67] K. Wu, H., Shi, X. Guo, Utilization of municipal solid waste incineration fly ash for sulfoaluminate cement clinker production, Waste Management 31 (9–10) (2011) 2001–2008.
[68] Y.C. Ersan, S. Gulcimen, T.N. Imis, O., Saygin, N. Uzal, Life cycle assessment of lightweight concrete containing recycled plastics and fly ash, European Journal of Environmental and Civil Engineering 26 (7) (2022) 2722–2735.
[69] T. Zhang, Z. Zhao, Optimal use of MSWI bottom ash in concrete, International Journal of Concrete Structures and Materials 8 (2014) 173–182.
[70] K.H. Chuang, C.H. Lu, J.C., Chen, M.Y. Wey, Reuse of bottom ash and fly ash from mechanical-bed and fluidized-bed municipal incinerators in manufacturing lightweight aggregates, Ceramics International 44 (11) (2018) 12691–12696.
[71] M. Margallo, M.B.M. Taddei, A. Hernández-Pellón, R., Aldaco, A. Irabien, Environmental sustainability assessment of the management of municipal solid waste incineration residues: a review of the current situation, Clean Technologies and Environmental Policy 17 (2015) 1333–1353.
[72] L.D. Poulikakos, C. Papadaskalopoulou, B. Hofko, F. Gschösser, A.C. Falchetto, M. Bueno, et al., Harvesting the unexplored potential of European waste materials for road construction, Resources, Conservation and Recycling 116 (2017) 32–44.
[73] S. Ascher, I. Watson, X., Wang, S. You, Township-based bioenergy systems for distributed energy supply and efficient household waste re-utilisation: techno-economic and environmental feasibility, Energy 181 (2019) 455–467.
[74] A. Roberto, E. Romeo, A., Montepara, R. Roncella, Effect of fillers and their fractional voids on fundamental fracture properties of asphalt mixtures and mastics, Road Materials and Pavement Design 21 (1) (2020) 25–41.
[75] H.F. Hassan, K. Al-Shamsi, Characterisation of asphalt mixes containing MSW ash using the dynamic modulus| E*| test, International Journal of Pavement Engineering 11 (6) (2010) 575–582.
[76] Liu, D., Li, L., and Cui, H. Utilization of Municipal Solid Waste Incinerator Bottom Ash Aggregate in Asphalt Mixture. Asphalt Pavements; Taylor & Francis Group: London, UK, 2014.
[77] J. Lederer, V., Trinkel, J. Fellner, Wide-scale utilization of MSWI fly ashes in cement production and its impact on average heavy metal contents in cements: the case of Austria, Waste Management 60 (2017) 247–258.
[78] M.R.M. Hasan, J.W. Chew, A. Jamshidi, X., Yang, M.O. Hamzah, Review of sustainability, pretreatment, and engineering considerations of asphalt modifiers from the industrial solid wastes, Journal of Traffic and Transportation Engineering (English Edition) 6 (3) (2019) 209–244.
[79] K. Yan, L., Li, D. Ge, Research on properties of bitumen mortar containing municipal solid waste incineration fly ash, Construction and Building Materials 218 (2019) 657–666.
[80] I. Garcia-Lodeiro, V. Carcelen-Taboada, A., Fernández-Jiménez, A. Palomo, Manufacture of hybrid cements with fly ash and bottom ash from a municipal solid waste incinerator, Construction and Building Materials 105 (2016) 218–226.
[81] H.M. Jafer, W. Atherton, M. Sadique, F., Ruddock, E. Loffill, Development of a new ternary blended cementitious binder produced from waste materials for use in soft soil stabilisation, Journal of Cleaner Production 172 (2018) 516–528.
[82] M.A. Al-Ghouti, M. Khan, M.S. Nasser, K., Al-Saad, O.E. Heng, Recent advances and applications of municipal solid wastes bottom and fly ashes: insights into sustainable management and conservation of resources, Environmental Technology & Innovation 21 (2021) 101267.

[83] H.L. Zhao, F. Liu, H.Q. Liu, L. Wang, R., Zhang, Y. Hao, Comparative life cycle assessment of two ceramsite production technologies for reusing municipal solid waste incinerator fly ash in China, Waste Management 113 (2020) 447–455.
[84] S.V. Vassilev, D. Baxter, L.K., Andersen, C.G. Vassileva, An overview of the composition and application of biomass ash. Part 1. Phase–mineral and chemical composition and classification, Fuel 105 (2013) 40–76.
[85] W. Xu, T.Y., Lo, S.A. Memon, Microstructure and reactivity of rich husk ash, Construction and Building Materials 29 (2012) 541–547.
[86] M.A. Munawar, A.H. Khoja, S.R. Naqvi, M.T. Mehran, M. Hassan, R. Liaquat, et al., Challenges and opportunities in biomass ash management and its utilization in novel applications, Renewable and Sustainable Energy Reviews 150 (2021) 111451.
[87] T.C. Herring, T., Nyomboi, J.N. Thuo, Ductility and cracking behavior of reinforced coconut shell concrete beams incorporated with coconut shell ash, Results in Engineering 14 (2022) 100401.
[88] T.P. da Costa, P. Quinteiro, L.A. Tarelho, L., Arroja, A.C. Dias, Environmental assessment of valorisation alternatives for woody biomass ash in construction materials, Resources, Conservation and Recycling 148 (2019) 67–79.
[89] R. Rajamma, L. Senff, M.J. Ribeiro, J.A. Labrincha, R.J. Ball, G.C. Allen, et al., Biomass fly ash effect on fresh and hardened state properties of cement based materials, Composites Part B: Engineering 77 (2015) 1–9.
[90] H.M. Hamada, B.S. Thomas, B. Tayeh, F.M. Yahaya, K., Muthusamy, J. Yang, Use of oil palm shell as an aggregate in cement concrete: a review, Construction and Building Materials 265 (2020) 120357.
[91] K. Muthusamy, M.H. Rasid, N.N. Isa, N.H. Hamdan, N.A.S. Jamil, A.M.A. Budiea, et al., Mechanical properties and acid resistance of oil palm shell lightweight aggregate concrete containing coal bottom ash, Materials Today: Proceedings 41 (2021) 47–50.
[92] B. Omil, F., Sánchez-Rodríguez, A. Merino, Effects of ash applications on soil status, nutrition, and growth of Pinus radiata d. don plantations, Recycling of Biomass Ashes (2011) 69–86.
[93] J. Qin, M.F. Hovmand, F. Ekelund, R. Rønn, S. Christensen, G.A. de Groot, et al., Wood ash application increases pH but does not harm the soil mesofauna, Environmental Pollution 224 (2017) 581–589.
[94] S.M. Kazmi, S. Abbas, M.A. Saleem, M.J., Munir, A. Khitab, Manufacturing of sustainable clay bricks: utilization of waste sugarcane bagasse and rice husk ashes, Construction and Building Materials 120 (2016) 29–41.
[95] B. Carrasco, N. Cruz, J. Terrados, F.A., Corpas, L. Pérez, An evaluation of bottom ash from plant biomass as a replacement for cement in building blocks, Fuel 118 (2014) 272–280.
[96] S.A. Tahami, M., Arabani, A.F. Mirhosseini, Usage of two biomass ashes as filler in hot mix asphalt, Construction and Building Materials 170 (2018) 547–556.
[97] Y. Xue, S. Wu, J. Cai, M., Zhou, J. Zha, Effects of two biomass ashes on asphalt binder: dynamic shear rheological characteristic analysis, Construction and Building Materials 56 (2014) 7–15.
[98] N.B. Singh, G., Nagpal, S. Agrawal, Water purification by using adsorbents: a review, Environmental Technology & Innovation 11 (2018) 187–240.
[99] B. Ruiz, R.P. Girón, I., Suárez-Ruiz, E. Fuente, From fly ash of forest biomass combustion (FBC) to micro-mesoporous silica adsorbent materials, Process Safety and Environmental Protection 105 (2017) 164–174.

[100] K.G. Santhosh, S.M., Subhani, A. Bahurudeen, Sustainable reuse of palm oil fuel ash in concrete, alkali-activated binders, soil stabilisation, bricks and adsorbent: a waste to wealth approach, Industrial Crops and Products 183 (2022) 114954.
[101] L. Tosti, A. van Zomeren, J.R. Pels, A., Damgaard, R.N. Comans, Life cycle assessment of the reuse of fly ash from biomass combustion as secondary cementitious material in cement products, Journal of Cleaner Production 245 (2020) 118937.
[102] A. Ketov, L. Rudakova, I. Vaisman, I. Ketov, V., Haritonovs, G. Sahmenko, Recycling of rice husks ash for the preparation of resistant, lightweight, and environment-friendly fired bricks, Construction and Building Materials 302 (2021) 124385.
[103] R.S. Padhi, R.K. Patra, B.B., Mukharjee, T. Dey, Influence of incorporation of rice husk ash and coarse recycled concrete aggregates on properties of concrete, Construction and Building Materials 173 (2018) 289–297.
[104] M. Cabrera, J. Rosales, J. Ayuso, J., Estaire, F. Agrela, Feasibility of using olive biomass bottom ash in the sub-bases of roads and rural paths, Construction and Building Materials 181 (2018) 266–275.
[105] M. Velay-Lizancos, M. Azenha, I., Martinez-Lage, P. Vázquez-Burgo, Addition of biomass ash in concrete: effects on e-modulus, electrical conductivity at early ages and their correlation, Construction and Building Materials 157 (2017) 1126–1132.
[106] P.K. Akarsh, G.O. Ganesh, S., Marathe, R. Rai, Incorporation of sugarcane bagasse ash to investigate the mechanical behavior of stone mastic asphalt, Construction and Building Materials 353 (2022) 129089.
[107] N. Maeda, T. Katakura, T. Fukasawa, A.N. Huang, T., Kawano, K. Fukui, Morphology of woody biomass combustion ash and enrichment of potassium components by particle size classification, Fuel Processing Technology 156 (2017) 1–8.
[108] H. Wu, Z. Lu, Study on double dovetail tenon and mortise combination of corn stalk biomass bricks, Construction and Building Materials 370 (2023) 130651.
[109] I. Carević, A. Baričević, N., Štirmer, J.Š. Bajto, Correlation between physical and chemical properties of wood biomass ash and cement composites performances, Construction and Building Materials 256 (2020) 119450.
[110] T.D. Gunawan, E., Munawar, S. Muchtar, Preparation and characterization of chemically activated adsorbent from the combination of coconut shell and fly ash, Materials Today: Proceedings 63 (2022) S40–S45.

Emerging innovative techniques for ash management

4

Abstract

This chapter offers a brief introduction to innovative and high added-value methods for the management of industrial solid ash. It also presents the influence of the complex structure of industrial ash and the excess amount of production on the ecological balance. Then, the innovative methods that contribute to the high commercial value and real potential of these industrial ashes are explained in detail. The chapter concludes by noting the benefits of innovative uses of industrial solid ash.

4.1 Introduction

Industrial solid ashes (e.g., coal ash, MSW ash, and biomass ash), which are formed during the burning of industrial wastes/coals to obtain energy and heat in thermal power plants, pose a major danger to the global order day by day due to its complex structure or abundant production. The failure to properly dispose of these ashes may cause water–soil pollution, disrupting the natural balance and causing environmental hazards. Considering the increasing strict disposal regulations, depletion of landfills and disposal costs, as well as environmental effects, there is also a need for using more economical/environmentally friendly technologies for industrial ashes. In comparison with the storage and/or disposal of industrial ash, adopting the reuse option of it can create significant opportunities for economic development and reduction of ecological hazards. As other industrial ashes, about 25% (one forth of production) of coal ash is actively consumed as alternative products in different sectors [1,2]. However, the imbalance between the production of industrial ash and the current usage amount makes it crucial to develop the scope of use and product base of these wastes with innovative methods. Therefore there is an inevitable need to transform problematic ashes into valuable products to ensure environmental protection and industrial development.

Currently, industrial ash has traditionally been used in significant quantities in concrete and cement for various applications in construction industry. In addition, these ashes are used as a raw material source in many sectors such as ceramics, glass, road structures, adsorbent, fills, meso-microporous materials (zeolites), and soil improvement applications due to their multicomponent

Machine Learning Applications in Industrial Solid Ash. DOI: https://doi.org/10.1016/B978-0-443-15524-6.00006-6
© 2024 Elsevier Inc. All rights reserved.

content and large surface areas [3–5]. However, economic/commercial value of these uses does not contribute to the real potential use of industrial ash and does not help reduce its consumption. For this reason, many thoughts and studies support the idea of converting industrial ashes into products with higher added value and using them in new areas. Now, thoughtful efforts are being made to embrace the transformation of industrial ashes into innovative materials compared with traditional uses. One of these materials is geopolymers, which are a modern and environmentally friendly alternative to cement. In addition, because of the innovative recycling of industrial ash, it is aimed to be used in many areas such as aerogel as an insulating material, carbon nanotubes as a thermal energy source, and the recovery of rare earth/radioactive elements. Fig. 4.1 shows some traditional and innovative uses for industrial ashes today.

The aforementioned usage areas have triggered the emergence of innovative, environmentalist, and sustainable paradigms that will leave the traditional uses of industrial ash behind. These innovative uses of industrial solid ash will close a significant economic gap today and in the future and contribute to environmental well-being. In addition, with the development of technology, there is no doubt that new methods will be found that will pointedly increase the commercial value of industrial ashes. Nevertheless, today, neither much consideration has been given to the usage and production possibilities of these methods, nor a comprehensive overview of these possibilities has been presented. Today, there is a need to examine and research the technologies related to the usage areas of industrial ash in detail and the benefits–use possibilities should be discussed in detail. Some innovative uses of industrial solid ashes are presented as follows.

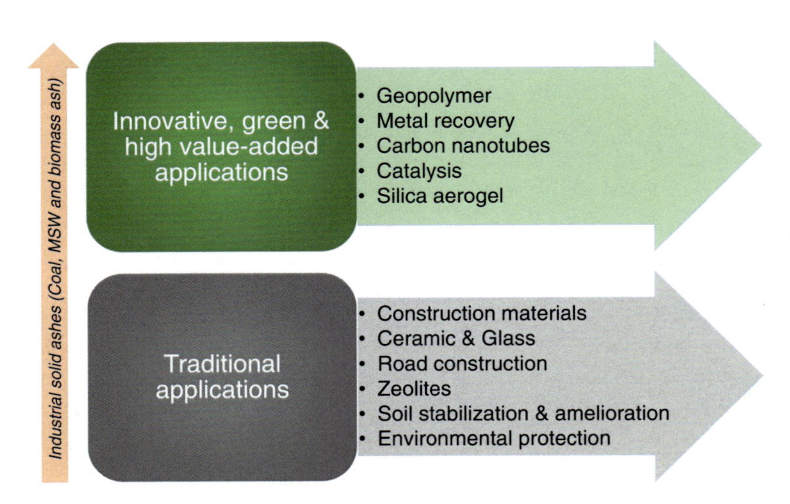

Figure 4.1 Some traditional and innovative uses of industrial ashes.

4.2 Geopolymer

Geopolymer cements are thought to be new fashion products that have an important role to change cement and lessen potential total heating and energy depletion in concrete production. The geopolymerization method can be carried out with the help of materials rich in alumino-silicate (aluminum oxide, Al_2O_3 and silicon dioxide, SiO_2). In addition, materials rich in Al_2O_3 and SiO_2 can increase the strength performance of concretes by supporting the creation of sodium alumino-silicate hydrate (NASH) and CSH gels. Thus the plenty of industrial solid ash and its richness in reactive oxides allow it to be evaluated as a potential raw material source for geopolymer production [6,7]. This method can be considered as a new field of discovery in industrial solid ash research and can act as a vital agent for ecological imbalances. Geopolymers, which are currently described as a new inorganic material compared with organic polymers, have a 3D structure with features such as high strength, good thermal resistance, resistance to sulfate attacks/deterioration, and permeability. Thanks to these properties, geopolymers increase the durability/performance of concrete more than cement. Using geopolymers as a cement alternative tends to change the fate of the construction industry in different areas such as aggregates, coating products, composites, insulation, and fiber-textile products [8].

Geopolymers produced using CFA and CBA are generally produced with the help of aluminosilicate powder and hydroxide-silicate solution. The highly reactive Al_2O_3 and SiO_2 in the coal ash dissolve quickly and contribute to the polymerization process. The main hydration products in the produced concrete are aluminosilicate hydrate gels with different Si and Al ratios, thus increasing the performance of the concrete. In addition, the strength development in concrete mostly varies as said by the volume, fineness, activator grade, and hardening conditions of the CFA and CBA grains. Although the SiO_2/Al_2O_3 content is higher in geopolymers containing CBA, it can be said that geopolymers with CFA perform better. This situation, that is, the lower reactivity of CBA, can be attributed to the size of the grain size and its porous structure. Therefore the reactivity of CBA can be removed by grinding and/or curing at high temperatures before it is used to prepare the geopolymer [9].

Last works have publicized that geopolymers synthesized using MSW ash could be employed as concrete [10]. MSW ash–based geopolymers are thought to be a capable cement product in construction implementations, as well as reduce the environmental impact of residual pollutants. For this reason, it can be said that the most key property of the geopolymer shaped with MSW burning ash is the immobilization of heavy metals. MSW ash can be preferred as a potential source in geopolymer making owing to its chemical content (CaO, SiO_2, and Al_2O_3), but it should not be ignored that the structure and physicochemical properties (e.g., Si/Al ratio, grain diameter range, phase structure, and formless content) of ashes directly affect the behavior of geopolymer. There are many studies in the literature about MSW ash–based geopolymers for reducing toxic leaching [11,12] and using sustainable building material instead of cement [13,14].

Geopolymer is an innovative method applied to reduce biomass solid ash and realize viable agronomic growth. While agricultural products such as rice husk ash/palm oil ash are commonly used for this approach, some biomass ashes such as corn cob, sawdust, and coconut shell are also evaluated for this purpose. Due to its high silica content and activity, rice husk ash is utilized as a clean alternative in the making of geopolymers. Thanks to the fill effect of rice husk ash nanosilica grains, it can improve geopolymers' micropores and the gel stage. Engaging it in concrete could also lessen the depletion of nanosilica. Rice husk ash with a high specific surface is broadly employed within geopolymer concrete [15]. In addition, palm oil fuel ash containing a great quantity of SiO_2 particles is also used as an extra cement-based product in the making of geopolymers. Palm oil fuel ash could rise the geopolymers' strength and promote hydration, effectively reducing the drying shrinkage rate [16]. However, this ash may reduce the workability of geopolymers owing to the unburned C content [17].

4.3 Recovery of REEs

REEs (rare earth elements) are vital critical raw materials that are in high demand for current and sustainable technologies [18]. These elements are used in a wide range of implementations (e.g., healthcare, defense, transportation, and communication technologies). The quick progress of techniques has recently increased the need for precious metals and REEs. Therefore the recovery of these elements from different industrial ashes is very important for sustainable development.

Although metal recovery from coal and coal ash dates to the past years, developing technology and needs have made this process more efficient. CFA and CBA contain many REEs that are left untreated and have high economic value. It is known that coal ash covers more than 300 organic—inorganic, radioactive, metallic, and nonmetallic elements. Generally, REE concentrations in the structure of coal ash are found in varying amounts between 270 and 1480 ppm [19]. Considering economically, the occurrence of some metals such as Al, Mg as well as the presence of Ge, U, Ga, Se, and V as critical elements has high commercial value. Furthermore, the existence of these elements makes CFA and CBA an important alternative when natural resources are considered. However, no matter how valuable the coal ash contains, the extraction methods of these components are quite complex and limit their potential use. Consequently, power plants are subject to strict regulations by various institutions (e.g., Atomic Energy Department, AERB and Natural Environmental Research Council, NERC) to analyze and limit the emissions of coal ash.

In general, REE recovery for waste materials such as coal ash is carried out in several steps. Although there are different processing methods for different raw materials, after crushing-grinding processes, materials are separated by methods such as flotation, gravity, and magnetic for concentrate making. These essences are treated into unpolluted metals with the aid of hydrometallurgical/pyrometallurgical processes. Finally, metal alloys with high purity are created by applying the reduction method [20].

Recent studies indicate that a major quantity of vital raw material is found on MSW ashes [21]. The low concentration of REEs in MSW ash necessitates the development of existing methods for the recovery of these elements. New cost-effective methods for REE recovery contribute to the management and economy of hazardous materials contained in waste. Recovery of MSW-FA rich in metal and organic pollutants can be difficult due to MSW-FA's origin and mineralogical chemical composition. MSW-BA generally contains glass, ceramics, ferrous/nonferrous metals, and ferrous materials are 7%–15% by weight of base ash [22]. The nonferrous products are merely found in a variety of 1–2 wt.% [23]. The quantity of metal constituents that could be detached from MSW-BA depends on the composition of waste. Fe is the most abundant metal that can normally be separated by a magnetic body in MSW-BA, and the regaining success of Fe can reach up to 80%. High-economic-value metals originated in MSW-BA are generally linked with slighter grain diameter. Hence, acid washing stands out as the most effective method for REE recovery [24].

Although biomass bottom ash covers low trace elements, it has valuable technical and environmental contributions thanks to its sustainable usage. Studies on rare earth elements in bottom ash have been dealt with depending on the behavior of these elements in thermo/chemical transformation of diverse biomass changes. The primary benefits of REE recovery are undoubtedly reducing the storage of these ashes and the increased risk of leaching. Low attentions (10–100 ppm) of biomass fly ash produced in large quantities can be cited as the reason why they are not used adequately as an REE source [25]. However, some studies also show that there are high concentrations of REEs. For this reason, it is vital to determine the properties of different biomass residues and to study the rare earth element recovery in terms of the evaluation of these wastes.

4.4 Carbon nanotubes

Carbon nanotubes (CNTs) are nanometer-scale graphene sheets that are considered a new class of materials and have unique characteristic properties (e.g., mechanical, optical, thermal, electrical, and electronic). In addition to these features, the morphology (diameter and helical structure) and orientation (vertical and horizontal state) of CNTs have many features such as fuel cell, battery, metal, polymer, composite material, catalyst, hydrogenation, field-emission indicator, microscope probe tip, and field emitters. However, the negative properties of CNTs (e.g., toxicity, inflammation, graphite content, and spine-like structure) lead to the portrayal of these materials as biologically harmful, limiting their commercial/economic potential [26].

CNTs were first recognized with materials called precarbon filaments or carbon-based nanotubes. Then, as a new product compared with carbon filaments, catalytic CNTs started to be evaluated in different sectors, attracting attention as a new alternative to carbon fibers due to their superior performance. Recently, focusing on the shapes of carbon nanotubes, different shaped fullerenes such as SWNT (single-walled), MWNT (multiwalled), PSWNT (plasma-torch single-walled) have begun to be produced. Among these types, the most preferred ones are SWNT and MWNT, which can be synthesized according to product needs. In the late 1900s, the groundbreaking discovery of molecular carbon fullerenes for CNTs took place [27]. At this stage, coal ashes have begun to be seen as an alternative source. Today, the production of CNTs mostly uses CFA compared with CBA. Generally, various methods and carbon-rich CFA-derived catalysts are used to synthesize CNT, CNT strip, carbon nanofibers, carbon nanomaterials, epoxy composites based on carbon nanotubes.

Another type of material used to obtain carbon nanotubes is MSW ash. For example, soft packing and PET ash have the feasible to be pyrolysis oil/MWCN for electrocatalytic [28], as the dominant pyrolysis products obtained from waste plastics are sources for CNT synthesis. The fact that the yield and quality of CNTs can be affected by many variables such as raw material type, pyrolysis conditions, and catalytic reaction conditions should not be overlooked.

Porous carbons are a material consisting of various types and physical formations such as activated carbon and graphite (high surface). Several porous C products have been employed in the literature to remove organic chemicals and heavy metals from wastewater. Biomass is employed as a natural C source in the fusion of this porous carbon, which has key contributions [29]. The biological texture of biomass properties is vital in the creation of the leaky assembly of carbon [30].

4.5 Catalysis

Catalysts usually used industrially consist of metals/metal oxides. Especially metal oxide catalysts can catalyze different acid–base reactions, especially oxidation. Coal ash (CFA and CBA), a valuable type of waste, has recently received a lot of attention for use as a catalyst to increase productivity, reduce costs, and conserve metal resources. CFA and CBA contain oxides (e.g., Al_2O_3, S_iO_2, CaO, Fe_2O_3, MgO and K_2O), making it an ideal catalyst and increasing its commercial value. In addition, the mineralogical structure of these ash meets the thermal stability and mechanical properties required for its use as a catalyst [4,31].

Heterogeneous catalysts produced using CFA and CBA are environmentally friendly and cost-effective. They are also easily recovered by reaction compared with homogeneous catalysts (e.g., acid/base). The activity of these catalysts is controlled by the interaction between the active ingredients and the support materials. The support materials of different catalysts are mostly SiO_2, Al_2O_3, TiO_2, and MgO metal oxides. The high alumina-silicate content of CFA and CBA makes them a suitable material for catalyst support because of their high stability in reactions. Using coal ash as a substance has been proven in studies. For example, ash could be employed as a substance for reactions in compact stages, gas, and liquid, providing better catalytic activity in H_2 production, $deSO_x$-$deNO_x$, and hydrocarbon oxidation [1,32].

4.6 Silica aerogel

Aerogels are a material that has attracted great interest due to its great surface, great pore volumes, low density/low thermal conductivity, flexibility, and durability. It was first developed in the early 1900s because of the change between the liquid phase and the gas phase [33]. Many materials have been used to produce aerogels, such as alumina, ferric oxide, tungsten oxide, tin oxide, cellulose, gelatin, egg white, and rubber. However, the complex and costly procedures involved in its manufacture have long slowed the development of this material. The synthesis process of aerogels was then simplified under sol–gel transition and supercritical conditions; so, aerogels started to become popular again. The first step in this process is to use TMOS (tetramethoxysilane), TEOS (tetraethylorthosilicate), and polyethoxydisiloxanes to produce silica aerogels, although it increases the synthesis cost. Small amounts of aerogels prepared with these substances are subjected to a supercritical dry route (prevention of capillary stress/drying shrinkage) to attain desirable features [26,34].

Recently, ashes (coals) have been recycled and used to produce silica aerogel in an environmentally friendly way. Owing to the high Si-Al content and excellent insulation performance of coal ash, silica aerogels can be considered as an important source. Silica-derived materials such as coal ash offer economical/practical solutions for the large-scale making of aerogels. With the use of coal ash, silica aerogels continue to be used in diverse fields (e.g., electronic devices, thermal insulation, optical lasers, sensors, and devices,), especially adsorption and catalyst.

References

[1] Z.T. Yao, X.S. Ji, P.K. Sarker, J.H. Tang, L.Q. Ge, M.S. Xia, et al., A comprehensive review on the applications of coal fly ash, Earth-Science Reviews 141 (2015) 105–121.

[2] A. Bhatt, S. Priyadarshini, A.A. Mohanakrishnan, A. Abri, M. Sattler, S. Techapaphawit, Physical, chemical, and geotechnical properties of coal fly ash: a global review, Case Studies in Construction Materials 11 (2019) e00263.

[3] M.A. Munawar, A.H. Khoja, S.R. Naqvi, M.T. Mehran, M. Hassan, R. Liaquat, et al., Challenges and opportunities in biomass ash management and its utilization in novel applications, Renewable and Sustainable Energy Reviews 150 (2021) 111451.

[4] S.M.H. Asl, A. Ghadi, M.S. Baei, H. Javadian, M. Maghsudi, H. Kazemian, Porous catalysts fabricated from coal fly ash as cost-effective alternatives for industrial applications: a review, Fuel 217 (2018) 320–342.

[5] Y. Zhang, L. Wang, L. Chen, B. Ma, Y. Zhang, W. Ni, et al., Treatment of municipal solid waste incineration fly ash: state-of-the-art technologies and future perspectives, Journal of Hazardous Materials 411 (2021) 125132.

[6] K.K. Le Ping, C.B. Cheah, J.J. Liew, R. Siddique, W. Tangchirapat, M.A.B.M. Johari, Coal bottom ash as constituent binder and aggregate replacement in cementitious and geopolymer composites: a review, Journal of Building Engineering 52 (2022) 104369.

[7] M. Sumesh, U.J. Alengaram, M.Z. Jumaat, K.H. Mo, M.F. Alnahhal, Incorporation of nano-materials in cement composite and geopolymer based paste and mortar—a review, Construction and Building Materials 148 (2017) 62—84.

[8] Z. Li, M.E. Fei, C. Huyan, X. Shi, Nano-engineered, fly ash-based geopolymer composites: an overview, Resources, Conservation and Recycling 168 (2021) 105334.

[9] S. Oruji, N.A. Brake, R.K. Guduru, L. Nalluri, Ö. Günaydın-Şen, K. Kharel, et al., Mitigation of ASR expansion in concrete using ultra-fine coal bottom ash, Construction and Building Materials 202 (2019) 814—824.

[10] N. Shao, X. Wei, M. Monasterio, Z., Dong, Z. Zhang, Performance and mechanism of mold-pressing alkali-activated material from MSWI fly ash for its heavy metals solidification, Waste Management 126 (2021) 747—753.

[11] M. Jin, Z. Zheng, Y. Sun, L., Chen, Z. Jin, Resistance of metakaolin-MSWI fly ash based geopolymer to acid and alkaline environments, Journal of Non-Crystalline Solids 450 (2016) 116—122.

[12] K. Shiota, T. Nakamura, M. Takaoka, K. Nitta, K. Oshita, T. Fujimori, et al., Chemical kinetics of Cs species in an alkali-activated municipal solid waste incineration fly ash and pyrophyllite-based system using Cs K-edge in situ X-ray absorption fine structure analysis, Spectrochimica Acta. Part B: Atomic Spectroscopy 131 (2017) 32—39.

[13] J. Xie, W. Chen, J. Wang, C. Fang, B., Zhang, F. Liu, Coupling effects of recycled aggregate and GGBS/metakaolin on physicochemical properties of geopolymer concrete, Construction and Building Materials 226 (2019) 345—359.

[14] Z. Zhang, Y. Zhu, T. Yang, L. Li, H., Zhu, H. Wang, Conversion of local industrial wastes into greener cement through geopolymer technology: a case study of high-magnesium nickel slag, Journal of cleaner production 141 (2017) 463—471.

[15] E.M. Raisi, J.V., Amiri, M.R. Davoodi, Mechanical performance of self-compacting concrete incorporating rice husk ash, Construction and Building Materials 177 (2018) 148—157.

[16] P.C. Lau, D.C.L., Teo, M.A. Mannan, Characteristics of lightweight aggregate produced from lime-treated sewage sludge and palm oil fuel ash, Construction and Building Materials 152 (2017) 558—567.

[17] H.M. Hamada, G.A. Jokhio, F.M. Yahaya, A.M., Humada, Y. Gul, The present state of the use of palm oil fuel ash (POFA) in concrete, Construction and Building Materials 175 (2018) 26—40.

[18] European Commission, European Commission Critical Raw Materials Resilience: Charting a Path Towards Greater Security and Sustainability 1, Brussels, 2020.

[19] S. Das, G. Gaustad, A. Sekar, E. Williams, Techno-economic analysis of supercritical extraction of rare earth elements from coal ash, Journal of Cleaner Production 189 (2018) 539—551.

[20] D. Talan, Q. Huang, A review study of rare earth, cobalt, lithium, and manganese in coal-based sources and process development for their recovery, Minerals Engineering 189 (2022) 107897.

[21] V. Funari, S.N.H. Bokhari, L. Vigliotti, T., Meisel, R. Braga, The rare earth elements in municipal solid waste incinerators ash and promising tools for their prospecting, Journal of Hazardous Materials 301 (2016) 471—479.

[22] J. Mehr, M. Haupt, S. Skutan, L. Morf, L.R. Adrianto, G. Weibel, et al., The environmental performance of enhanced metal recovery from dry municipal solid waste incineration bottom ash, Waste Management 119 (2021) 330—341.

[23] M. Reig, X. Vecino, C. Valderrama, I., Sirés, J.L. Cortina, Waste-to-energy bottom ash management: copper recovery by electrowinning, Separation and Purification Technology (2023) 123256.

[24] M.A. Al-Ghouti, M. Khan, M.S. Nasser, K., Al-Saad, O.E. Heng, Recent advances and applications of municipal solid wastes bottom and fly ashes: insights into sustainable management and conservation of resources, Environmental Technology & Innovation 21 (2021) 101267.

[25] S.E. Perämäki, A.J., Tiihonen, A.O. Väisänen, Occurrence and recovery potential of rare earth elements in finnish peat and biomass combustion fly ash, Journal of Geochemical Exploration 201 (2019) 71–78.

[26] A.R. Gollakota, V. Volli, C.M. Shu, Progressive utilisation prospects of coal fly ash: a review, Science of the Total Environment 672 (2019) 951–989.

[27] S. Iijima, Helical microtubules of graphitic carbon, Nature 354 (6348) (1991) 56–58.

[28] A. Veksha, K. Yin, J.G.S. Moo, W.D. Oh, A. Ahamed, W.Q. Chen, et al., Processing of flexible plastic packaging waste into pyrolysis oil and multi-walled carbon nanotubes for electrocatalytic oxygen reduction, Journal of Hazardous Materials 387 (2020) 121256.

[29] C.J. Thambiliyagodage, S. Ulrich, P.T., Araujo, M.G. Bakker, Catalytic graphitization in nanocast carbon monoliths by iron, cobalt and nickel nanoparticles, Carbon 134 (2018) 452–463.

[30] L. Wang, X. Hu, Recent advances in porous carbon materials for electrochemical energy storage, Chemistry–An Asian Journal 13 (12) (2018) 1518–1529.

[31] M. Shahbaz, S. Yusup, T. Al-Ansari, A. Inayat, M. Inayat, H. Zeb, et al., Characterization and reactivity study of coal bottom ash for utilization in biomass gasification as an adsorbent/catalyst for cleaner fuel production, Energy & Fuels 33 (11) (2019) 11318–11327.

[32] C. Belviso, State-of-the-art applications of fly ash from coal and biomass: a focus on zeolite synthesis processes and issues, Progress in Energy and Combustion Science 65 (2018) 109–135.

[33] C.M. Almeida, M.E. Ghica, L. Duraes, An overview on alumina-silica-based aerogels, Advances in Colloid and Interface Science 282 (2020) 102189.

[34] Y.R. Lee, J.T. Soe, S. Zhang, J.W. Ahn, M.B. Park, W.S. Ahn, Synthesis of nanoporous materials via recycling coal fly ash and other solid wastes: a mini review, Chemical Engineering Journal 317 (2017) 821–843.

Legal framework for ashes

Abstract

This chapter provides a brief introduction to the management and assessment of industrial solid ash within the legal framework. It also reveals the impact and importance of the resourceful management of built-up ash on the environment and human health. Then, key legislation and regulations around the world for the management of these ashes are examined in detail. Besides, the effective and sustainable management of solid ash necessitates addressing many legislations and regulations covering different processes from the production of these ashes to their disposal. Legislation and regulations related to industrial solid ashes actually aim at the safe disposal of these ashes, which minimizes or prevents their harm to the environment. This chapter deals with the safe, sustainable, and environmentally friendly ways of industrial solid ashes by considering their legalities and rules.

5.1 Introduction

Industrial solid ashes are by-products of industrial processes such as fossil fuel use, waste incineration, and metal production. Due to their structural characteristics, they are partially included in the hazardous waste category, showing that these ashes can cause many environmental problems. For example, these ashes contain various pollutants including heavy metals, arsenic, and radioactive materials that can pollute the soil, water, and air, causing them to be seen as a potential threat. Therefore, improper disposal and management of industrial ash can lead to groundwater pollution, air pollution, and other ecological problems [1]. In addition, these pollutants can pose a significant risk to social environment, chiefly when they move in the nutrition chain by the polluted soil or water. Thus it is important that industrial solid ash is properly managed, regulated, and recycled to ensure that it does not harm the environment and human health [2].

Sustainable and safe management of industrial solid ashes includes diverse processes from production to disposal of these ashes. These processes include operations such as wastes' transportation/storage/recycling/disposal. Landfill is the best ideal method, but it leads to environmental pollution such as groundwater pollution and air pollution from landfill gas emissions and other effects. In addition, industrial solid ash recycling and/or reuse could lessen the quantity of the waste created and provide many economic benefits. There are many legislations and regulations for the management of industrial solid ashes. Overall, the legislation and regulations related to industrial solid ashes aim to ensure the safe storage, transportation, and disposal of these ashes without harming the environment. In this way, it is aimed to minimize the

Machine Learning Applications in Industrial Solid Ash. DOI: https://doi.org/10.1016/B978-0-443-15524-6.00003-0
© 2024 Elsevier Inc. All rights reserved.

environmental effects of wastes and to prevent wastes from harming human health. The management of industrial solid ashes should be carried out in cooperation with waste makers, municipalities, local governments, ecological protection organizations, and other interested parties [3–5]. In this manner, while the environmental effects of wastes are minimized, resource utilization can be optimized.

Regulations and regulations on industrial solid ashes vary by country and region. Some countries have strict regulations on the management of industrial solid ashes, while in others the regulations are more lenient. Let us say, EPA puts principles for managing hazardous wastes such as coal ash. It limits disposal of waste with strict rules such that facilities must monitor groundwater, maintain appropriate shutdown and postclosure procedures, and meet certain structural and engineering desires [6]. It also governs requests for managing waste out of industrial processes, including solid ash, under the EU Industrial Emissions Directive (IED). IED requires member states to establish permit conditions for industrial facilities that generate waste, including the usage of leading existing practices to manage waste, as well as waste prevention and minimization requests [7]. In China, the Hazardous Waste Management Legislation emphasizes that facilities should establish a hazardous waste management system, carry out risk assessments, and use right storage and disposal methods.

The management, regulation, and legislation of industrial solid ashes are complex issues that require collaboration between governments, industries, and stakeholders. As the world continues to industrialize, there will be a corresponding increase in the amount of industrial solid ashes produced. It is therefore crucial to have a robust regulatory framework globally to ensure the safe management of industrial solid ashes. Industrial solid ash regulation and legislation can help establish minimum standards and ensure compliance with best practices. However, the effectiveness of these regulations depends on their implementation and the industry's willingness to comply with them. The regulations and legislation discussed in this section provide a snapshot of how countries around the world are addressing this issue. While there are differences in approaches and practice, the ultimate goal is to manage industrial solid ash safely and sustainably.

5.2 Review of coal fly/bottom ash regulations

Legislation on coal burning products, also known as coal ash, differs from country to country, and these legislations are inspected by different institutions in countries. The basis of the legislation on coal ash is based on the classification of these ash. In addition, these regulations allow countries to consider coal ash as hazardous or nonhazardous waste. Most countries do not qualify coal ash as hazardous waste, so the storage/disposal of coal ash is limited by certain regulations/rules. The aim of the states' characterization of coal ash as nonhazardous waste can be thought of as boosting the coal usage and escaping needless restrictions from a sectoral point of view.

Until today, the best widespread universal ecological pact for hazardous and harmless wastes, especially coal ash, is the *Basel Pact on the Regulator of Transboundary Activities of Risky Wastes and Their Discarding*. This pact, executed by UN/OECD, entered into power in 1992 and has common participation with 175 members (as of 2011). It also aims to abate human health and environmental damage by eliminating the negative conditions related to wastes' management/cross-border measure. It covers the management of toxic, explosive-corrosive, flammable, eco-toxic, and infectious wastes in general terms. They have obligations such as minimizing the amount of waste, treatment/disposal at the closest location where they are made and preventing/reducing the source of waste generation. In addition to the Basel pact, there are many industrial rules and limitations due to the potential problems of coal ash in many countries where coal making is intense [8]. Below is a brief description of the regulation in some nations where coal/coal ash making is intense (e.g., China, India, United States, EU, Australia).

5.2.1 Legislations for Chinese coal ashes

In general, Chinese coal ashes are categorized as manufacturing solid waste, not hazardous waste, due to its potential benefits. The control of coal ash in the country is provided by both NDRC (National Development and Reform Commission) and MEE (Ministry of Ecology and Environment). MEE continues its activities by replacing MEP (Ministry of Ecological Safety) as of 2018. NDRC reassures firms to consider coal ashes as a reuse source. It also oversees the development of clear rules and plans for coal ash's disposal, decreasing its environmental impacts over a period. MEP, conversely, carries out studies on reducing the environmental effects of coal ash and protecting the ecological balance. This institution is responsible for instituting and enforcing regulations for the preservation of ecological environment in urban/rural areas, the rule of pollutant emissions, and the decrease of ecological pollution (air, water, soil) [9–11].

China now spreads *Regulation on Stoppage and Govern of Ecological Trash by Industrial Waste*, which is the most relevant policy on coal ash and has been in effect since 2005. Although this law includes various regulations on the prevention of environmental effects of ash, MEE is responsible for its implementation and control. In this law, solid wastes together with coal ash are divided into classes I and II. Coal ash is specified as class II, and it is emphasized that coal ash that is not reused/reused can be stored [10]. Moreover, *Principles for Contamination Govern on the Storage/Disposal Areas for Overall Built-up Waste* remains valid for these ashes. The standard includes instructions for the selection of storage areas for coal wastes and the prevention of leakages (e.g., dust and leakage) in the storage areas. Other existing local/regional laws for environmental protection and resource use in China are keenly used for coal ash management. However, almost all of them consist of basic laws, their contents are quite general, and their execution in practice is quite difficult. Some problematic standards and laws for ashes are exerted on prevention of leakage from this ash. The most common and relevant among those used is *Contamination Govern on the Storage/*

Disposal Areas for Overall Built-up Waste. There also exist over 20 state/business codes in practice or in making for coal ash. Although some of the specified standards are mandatory, they consist of voluntary standards of institutions and industries for incentive purposes. Therefore, the excess and content of the standards are limited for coal-fired power plants if they are not implemented or supervised [8,12].

5.2.2 India

Indian power generation is highly based on coal and is expected to remain so soon. Although the quality and calorific rate of Indian coals are low, ash content (30%–45%) is quite high [13]. Though the produced coal ashes are not considered hazardous waste by the Indian government, the disposal and storage of these ashes have become a major challenge. Management of coal ash is also a major concern due to its environmental impacts (e.g., air and water pollution) and disposal conditions. Hence, India's EFCC (Ministry of Environment, Forest and Climate Change) circulated a communiqué in 2000 to address the environmental impacts on the management of coal ash. In this communiqué, targets are set for the gradual 100% use of ash from coal/lignite thermic power facilities. The targets set in 1999 were revised by this ministry in 2003 and 2009. Overall, the goals of this communiqué are to keep the environs-top soil, to prevent the discharge of the ashes created in thermic power facilities to land, and to encourage employing coal ash in building/construction activities. After this communiqué was published in India, positive results were obtained on the use and management of coal ash. For example, the use of coal ash increased from 13.5% to 57.6% in 2013 and 2014, but the targeted 100% usage amount could not be reached.

EFCC held a Monitoring Committee meeting in 2014 to review the use status of coal ash. The content of this meeting has been formed based on the provisions covered by the coal ash use notification. Within the scope of the meeting, institutions such as Development Department, Central Public Works Department, Indian National Highways Authority were asked to use ashes in structure practices. Studies were also requested from Central Pollution Control Board to evaluate the ecological effects connected to the usage of coal in abandoned mines. In addition to the requirement for coal ash's reuse, EFCC has imposed an obligation to pay for the transport/storage of ash out of thermic power facilities, depending on the proximity (300 km radius) to government projects such as buildings, roads, dams, and earth fills. There are permits for ash using in agricultural areas and soils [12].

5.2.3 United States

The regulations/guidelines for coal ash in United States are generally developed by EPA. The EPA approved the final regulation for *Discarding of Coal Burning Wastes from Electric Utilities* in 2014 and published it in the Federal Register in 2015. The entrance into power of this rule was triggered by leakage of large amounts of coal ash (over 4 million m^3) because of the dam accident in New York. The increase in the release of various metals, especially methyl mercury, in the research within this region has revealed the environmental risks that may be caused by the improper disposal and storage of coal ash [14]. The 2015 coal ash regulation addresses many issues such as storage, facility location, safety, groundwater protection, and air transfer of dust from ash. Moreover, this arrangement has improved various aspects such as design, monitoring, operation, and closure for surface dams and storage areas. While this regulation covers existing, old, and new facilities, closed/powerless facilities are excluded from this regulation [6].

Discarding of Coal Burning Wastes from Electric Utilities (2015) sets out practical needs for coal ash fills and surface dams below subheading D of *Resource Conservation and Recovery Act (RCRA)*, United States' first law to regulate solid waste. RCRA is United States' key law leading the disposal of solid/risky waste. In 1976, RCRA was adopted to solve the rising difficulties of the involuntary growth of municipal/industrial waste. RCRA was enacted by effectively replacing the *Solid Waste Disposal Act* of 1965. RCRA's main objectives are to guard the environment/human health from waste disposal, to lessen the sum of waste, to manage waste in a more environmentally responsible way, and to protect energy and natural resources. In line with the stated objectives, RCRA has recognized three diverse plans, namely interrelated solid waste, risky waste, and underground storage tank (UST) programs. Solid waste program under RCRA's Subtitle D reassures nations to improve policies to better treat harmless industrial/municipal wastes. It creates principles for municipalities' waste storage/disposal points and stops these wastes from being discharged in open areas. Risky waste program below RCRA's Subtitle C starts a scheme for the final disposal of hazardous waste from its generation. The UST program under RCRA's Subtitle I oversees/regulates the conditions for underground storing of dangerous goods and petroleum-containing products. EPA classified coal ash as "special wastes" under RCRA's Subtitle C because of the environmental and health problems it causes. But, in 1980, the US Congress passed the Solid Waste Disposal Act Amendments, exempting special wastes such as coal ash from regulation until further risk assessment. Coal ash was removed from Subtitle C of RCRA by EPA after detailed examination and included in Subtitle D of RCRA. As a result of this situation, coal ash has been separated from federal government regulations and subjected to State-based regulations in United States [15–17].

Along with the regulations stated earlier, there are many works on regulation of ash in United States. Although these studies have a long history, they have a complex structure. Fig. 5.1 summarizes the evolution of rules for US coal ashes.

5.2.4 European Union

In recent years, coal ash has received major attention in the European Union (EU) due to its potential negative impact on both human health/atmosphere. To solve this worry, EU developed a set of regulations aimed at regulating the treatment, disposal, and use of ash throughout the region. EU regulations on the treatment of the ash gained momentum after a series of major coal ash spills in the United States in 2014. Regulations set out various requirements for the secure control/disposal of

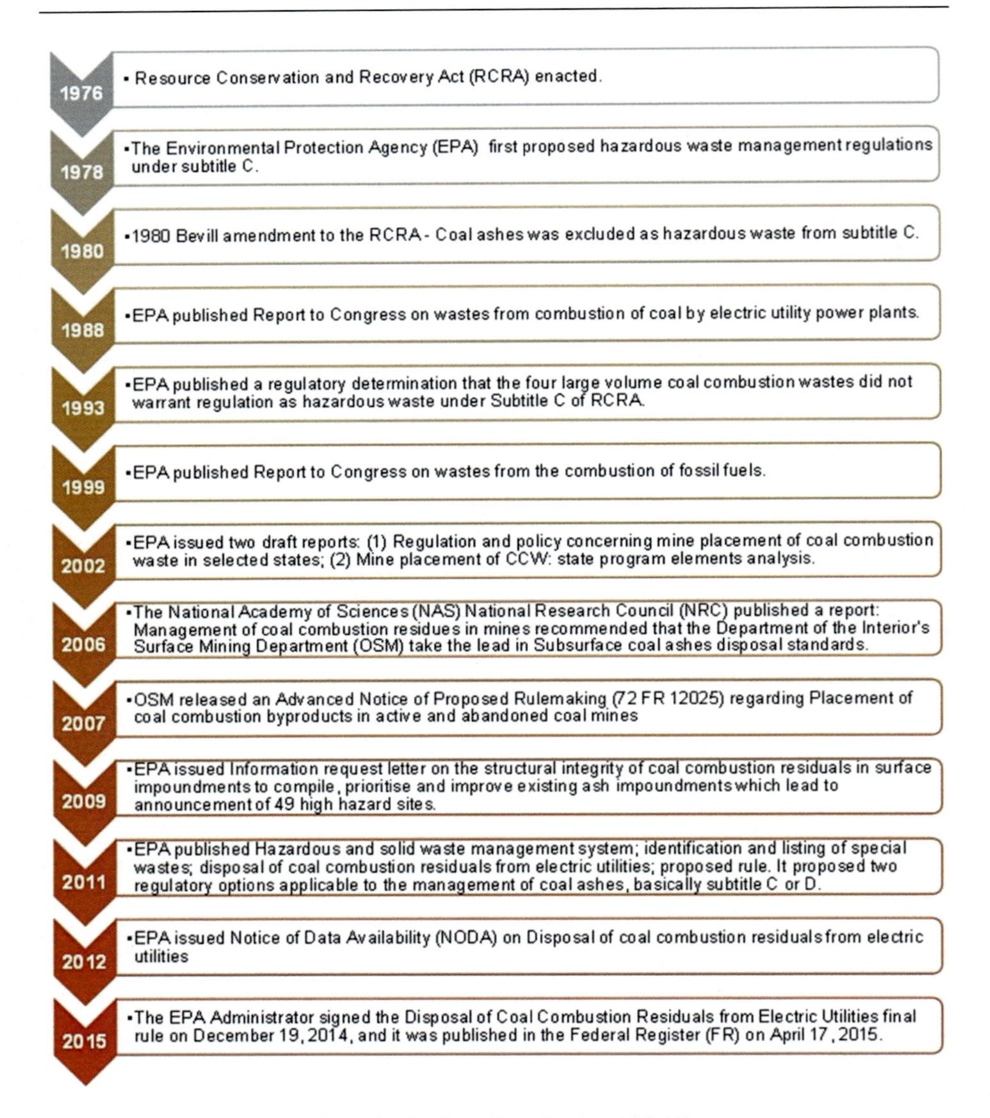

Figure 5.1 Historical evolution of rules for US coal ashes [6,8,18].

ash, including monitoring, reporting, and containment standards. It also identifies a set of criteria that must be met to determine whether coal ash is risky, which has implications for how they are managed and disposed of. In addition to the general coal rules, EU has established specific regulations for ash using in building products. These regulations outline ash's permitted uses in construction materials such as concrete, asphalt, and masonry products. Regulations also set limits on the amount of certain hazardous elements, such as mercury and cadmium, that may be present in ash-based products.

EU has some regulations governing coal ash management, including IED (Industrial Emissions Directive), Waste Framework Directive, and Landfill Directive. These rules set out supplies for coal ash's delivery, storage and disposal, and monitoring and reporting of its environmental impacts. Under IED, operators of large combustion plants, including coal-fired power plants, are required to employ principal existing systems to minimize emissions and cut release of pollutants such as metals and other harmful substances. IED also sets requirements for monitoring and reporting emissions, including those from coal ash storage and disposal facilities [19]. The Waste Framework Directive sets rules for managing risky wastes, including coal ash, and aims to promote using viable waste treatment ways such as reuse/recycling/recovery. The Directive requires that hazardous waste be managed and stored safely, with a focus on minimizing possible hazards to community and atmosphere [20]. Landfill Directive, conversely, sets principles for waste storage, covering ash, in landfills. It aims to cut the ecological effect of landfills by promoting the use of sustainable waste management practices and requiring landfill operators to take measures to prevent/minimize the release of contaminants such as metals and other harmful stuffs [21].

In addition to these regulations, EU has developed technical guidance on coal ash management. For example, the Best Available Techniques Reference (BREF) documents provide guidance on key existing methods to diminish releases from diverse sectors, covering coal power plants. BREFs are updated periodically to reflect advances in technology and changes in regulations [22]. EU also maintains a public register, the European Pollutant Release/Transfer Register (E-PRTR), which offers data on pollutant releases and transfers from industrial facilities, including coal power plants. E-PRTR goals to endorse clearness and accountability in ecological decision-making by providing public access to information on emissions and transfers of pollutants [23].

Along with the regulations mentioned earlier, almost all EU countries interpret and implement these regulations according to their own domestic legislation. Fig. 5.2 shows the regulation and management practices of some EU countries in general. As can be seen here, coal ash is not considered as hazardous waste by EU countries, despite its environmental and human health effects.

5.2.5 Australia

In Australia, coal ash is classified as industrial waste. While the regulations used in the country reflect those used in the United States/EU, there exists no federal regulation. Consideration is given to coal ash disposal and disposal at the state level (e.g., New South Wales, Queensland, and Victoria) for major coal mines and power plants. In addition, the management of coal ash on a state basis is based on general environmental regulations, apart from current regulations.

Regulations and rules for coal ash in Australia vary by state or territory. For this reason, coal ash landfills are inconsistent and do not meet comparable construction, management, and rehabilitation standards in other nations. For example, regulations regarding the management of storage and disposal of coal ash change from Nation

Figure 5.2 Several EU countries and their coal ash regulations [8].

to Nation and from coal thermic power facility to coal thermic power facility in each Nation. There are implementation management standards for coal ash disposal at the state or national level. Operators in thermal power plants do not have rehabilitation and postclosure plans for power plant shutdown. Within the scope of information access and freedom, access to groundwater monitoring, access to coal ash management, and information access to coal ash storage are extremely limited. In addition, Victoria is the only state with nationally recognized terms and regulations to provide financial security for ash deposits [24].

5.3 MSW/biomass ash regulations

Increasing population/living standards greatly increase the product and energy consumption in the world, and this triggers the generation of waste, which has been explained in detail in the previous sections. In this context, it is well known that waste management has an important place for countries, and the solutions to be implemented should not only have environmental concerns but also cost-effective, social, and sustainable purposes. Thus waste management should be evaluated from diverse views (e.g., political, economic, environmental, social, technological developments, education). Waste management aims to successfully carry out multistage processes such as waste collection, treatment, and disposal due to diverse countries determining their own strategies, creating national plans, and implementing the basic legal policies. For this purpose, regime rules, tolls, and provision programs have a key character. Table 5.1 shows the official legal instruments for solid waste management of some countries. From this section, information about the waste management of nations/unions in world's diverse regions will be presented, and municipal solid waste and biomass ashes will be mentioned in these regulations.

5.3.1 Legislations for ashes in China

The plain rule of Chinese MSW/biomass ashes is based on the Stoppage and Govern of Ecological Effluence by Built-In Waste, which was adopted in 1996 and reviewed in 2021. Due to wastes' increasing amount, many Chinese national policies and rules have been established to further use them in diverse industries as a raw material [26]. Much care is taken to MSW's classification in China, and the first administration rule on MSW treatment is "Temporary Trials for the Sorting and Running of City Wastes in China (Guangzhou)."

Although regulations for the classification and disposal of MSW date back to earlier, the inclusion of fly ash formed through burning of MSW wastes into National Risky Waste List took place in 2016. With the law (China's National Ecological Protection Standards) that came into force in August 2020, the environmental management of MSW-FA has been standardized, and the quality of the ecological environment has been improved by preventing ecological environmental

Table 5.1 Some legal instruments for solid waste management of different countries [25].

Country	Legitimate tools
Denmark	Ecological safety act
Japan	Waste management/public cleansing law
Singapore	Ecological security/management act
Germany	Basic law
Taiwan	Waste disposal act
South Korea	Waste control act
Netherlands	Soil protection act

pollution. This law sets general requests for MSW-FA gathering, storing, transport, treatment, and discarding of and controls pollution. Various regulations mandate the disposal of MSW residues by appropriate pretreatment in China.

5.3.2 Legislations for ashes in the European Union

EU has some laws covering authorized regulations and implementation procedures on managing options of wastes. While they must be implemented by EU nations, affiliated nations could create proper legislation depending on different conditions, provided that they do not contradict the general understanding. The legal regulations and rules created are responsible for reducing the effects of wastes on environment/human health.

As burning of European wastes is legitimately banned for entire nations, national law [27] has been harmonized with Ruling 2010/75/EU on industrial emissions [28,29]. It is understood from this that if wastes' burning is run in other nations other than the EU, it must meet the least requests left in this ruling. The working settings that certify appropriate recycling of waste are examples of minimum requirements. The relevant directive requires not only the proper incineration of waste but also the reduction and recycling of plant wastes. Before recycling processes, it is needed to determine the characteristics of waste and its pollutant potential. Wastes from burning are also subject to Directive Conclusion 2014/955/EU [30] as they are considered as waste. With this commission decision, a Waste List has been created that classifies waste as risky or harmless and defines the types of wastes.

It is extremely important to develop applicable waste treatment policies for biomass ash in terms of the creation of sustainable energy. Ash gained as a result of the incineration of biomass is categorized as waste as stated by the List of EU Wastes [31]. The joint managing used for biomass ash is fill storage, causing some environmental problems. The European Landfill Directive [32] has increased penalties for discarding practices and has made it difficult for companies to seek different management ways instead of storage areas. EU has planned for the recycling of waste as a replacement source and has determined threshold values for fill's discarding. WFD (Waste Framework Directive) uses products identified as waste and provides several incentives for increased waste generation [33].

EU has commenced some events to give to round economy of waste reuse/recycling activities. For example, the Waste-end strategy [34] and the review of the European fertilizer directive [35]. Legislation about use of biomass ash in various fields (e.g., forestry, agriculture, or construction) may differ between European countries. For example, while national legislation is in force in Denmark, Finland, and Austria, Sweden offers approvals for using biomass ash in forestry/agriculture [36]. Employing biomass ash in topsoil depends on biomass ash's element content. In addition, recycling these ashes as a building product meets the endorsements of EU Directive 2008/98/EC and has major ecological aids correlated to the reduction of waste transported to landfills. Reusing biomass ash as a paltry cement substitution product is not permitted by ASTM C 618 and CEN EN 450−1, the standards governing fly ash usage as mineral additives in cement-based products such as concrete.

5.3.3 Legislations for ashes in the United States

The leaching properties of these ashes, alone or together, are the critical determinants of the environmental harmlessness and use of MSW residues (BA and FA). Each country evaluates the leaching potential of soluble salt and heavy metal presence of ashes as stated by the threshold limits measured by using their own leaching test standards. Therefore there is wide variation between some chemical constituents of MSW residues.

In the United States, bottom/fly ashes (BA/FA) are mixed together, and the blend ash is dumped into landfill. For ashes to be considered as harmless waste, TCLP (Toxicity Characteristic Leaching Procedure; SW-846 EPA Method 1311) should be provided by RCRA (Resource Conversation and Recovery Act). BA usually provides this procedure, while FA often fails. Thus the United States combines and disposes of these ashes to avoid high costs. TCLP test, which uses acetic acid to simply sort if ash is risky or harmless, regardless of on-site leaching status, was designed by the EPA. As the results of leaching tests using this procedure generally yield higher results than field conditions, an another leaching test method, The Artificial Rain Leaching Procedure (EPA Method 1312) was developed to better reflect field conditions and obtain accurate results.

RCRA is the law considered in the discarding of risky waste in United States. RCRA regulations regarding all stages of hazardous waste (e.g., identification, classification, making, management, and disposal) are set out in diverse sections of Federal Regulation codes. Title 40-Environmental Protection mainly regulates environmental regulations promulgated by the EPA.

References

[1] B. Qiu, C. Yang, Q. Shao, Y. Liu, H. Chu, Recent advances on industrial solid waste catalysts for improving the quality of bio-oil from biomass catalytic cracking: a review, Fuel 315 (2022) 123218.

[2] N.K. Soliman, A.F. Moustafa, Industrial solid waste for heavy metals adsorption features and challenges; a review, Journal of Materials Research and Technology 9 (5) (2020) 10235–10253.

[3] A.T. Hoang, P.S. Varbanov, S. Nižetić, R. Sirohi, A. Pandey, R. Luque, et al., Perspective review on municipal solid waste-to-energy route: characteristics, management strategy, and role in circular economy, Journal of Cleaner Production (2022) 131897.

[4] S. Kalisz, K. Kibort, J. Mioduska, M. Lieder, A. Małachowska, Waste management in the mining industry of metals ores, coal, oil and natural gas-a review, Journal of Environmental Management 304 (2022) 114239.

[5] J. Zhai, I.T. Burke, D.I. Stewart, Beneficial management of biomass combustion ashes, Renewable and Sustainable Energy Reviews 151 (2021) 111555.

[6] EPA, 2023. United States environmental protection agency, https://www.epa.gov/.

[7] EC, 2023. European Commission, https://commission.europa.eu/index_en. Environmental Justice Australia 2019. Unearthing Australia's toxic coal ash legacy (how the regulation of toxic coal ash waste is failing Australian communities). https://envirojustice.org.au/.

[8] X. Zhang, Management of coal combustion wastes, IEA Clean Coal Centre (2014) 2–68.
[9] Ministry of Commerce People's Republic of China (MOFCOM) 2007. Law of the people's republic of china on prevention and control of environmental pollution by solid waste. http://www.mofcom.gov.cn/.
[10] Ministry of Ecology and Environment (MEE) 2023. https://www.mee.gov.cn/.
[11] National Development and Reform Commission (NDRC) 2023. https://www.ndrc.gov.cn/.
[12] K.M. Zierold, C. Odoh, A review on fly ash from coal-fired power plants: chemical composition, regulations, and health evidence, Reviews on Environmental Health 35 (4) (2020) 401–418.
[13] H.P. Jambhulkar, S.M.S. Shaikh, M.S. Kumar, Fly ash toxicity, emerging issues and possible implications for its exploitation in agriculture; Indian scenario: a review, Chemosphere 213 (2018) 333–344.
[14] J.G. Smith, T.F. Baker, C.A. Murphy, R.T. Jett, Spatial and temporal trends in contaminant concentrations in hexagenia nymphs following a coal ash spill at the Tennessee Valley Authority's Kingston Fossil Plant, Environmental Toxicology and Chemistry 35 (5) (2016) 1159–1171.
[15] J.Y. Park, Assessing determinants of industrial waste reuse: The case of coal ash in the United States, Resources, Conservation and Recycling 92 (2014) 116–127.
[16] Resource Conservation and Recovery Act (RCRA) 1976. Summary of the Resource Conservation and Recovery Act. 42 U.S.C. §6901, https://www.epa.gov/.
[17] Environmental Protection Agency (EPA) Final Rule 2015, Hazardous and Solid Waste Management System; Disposal of Coal Combustion Residuals From Electric Utilities. Federal Register 80(74), 2015-00257.
[18] Luther, L.G. 2009. Managing coal combustion waste (CCW): Issues with disposal and use. Congressional Research Service, Library of Congress.
[19] Industrial Emission Directive (IED) 2010, Directive 2010/75/EU of the European Parliament and of the Council of 24 November 2010 on industrial emissions (integrated pollution prevention and control). http://data.europa.eu/eli/dir/2010/75/2011-01-06.
[20] Waste Framework Directive (WFD) 2008. Directive 2008/98/EC of the European Parliament and of the Council of 19 November 2008 on waste and repealing certain Directives. http://data.europa.eu/eli/dir/2008/98/2018-07-05.
[21] The Landfill Directive (LD) 1999. Council Directive 1999/31/EC of 26 April 1999 on the landfill of waste. http://data.europa.eu/eli/dir/1999/31/2018-07-04.
[22] Lecomte, T., Ferreria de La Fuente, J.F., Neuwahl, F., Canova, M., Pinasseau, A., Jankov, I., et al., 2017. Best available techniques (BAT) reference document for large combustion plants. Industrial emissions directive 2010/75/EU (Integrated pollution prevention and control) (No. JRC107769). Joint Research Centre (Seville site).
[23] The European Pollutant Release and Transfer Register (E-PRTR) 2023. https://industry.eea.europa.eu.
[24] Australia, E.J. Unearthing Australia's toxic coal ash legacy (report), 2019.
[25] A. Liu, F. Ren, W.Y. Lin, J.Y. Wang, A review of municipal solid waste environmental standards with a focus on incinerator residues, International Journal of Sustainable Built Environment 4 (2) (2015) 165–188.
[26] Y. Wei, J. Li, D. Shi, G. Liu, Y. Zhao, T. Shimaoka, Environmental challenges impeding the composting of biodegradable municipal solid waste: a critical review, Resources, Conservation and Recycling 122 (2017) 51–65.
[27] EU. Regulations, Directives and other acts. EU law. European Union, 2019.

[28] D. Blasenbauer, F. Huber, J. Lederer, M.J. Quina, D. Blanc-Biscarat, A. Bogush, et al., Legal situation and current practice of waste incineration bottom ash utilisation in Europe, Waste Management 102 (2020) 868–883.
[29] EU. Directive 2010/75/EU of the European Parliament and of the Council of 24 November 2010 on industrial emissions (integrated pollution prevention and control) Text with EEA relevance. European Parliament and the Council of the European Union, 2010.
[30] EU. Commission decision of 18 December 2014 amending Decision 2000/ 532/EC on the list of waste pursuant to Directive 2008/98/EC of the European Parliament and of the Council (Text with EEA relevance) (2014/955/EU), 2014/ 955/EU. European Commission, 2014.
[31] EC (European Commission). Commission Decision (2000/532/EC) of 3 May, 2000.
[32] European Council. Council Directive 1999/31/EC on the landfill of waste. Off. J. Eur. Communities L 182/1e19, 16 July, 1999.
[33] I. Costa, G., Massard, A. Agarwal, Waste management policies for industrial symbiosis development: case studies in European countries, Journal of Cleaner Production 18 (8) (2010) 815–822.
[34] European Commission. Report from the commission to the European parliament, the council, the European economic and social committee and the committee of the regions on the Thematic Strategy on the Prevention and Recycling of Waste SEC(2011) 70 final (No. COM(2011) 13 final), 2011.
[35] European Commission. Proposal for a regulation of the European parliament and of the council laying down rules on the making available on the market of CE marked fertilising products and amending Regulations (EC) No 1069/2009 and (EC) No 1107/2009 (Proposal No. COM(2016) 157 final), 2016.
[36] L. Tosti, A. van Zomeren, J.R. Pels, J.J., Dijkstra, R.N. Comans, Assessment of biomass ash applications in soil and cement mortars, Chemosphere 223 (2019) 425–437.

Background of machine learning

6

Abstract

This chapter provides a brief introduction to the background of machine learning. First, the development history of machine learning is described and a clear comparison between machine learning and other widely-used terms is present. The categories of machine learning types are then summarized, including supervised learning, unsupervised learning, and semisupervised learning. Some technologies commonly used in machine learning are introduced in depth, including but not limited to decision tree, random forest, support vector machine, artificial neural network, etc. Finally, the programming languages and widely-used open-source libraries commonly used in machine learning are briefly summarized at the end of this chapter.

6.1 History of machine learning

With the advent of the era of big data, artificial intelligence (AI), machine learning (ML), and deep learning (DL) have become increasingly popular. These terms appear frequently and are often used interchangeably, leading to the belief that they are synonyms; however, there are subtle and important differences between these three concepts (Fig. 6.1).

ML stems from the pursuit of AI. AI itself is a technical science that investigates theories, methods, technologies, and application systems to simulate and extend human intelligence. AI uses experience to acquire knowledge and skills and then applies that knowledge to new, unseen environments. AI can not only identify

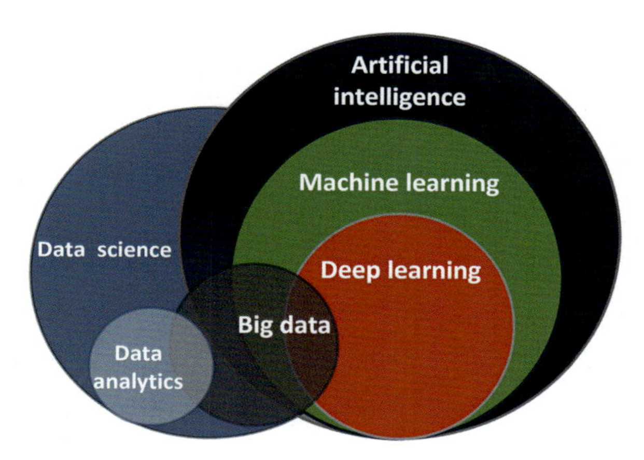

Figure 6.1 Differences and connections between ML and other terms.

Machine Learning Applications in Industrial Solid Ash. DOI: https://doi.org/10.1016/B978-0-443-15524-6.00015-7
© 2024 Elsevier Inc. All rights reserved.

important patterns in the massive data collected for data mining but can also be used to analyze and classify image data, an application area with a huge scope that has received extensive research attention. The core concept of ML, as an important branch of AI, is to learn patterns through experience. ML was originally used only as a training program for AI, but increasing emphasis on logical and knowledge-based approaches has led to ML being differentiated as a separate field that has begun to flourish in its own right. Researchers now less commonly use the "traditional" approach of writing software by hand, determining a particular set of instructions, and then asking the program to perform a particular task; instead, the approach of "training" machines to learn how to algorithmically perform tasks using huge volumes of data has become mainstream. The research focus has also shifted from the symbolic methods inherited from AI to methods and models borrowed from statistics, fuzzy logic, and probability theory [1]. Finally, DL, a subset of ML, is a technology that analyzes and learns from data by building neural networks to imitate the mechanisms of the human brain.

ML is also closely related to the fields of data mining, optimization, generalization, and statistics [2]. ML utilizes statistical methods in which algorithms are trained to perform categorization or prediction and discover key insights in data mining projects. Although ML intersects with many fields of data science, it has its own unique characteristics. For example, the field of statistics extracts population inferences from a sample, whereas ML finds generalizable prediction patterns. Although there is overlap and crossover with data mining, the goals of ML approaches are different. ML focuses on discovering known attributes from training data, whereas data mining focuses on mining the unknown attributes of data [3]. ML integrates data science to simulate and extend AI and has become an important method to enable advances in the field of cutting-edge technology.

ML has a long history, and the term ML was first formally coined by IBM's Arthur Samuel in 1952 [4]. By designing a checkers program that could teach itself, Arthur Samuel overturned the conventional wisdom that machines could not outdo humans, write code, or learn like humans. In terms of the foundations of ML, in the 18th century, Thomas Bayes developed a probabilistic mathematical theorem known as Bayes' theorem [5], which remains a central concept in some modern ML methods. In the early to mid-19th century, Adrien-Marie Legendre described the least squares method [6], and Warren McCulloch et al. developed a mathematical model that mimics the function of biological neurons [7]. Additionally, Alan Turing proposed a "learning machine" in 1950 that foreshadowed genetic algorithms [8], and Marvin Minsky et al. built the first learning neural network machine, SNARC, in 1951 [9]. All of these approaches laid the theoretical and methodological foundations for developing and applying ML. Subsequently, until the early 1960s, Frank Rosenblatt combined brain cell interactions with ML efforts to create the perceptron concept [10]. In 1967, the nearest neighbor algorithm was created, which formed the basis of simple pattern recognition [11]. As some limitations of the perceptron and neural network approaches have progressively been identified, the discovery and use of multilayers have opened a new path for neural network research. Approaches including backpropagation, artificial neural network

(ANN), convolutional neural network (CNN), and recursive neural network techniques have been proposed, and there has also been a recent boom in the use of deep ML [12]. ML was first commercialized in 1989 when Axcelis, Inc. released the first software package that applied genetic algorithms to personal computers [13]. Later, the MNIST database, the Torch ML library, the launch of the Kaggle website, the development of speech recognition programs, the face recognition leap, etc., have all become important milestones in the history of ML. ML is now being applied to new industries and technological challenges such as self-driving cars and exploring the Milky Way galaxy. A range of new concepts and technologies, including supervised and unsupervised learning, new algorithms for robots, the Internet of Things, analytical tools, and chatbots have been proposed, and ML has made many extraordinary achievements on this basis. Google's AlphaGo program became the first computer go program to beat professional human players by combining ML and tree search techniques [14]. AlphaFold2 uses DL algorithms to predict the 3D shapes of proteins from linear sequences, thus accelerating research developments in protein engineering and experimental structural biology [15]. ML shows strong adaptability in continuous learning. Its scalability and efficiency have been improved by combining ML with new computing technologies, and modern versions of the technology can solve many complex problems. A timeline of the major events in the history of ML is shown in Fig. 6.2.

6.2 Machine learning categories

There is a range of existing ML approaches that use different classification standards and methods. Broadly, ML can be divided into supervised learning, unsupervised learning, and reinforcement learning (RL) according to the amount and type of supervision (learning style) received during training, as shown in Fig. 6.3.

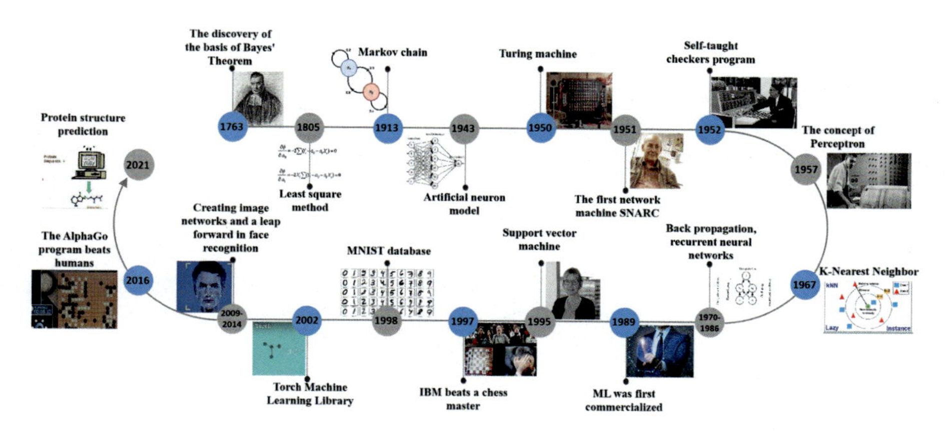

Figure 6.2 The timeline of big events in the history of ML.

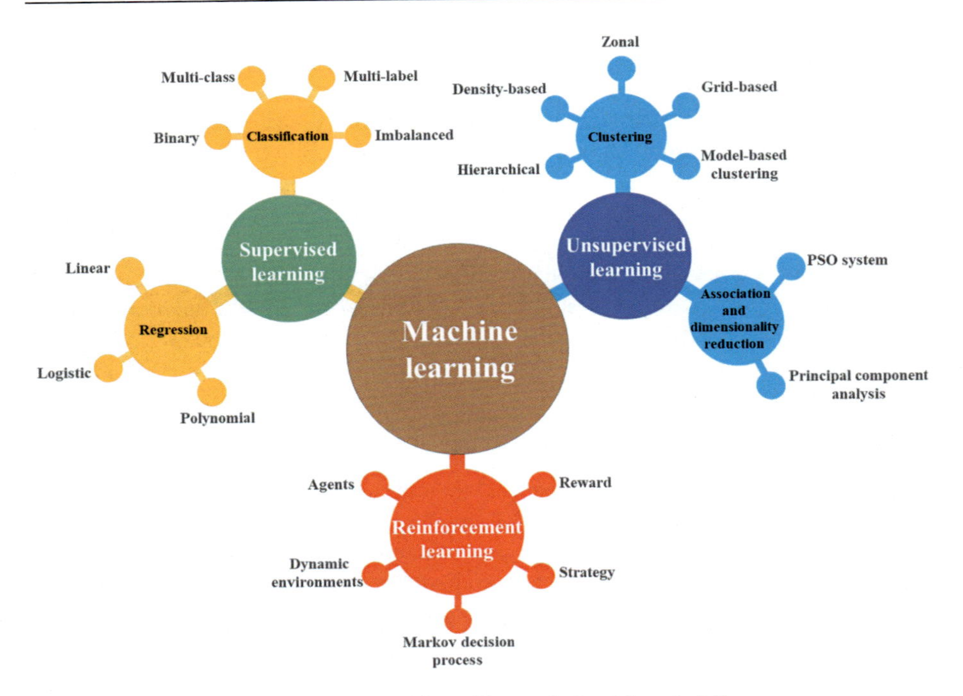

Figure 6.3 Classification and relationships of internal algorithms in ML.

6.2.1 Supervised learning

In supervised learning, a machine learns from a large sample of data and trains a mathematical model that can predict the corresponding outputs (labels) based on the model inputs (features) [16,17]. Thus the essence of supervised learning is to find mappings between features and labels. Supervised learning consists of three key elements: model, strategy, and algorithm. Among these, the model is a type of function system that summarizes the internal laws of the data. The strategy is the evaluation criterion used to select the optimal model, and the algorithm is the specific mathematical method used in the prediction.

Supervised learning problems can be divided into two categories: classification and regression [18]. Classification algorithms are used when the output is restricted to a finite set of values (discrete values), whereas regression algorithms are used when the output can take any value within a range (continuous values) [19]. Specifically, classification algorithms train a classification function (classifier) from a given manually labeled dataset and make predictions for new, unknown data, so as to map the new data to a certain class [20]. In contrast, regression algorithms are used to select a function curve, optimize its fit to the known data, and then use the function to map the unknown data to specific values within a range.

Classification problems can be divided into four types: binary classification, multiclass classification, multilabel classification, and imbalanced classification.

Binary classification describes classification tasks with two class labels, which are typically modeled using Bernoulli probability distributions for each data point [21]. Logistic regression and support vector machine (SVM) approaches are designed specifically for binary classification and thus cannot be applied to problems with more than two classes. Multiclass classification refers to classification tasks with more than two class labels such as face recognition [22]. Multiclass classification tasks are usually modeled using methods that predict the Multinoulli probability distribution for each data point. Algorithms designed for binary classification can also be used for multiclass problems; however, these algorithms can only be applied in two scenarios: (1) multiple binary classification models are made that compare each class with all the other classes (called one-vs-rest) or (2) one model is fitted to each pair of classes (called one-vs-one) [23,24]. Multilabel classification means predicting one or more class labels for each sample [25]. Multilabel classification tasks are typically modeled using models that predict multiple outputs, where each is predicted as a Bernoulli probability distribution. Classification algorithms used for binary or multiclass classification cannot be used directly for multilabel classification. However, a multilabel version of the standard classification algorithm or a separate classification algorithm can be used to predict the labels of each class. Unbalanced classification is a binary classification task type where the number of samples in each class is unevenly distributed; for example, fraud detection and outlier detection represent common unbalanced classification problems. Specialized techniques for changing the composition of samples in a training dataset by undersampling most classes or oversampling a few classes, as well as specialized modeling algorithms and alternative performance metrics, can be used to solve this type of problem.

There are three main regression analysis types: linear regression, logistic regression, and polynomial regression [26]. Linear regression is one of the most widely used regression techniques; in this approach, it is assumed that there is a linear relationship between the objective variable (Y) and the independent input parameter (X). Based on the number of dependent and independent variables, linear regression methods can be divided into simple linear regression and multiple linear regression. Logistic regression is a regression method used to solve classification problems. In this method, unlike other regression techniques where the goal is continuous, the dependent variables are discrete. Polynomial regression is used to identify nonlinear relationships between a dependent and an independent variable. Unlike linear regression, in which the relationship between variable pairs forms a straight line on a graph, polynomial regression optimally fits higher-order curves to the data points. Many ML techniques have been designed to solve multivariable and nonlinear regression problems, and the related regression techniques are described in detail in Section 6.3.

6.2.2 Unsupervised learning

Unsupervised learning is a type of training method that detects underlying structures or hidden patterns in unlabeled data without human intervention [27]. Unsupervised learning is not a response to feedback but instead an algorithm that automatically learns commonalities in data and then responds to the presence or absence of these commonalities when exposed to new data [28]. Unsupervised learning is often used for clustering, association, and dimensionality reduction tasks.

The clustering algorithm is a type of data mining technology and represents a typical unsupervised learning algorithm [29]. The clustering algorithm locates and distributes the sample data into a unified class based on the similarity between data objects [30]. The goal of cluster analysis is to achieve high intracluster similarity and low intercluster similarity, that is, the analysis aims to maximize the distance between clusters while minimizing the distance between samples in the cluster and the cluster's center. This configuration represents the optimal clustering effect. Clustering algorithms can be divided into five types: hierarchical, zonal, density-based, grid-based, and model-based clustering [31], each of which has different applications.

Association rule learning is a rule-based approach that focuses on identifying the relationships between features within a given dataset rather than making purposeful predictions [32]. Association rule learning identifies rules that have high support and confidence and takes effective rules that can really promote the improvement of the front and back items in the rules as evaluation criteria [33]. This approach has strong potential in applications involving identifying regularity between products in large-scale transaction data, such as those recorded by point-of-sale systems in supermarkets.

Dimensionality reduction is a method to transform high-dimensionality input samples into training samples with low dimensionality to improve data processing speeds and reduce memory usage while also preserving as much of the original information as possible [34]. Principal component analysis (PCA) is the most-used dimensionality reduction algorithm whose main concept is to map n-dimensional features to reconstructed k-dimensional orthogonal features. PCA can be implemented in two approaches: one is based on the eigenvalue decomposition covariance matrix, and the other is based on the singular value decomposition covariance matrix [35]. The PCA method is typically used for exploring and visualizing high-dimensional datasets as well as for data compression and preprocessing.

6.2.3 Semisupervised learning

Semisupervised learning, which occupies a region between supervised and unsupervised learning, is an algorithm that combines a small amount of labeled data with a large amount of unlabeled data during training [36], as shown in Fig. 6.4.

Semisupervised algorithms can be extended from finite training sets to infinite invisible sets only if the data structure remains unchanged. Therefore, at least one of the three assumptions (smoothness, cluster, or manifold) must be used to establish the relationship between the prediction samples and learning objectives [37].

Semisupervised learning can be broadly divided into five categories, as listed in Table 6.1. In addition, this technique can be divided into inductive learning and transduction learning based on the learning mode.

In real life, semisupervised learning is of great practical value given that it is difficult and expensive to acquire large, fully labeled training sets, whereas it is relatively cheap to acquire unlabeled data.

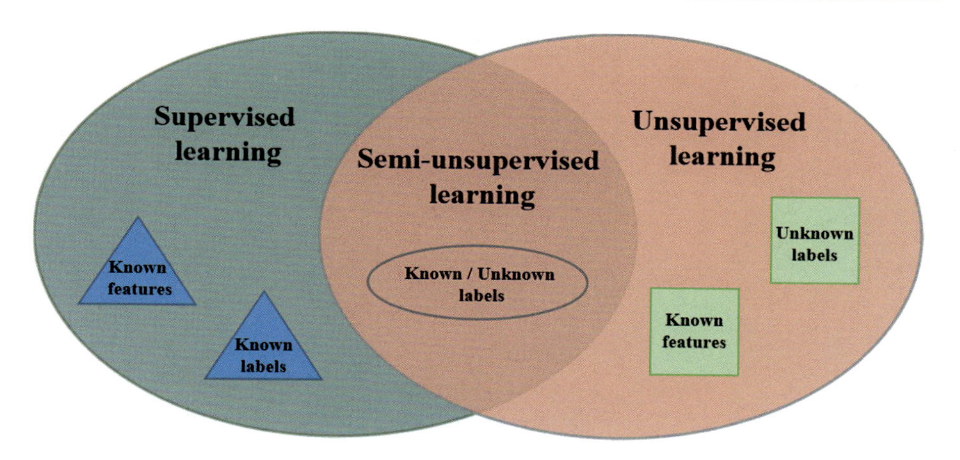

Figure 6.4 The relationships between semisupervised, supervised, and unsupervised learning.

Table 6.1 Classification of semisupervised learning approaches.

Serial number	Category	Description
1	Entropy minimization	Models are encouraged to predict high-confidence outcomes
2	Consistency regularization	The feature/prediction results should not change significantly after the data are disturbed, so the model can be trained by constraining the features corresponding to data disturbance.
3	Generative model	Good discriminators for classification are learned from unlabeled data in conjunction with Generative Adversarial Networks or Variational Autoencoder.
4	Proxy-label method	Unlabeled data are labeled by a pretraining model, and then the model is trained using those data.
5	Graph-based methods	Both labeled and unlabeled data are regarded as graph nodes. The labels of the labeled nodes are then propagated to unlabeled nodes using the similarity between nodes.

6.2.4 Reinforcement learning

RL is a branch of ML that does not rely on static datasets but learns to perform tasks through repeated trial-and-error interactions between software agents and dynamic environments [38] (Fig. 6.5). Unlike supervised learning, RL does not require labeled input/output pairs or explicit correction of suboptimal operations but instead focuses on finding a balance between exploration and development. In

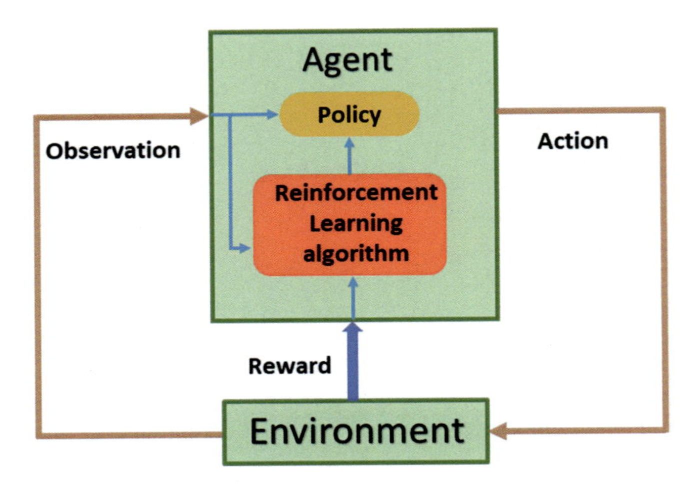

Figure 6.5 Key elements and schematic diagram of RL.

addition, although RL involves a similar experience to unsupervised learning in terms of taking correct actions, the purpose of RL is to make an agent learn an optimal or near-optimal strategy to maximize a task's reward index.

The Markov decision process (MDP) is an idealized form of RL in mathematics. Almost all RL problems can be solved by transforming them into the MDP [39]. The general workflow for using RL training agents comprises the following steps:

1. Create an environment—this step defines the environment in which the RF agent runs, including the interface between the agent and the environment.
2. Define the reward—specify the reward signal that the agent uses to measure its performance against the task goal and how to calculate this signal from the environment.
3. Create an agent—choose an approach to represent the desired strategy and an appropriate training algorithm.
4. Train and validate agents—set the stop condition option and train the agent's adjustment strategy. Ensure that the design strategy is validated at the end of the training step. If necessary, reexamine the agent's design choices, such as reward signals and strategic architecture, before retraining.
5. Strategy deployment

Based on whether an agent can fully understand or learn the model of its environment, RL algorithms can be divided into two categories: model-based and model-free. Among these, model-based approaches have some understanding of their environment and can be planned in advance. However, when these models are inconsistent with the environments in which they are deployed, their performance in real-use scenarios tends to be slightly worse. Although model-free approaches are less efficient, they are easier to implement and adjust to real scenarios and thus have greater development value.

6.3 Deep learning

As an important branch of ML, DL techniques seek to imitate the behavior of the human brain through a combination of data input, weight, and bias [40], as shown in Fig. 6.6. DL differs from classical ML in terms of the data types and learning methods it uses. DL can process unstructured data and achieve optimal accuracy by training models using large volumes of labeled data and neural network architectures containing multiple layers [41]. Additionally, deep neural network architectures can automatically extract features, thereby eliminating some reliance on human experts.

The core concept of DL is a deep neural network containing three or more layers, which are divided into three types: input layers, hidden layers, and output layers [42]. In any DL model, there is only one input layer and one output layer; however, to enhance the model's expression ability, the hidden layer in the middle of the network can contain multiple layers. The neuronal nodes between the layers of the DL neural network are fully connected, and each node is built based on the previous layer, thus optimizing the prediction or classification.

Deep neural networks are mostly feedforward neural networks, in which the output of the next layer is calculated through the forward propagation process. These networks also make use of backpropagation. In this process, algorithms such as gradient descent are used to calculate the error in the prediction, and the weight and bias of the function are then adjusted by moving backward through the network's layers to train the model [43]. The effective combination of forward and backpropagation approaches enables the neural network to make predictions and correct any errors, thus making the algorithm progressively more accurate. Some studies have also considered sequence characteristics and extended deep neural networks to recurrent neural networks (RNNs) to solve problems in fields such as speech

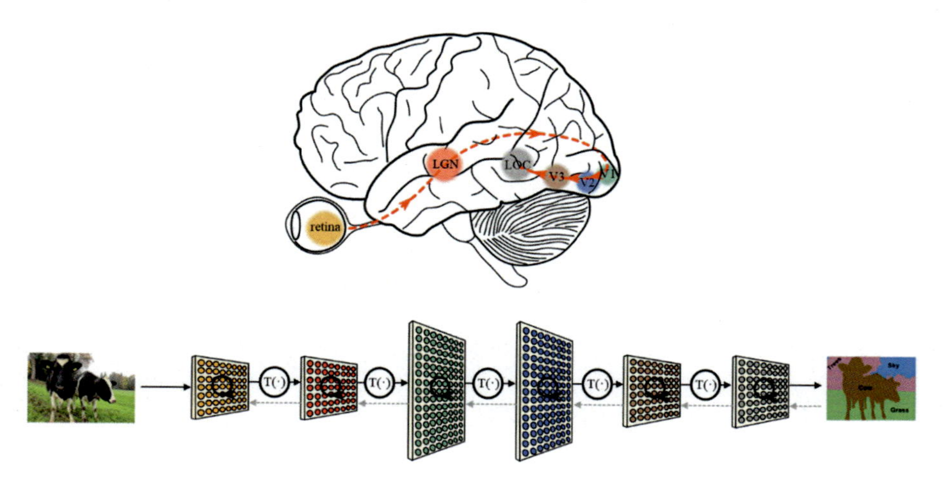

Figure 6.6 A behavior of imitating the human brain.

recognition, machine translation, and timing analysis [44]. The principles and algorithms of RNNs will be introduced in detail in Section 6.3.

6.4 Introduction to machine learning techniques

As discussed in Section 6.2, ML techniques can be divided into five types: supervised learning, unsupervised learning, semisupervised learning, RL, and DL. Each type contains a wide variety of ML algorithms; however, there is no single optimal ML technique or algorithm that is suitable for all cases. Therefore, it is necessary to select the appropriate ML algorithm for different problems and data types. Fig. 6.7 briefly summarizes a range of different ML algorithm types. The most common and popular ML algorithms are described in detail in Sections 6.3.1–6.3.6.

6.4.1 Decision tree

The decision tree approach is a nonparametric supervised learning method. This method can summarize decision rules from a series of data with features and labels and then present these rules in a dendrogram structure to solve classification and regression problems [45], as shown in Fig. 6.8.

As one of the best learning algorithm types, the outputs of decision trees are easy to read and interpret. In addition, they have fast run speeds, which can allow large datasets to be processed relatively quickly. However, the disadvantages of decision trees are that they may be insufficiently stable for reliable prediction, model overfitting can easily occur, and it can be difficult to express more complex concepts in these models.

A decision tree usually comprises four elements: root nodes, branches, internal nodes, and leaf nodes [46]. The decision tree starts at the root node where the original problem is located without any incoming branches. The outgoing branches from the root are then linked to the model's internal nodes, which all correspond to a feature test. These internal decision nodes are ultimately all evaluated to obtain a series of leaf nodes that describe the results of the dataset.

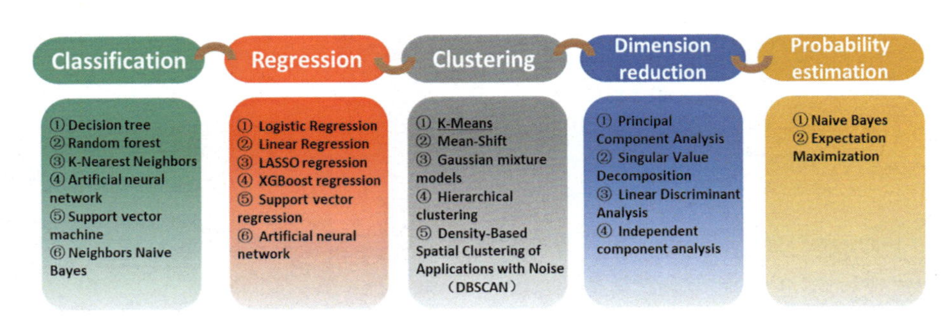

Figure 6.7 Summary of common ML algorithms for different problem types.

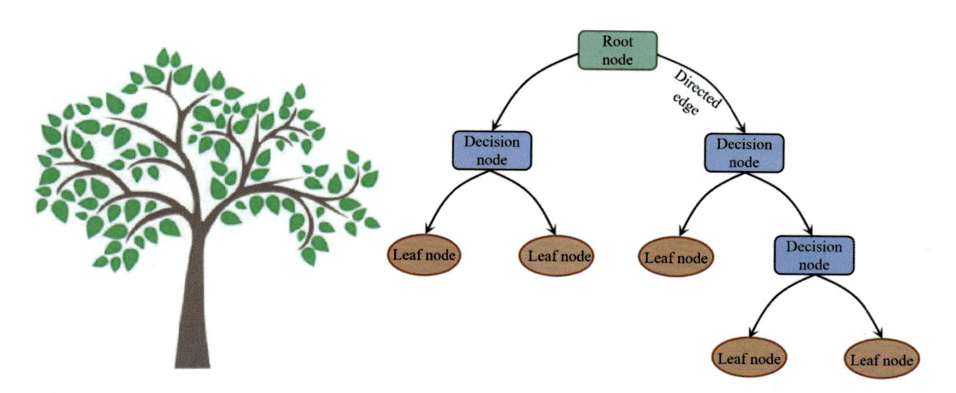

Figure 6.8 Schematic diagram of decision tree.

The basic process of the decision tree is summarized as follows: (1) calculate the impurity index of all features; (2) select the feature with the best (i.e., lowest) impurity index to branch; (3) under the branch of the first feature, calculate the impurity index of all features; (4) select the feature with the best impurity index and continue branching until either no more features are available or the overall impurity index is optimal, at which point the decision tree stops growing. Therefore, identifying the best nodes and branches from the data and optimizing the impurity index represent the core problems to be solved by the decision tree algorithm.

Information entropy, information gain, and the Gini coefficient are popular splitting criteria for decision tree models and are used to select optimal split features [47]. The calculation formulas for these parameters are as follows:

$$Entropy(y) = -\sum_{i=0}^{c-1} p(i)\log_2 p(i) \tag{6.1}$$

$$InformationGain = Entropy_{parent} - Entropy_{children} \tag{6.2}$$

$$Gini = 1 - \sum_{i=1}^{j} p(i)^2 \tag{6.3}$$

Where $p(i)$ is the probability of the i-th outcome, and the summation is taken over all possible outcomes. $Entropy_{parent}$ is the entropy of the parent node, and $Entropy_{children}$ represents the average entropy of the child nodes that follow this variable. j represents the total classes in the target variable. $p(i)$ represents the probability of each class appearing.

Entropy is a measure used to represent the amount of information in a dataset, that is, a measure to represent the uncertainty of random variables. To select the best feature for segmentation and determine the optimal decision tree, the attribute with the lowest entropy (and thus the lowest impurity) should be used [48]. The information gain parameter describes the entropy difference before and after the splitting of a given attribute. In this case, the attribute with the largest information gain should be selected to produce the optimum split [49]. The interpretation of the Gini coefficient is similar to that of the information entropy.

ID3, C4.5, C5.0, and CART are basic generating algorithms for decision tree models [50]. Among these, ID3 uses entropy and information gain to evaluate the decision tree's segmentation points. However, this model tends toward overfitting, because ID3 trees tend to select categorical variables with more values during the growth process. To mitigate this issue, C4.5 uses information gain or gain ratio as an indicator to evaluate splitting. Additionally, C4.5 compensates for the inability of the ID3 model to handle continuous values of feature attributes [51]. C5.0 usually uses Gini impurity to identify desirable attributes to be split; this approach improves on the limitations of C4.5, which cannot be applied to large datasets, in addition to improving execution efficiency and saving memory. CART takes the binary tree as its logical structure and also uses the Gini index to select partition attributes, thus representing an upgraded version of the decision tree [52]. Furthermore, many powerful models such as the random forest (RF) technique are built based on these basic generation algorithms.

6.4.2 *Random forest*

The RF method is a powerful supervised ML algorithm. The RF approach uses Bagging and feature randomness to expand and combine multiple decision trees to create a "forest" in which there is no correlation between different decision trees [53], as shown in Fig. 6.9. Bagging is a parallelization method used to organize

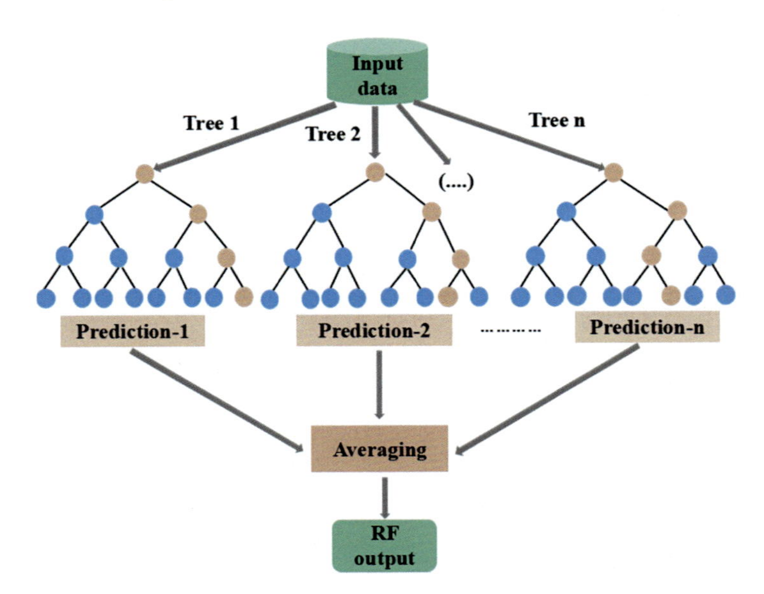

Figure 6.9 Parallel learning and prediction principles of RF.

multiple learners in parallel. Its core concept involves constructing multiple independent estimators and then the average or majority voting principle is applied to their predictions to determine the results of the integrated estimators.

The main steps of RF construction are as follows [54]:

1. To train the decision tree, N samples are randomly selected, one of which is selected each time [55].
2. When the nodes of the decision tree are split, m feature subsets (m << M) are randomly selected from M features of the sample. An optimal attribute is then selected from the m features as the splitting attribute of the node using decision strategies such as information gain, as described in Section 6.3.1.
3. Each decision tree grows to the maximum extent until it can no longer split. In addition, the decision tree has no pruning process.
4. Steps 1−3 are repeated to build numerous decision trees to form an RF.

The RF technique is flexible and easy to use and is mainly applied for solving regression or classification problems. When using RF for classification, each tree gives a classification or "vote," and the RF selects the classification with the most "votes." When RF is used for regression, the forest will select the average of the output from all of the trees [56].

In the RF generation process, an unbiased estimate of the out-of-bag (OOB) error, that is, the internally generated error, can be obtained without cross-validation to achieve the minimum variance [57]. In addition, the RF method can calculate the relative importance of a single feature variable to the prediction [58]; this allows the deletion of features that do not contribute sufficiently to the prediction process or have no contribution to prevent model overfitting.

As a predictive modeling method, RF has the advantages of rapid training speed, simple implementation, high accuracy, and strong overfitting resistance; thus, it is suitable for use as a benchmark model in many fields.

6.4.3 Extreme Gradient Boosting

Extreme Gradient Boosting (XGBoost) is a distributed ML algorithm optimized within the Gradient Boosting framework. It provides parallel tree lifting to rapidly and accurately solve a variety of data science problems such as classification and regression [59].

The Gradient Boosting Decision Tree (GBDT) technique, which forms the basis of XGBoost, is an additive model based on the Boosting ensemble. This approach seeks to minimize the loss function by adding a weak learner using the Gradient descent optimization algorithm [60]. GBDT trains an ensemble of shallow decision trees iteratively, where each iteration uses the error residuals of the previous model to fit the next model. The resulting final prediction is the weighted sum of all tree predictions.

To improve the model's effect and performance, the XGBoost approach is optimized and improved based on GBDT. The objective function is expanded using a second-order Taylor formula to optimize the loss function, and an L2 regularization term is used to simplify the model and avoid overfitting. In addition, XGBoost uses a block storage structure, adding characteristics such as automatic handling of missing values. The key principles of XGBoost are as follows:

1. Construct the objective function consisting of the training loss $l(y_i, \hat{y}_i)$ and regularization term $\Omega(f_k)$ [61]:

$$\mathrm{Obj} = \sum_{i=1}^{n} 1(y_i,\ \hat{y}_i) + \sum_{k=1}^{k} \Omega(f_k) \tag{6.4}$$

Where i is the i-th sample, and there are K trees, $\hat{y}_i$ is the predicted value of the i-th sample, and $\sum_{k=1}^{k} \Omega(f_k)$ represents the complexity of the tree.

2. Learn the t-th tree. Assuming that the tree model we want to train in the t-th iteration is$f_t()$, then $\hat{y}_i^{(t)} = \sum_{k=1}^{t} f_k(x_i) = \hat{y}_i^{(t-1)} + f_t(x_i)$. This can be placed in the objective function and arranged as follows:

$$Obj^{(t)} = \sum_{i=1}^{n} l(y_i, \hat{y}_i^{(t-1)} + f_t(x_i)) + \Omega(f_t) + constant$$

Where $\hat{y}_i^{(t)}$ is the prediction result of sample i after iteration t, $\hat{y}_i^{(t-1)}$ is the prediction result of the previous t-1 tree, $f_t(x_i)$ is the function of the t tree, and $\Omega(f_t)$ is the structural complexity of the t-th tree. Note that given the structure of the t-1 trees has been determined, the sum of the complexity of the t-1 trees can be represented by a constant.

3. Expand the objective function using a second-order Taylor formula and also expand the regularization term [62]. The constant term is removed and the coefficients of the first and second terms are combined; thus, the final objective function can be obtained as follows:

$$Obj^{(t)} \sum_{i=1}^{n} [g_i f_t(x_i) + \frac{1}{2} h_i f_t^2(x_i)] + \Omega(f_t) \tag{6.5}$$

Where $g_i = \partial_{\hat{y}^{(t-1)}} l(y_i, \hat{y}^{(t-1)})$, $h_i = \partial^2_{\hat{y}^{(t-1)}} l(y_i, \hat{y}^{(t-1)})$ are the first and second partial derivatives of the loss function with respect to $\hat{y}^{(t-1)}$. $f_t(x_i)$ is the t-th tree, which is equivalent to Δx.

4. Redefine the weight vector ω and the mapping relation q of the leaf nodes of a tree, and then define the complexity Ω of the tree.

$$f_t(x) = \omega_{q(x)} \tag{6.6}$$

$$\Omega(f_t) = \Upsilon T + \frac{1}{2} \lambda \sum_{j=1}^{T} \omega_j^2 \tag{6.7}$$

Where ΥT indicates the number of leaf nodes, and $\frac{1}{2} \lambda \sum_{j=1}^{T} \omega_j^2$ indicates the L2 norm of the weight vector of the leaf nodes.

5. Substitute Eqs. (6.6) and (6.7) into Eq. (6.5) and simplify them to obtain the final objective function [63]:

$$Obj^{(t)} = \sum_{j=1}^{T} [G_j \omega_j + \frac{1}{2} (H_j + \lambda) \omega_j{}^2] + \Upsilon T \tag{6.8}$$

Where G_j is the sum of the first partial derivatives of the samples contained in leaf node j, and H_j is the sum of the second partial derivatives of the samples contained in leaf node j, which is a constant.

6. Apply the maximum value formula of the quadratic function with one variable to identify the optimal solution of the objective function.
7. Based on the generation strategy of the decision tree, identify the split node that makes the loss function decrease the fastest and continue splitting until the tree reaches its maximum depth and stops growing.

XGBoost generally uses a greedy algorithm to split the tree nodes [64]. However, when the amount of data is too large, the greedy algorithm will become highly inefficient. Therefore, a range of schemes has been proposed to accelerate the identification of the optimal split point, such as feature presorting + cache and partial point approximation. The parallel lookup method used by XGBoost can adopt multiple threads to calculate the optimal splitting point of each feature in parallel because each feature has been prestored as a block structure. This approach not only greatly improves the node splitting speed but also significantly increases the ease of use of large-scale training sets [65].

6.4.4 Support vector machine

The SVM method is a supervised ML algorithm for analyzing data and recognizing patterns [66]. SVM was first introduced in the 1960s and further improved in 1990 [67]. SVM is derived from statistical learning theory, supports regression and classification tasks, and can handle multiple continuous and categorical variables.

The basic SWM model comprises a linear classifier with the largest interval defined in the feature space. For linearly separable data, the SVM learns linear separation through hard margin maximization, as shown in Fig. 6.10. Specifically, the SVM finds the optimal hyperplane in N-dimensional space that maximizes the margin between the hyperplane and the nearest data point, which can clearly classify data points [68]. The dimensionality of the hyperplane depends on the number of

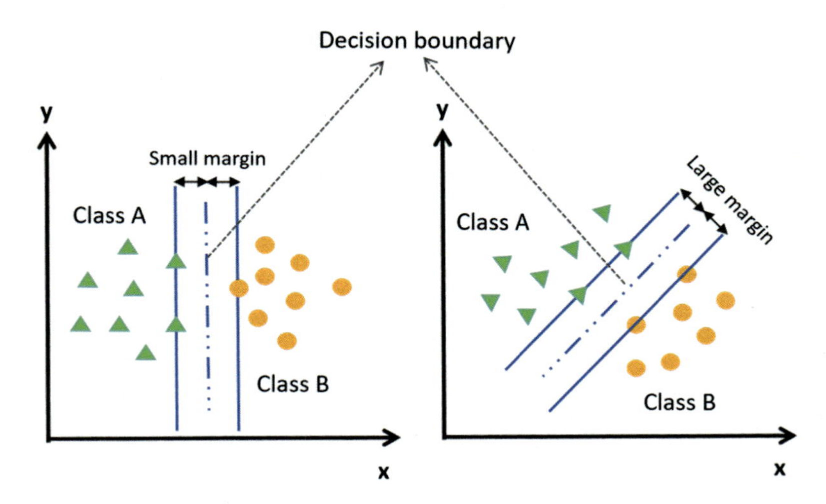

Figure 6.10 SVM learning strategy based on a separating hyperplane.

features. If the dimension of the input elements is 2, the hyperplane forms a line. However, if the dimension of the input elements is 3, the hyperplane will become a 2D surface. In this case, the problem under study is transformed into high-dimensional features and has linear inseparability.

When the training data are not linearly separable, the SVM can become nonlinear using kernel trick and soft interval maximization approaches [69]. This is equivalent to implicitly teaching the linear SVM in high-dimensional feature space (Fig. 6.11).

As an important branch of SVMs, the support vector regression (SVR) method is used to predict the regressor values of continuous ordered variables [70]. In contrast to SVM classification, the core concept of SVR is to identify a regression plane such that all the data of a set are closest to the plane. Instead of using the curve as a decision boundary, the SVR instead uses it to find a match between the vector and the position of the curve.

Unlike traditional regression methods, SVR assumes that as long as the deviation between f(x) and y is not too large, the prediction can be considered correct and there is thus no need to calculate the loss [71]. Consequently, after setting a threshold value of α, the loss is only calculated when $|f(x) - y| > \alpha$.

Both SVM classification and SVR approaches have been shown to perform well when applied to various practical problems. SVMs are widely used in handwriting recognition, digit recognition, and face recognition; however, this approach is also important in text and hypertext classification, as SVMs greatly reduce the requirements for markup training instances in standard induction and transformation settings. Meanwhile, the SVM method is also used to perform image classification and image segmentation [72], in addition to its widespread usage in cutting-edge biological science research.

6.4.5 Convolutional neural network

The CNN approach is a type of DL neural network designed to process structured data arrays such as images. The CNN method simplifies complex problems by reducing the dimension of many parameters to a few parameters; in addition, this technique preserves the original features of images in a vision-like way, providing an extensible method for image classification and object recognition tasks [73].

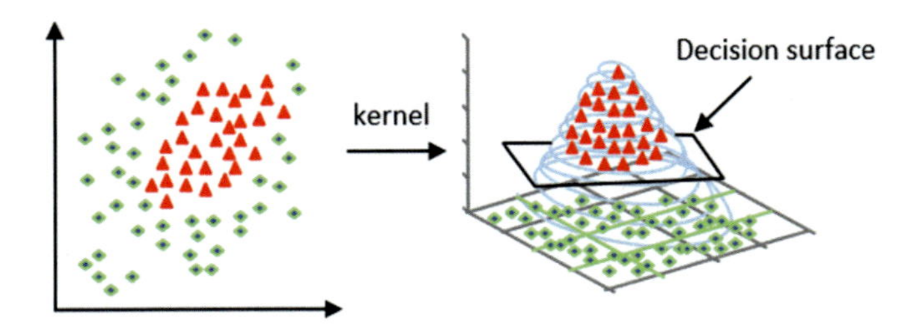

Figure 6.11 Kernel principle of SVM for solving linearly inseparable problems.

The CNN is a feedforward neural network, usually containing up to 20 or 30 layers, mainly composed of a data input layer, convolution layers, Rectified Linear Unit (ReLU) layers, pooling layers, and fully connected layer [74], as shown in Fig. 6.12.

The data input layer mainly preprocesses the original image data through steps such as mean removal and normalization. As the network's first layer, the convolutional layer is the core constructing block of the CNN and is mainly responsible for extracting the local features in images [75]. An input image consists of a 3D pixel matrix with three dimensions: height, width, and depth. When dealing with high-dimensional inputs, each output value in the factor map need not be connected to each pixel value in the input image but instead only to the receptive field to which the filter is being applied [76]. The filter is a 2D weight array, usually a 3×3 matrix, which represents part of the image and determines the size of the receptive field, that is, the size of the region mapped on the original image by the pixels on the feature map output by each layer of the CNN. When filtering the image, the dot product of the filter is calculated with each equally large area of the image; the filter is then slid across all areas of the image from left to right and from top to bottom, and the product after dot multiplication is summed to obtain a new filtered image [77]. When a certain area of the image is similar to the feature detected by the filter, the filter will be activated when passing through the area and return a large positive value. Conversely, the filter will return zero or a smaller value when the area is not sufficiently similar.

The parameters of the convolutional layer are composed of learnable filters (kernels). To achieve parameter sharing, the weight of the filter should remain constant as it moves through the image. As the number of filters, stride, and zero-padding represent three hyperparameters that control the number and arrangement of neurons in the output data, these parameters must be set before the neural network is trained. The number of filters affects the depth of the output, the stride is the distance or number of pixels that the kernel moves on the

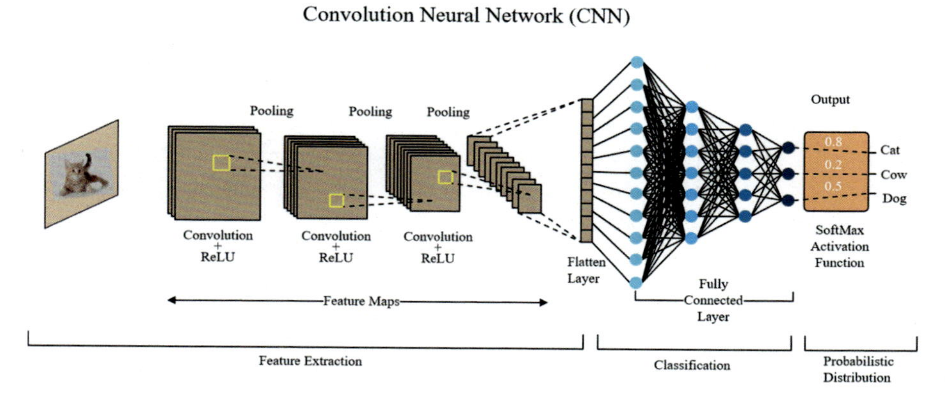

Figure 6.12 Structure and principle of CNN.

input matrix, and zero-padding is usually used when the filter is not suitable for the input image, with three main approaches including valid padding, same padding, and full padding [78].

The ReLU layer uses an activation function (modified linear unit) to transform the output results into nonlinear mapping; this step is often integrated with the convolution layer [79]. The pooling layer is sandwiched between continuous convolutional layers. Gradually reducing the spatial size of the data volume allows the number of parameters in the network to be decreased, thus reducing computing resource needs and effectively controlling model overfitting [80]. Similar to the convolutional layer structure, each neuron in the pooling layer is connected only to a local region of the input data, and the kernel applies an aggregation function to the receptive field and populates the output array. Note that the filter that moves through the entire input does not have any weight.

The pooling layer can perform two pooling operations: max pooling and average pooling [81]. As shown in Fig. 6.13, max pooling involves "scanning" the values in the neighborhood of a 4×4 feature map using a 2×2 filter with a step of 2; the maximum value is then selected and output to the next layer. In this process, the height and width of the feature map are halved, and the number of channels remains unchanged. Average pooling refers to using a similar filter approach to change the maximum value of an area to the average value of this area. Of the two techniques, max pooling is more widely used.

The fully connected layer is the final layer of the CNN network structure, and each neuron in the fully connected layer is fully connected with all the neurons in the previous layer by weight [82]. The classification task is performed using the features extracted from the previous network using its various filters.

Different combinations and collocations of network layers form different CNN architectures. This type of network is widely used in image analysis but has also achieved a series of successes in natural language processing and drug discovery.

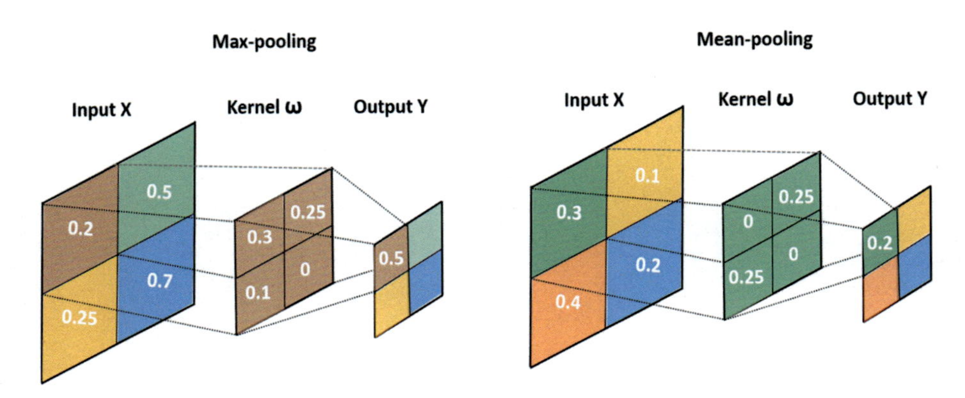

Figure 6.13 Two pooling methods for feature dimension reduction.

6.4.6 Recurrent neural network

The recurrent neural network (RNN) approach is a neural network type dedicated to processing and predicting sequence data. This network type originated from the Hopfield network proposed by Saratha Sathasivam in 1982 [83]. In contrast to traditional feedforward neural networks that receive specific inputs to obtain their output, RNNs memorize previous information and apply it to calculate the output of the following node [84]. Thus the nodes between the hidden layers of the RNN are connected, and the input of the hidden layer includes not only the output of the input layer but also the previous output of the hidden layer [85].

RNNs consist of artificial neurons and one or more feedback loops, including one-to-one, one-to-many, many-to-one, and many-to-many structures [86]. The typical RNN structure is expanded in the time dimension as shown in Fig. 6.14.Where $x_t$and o_t are vectors representing the input data and output data, respectively. s_t is the output of the hidden layer and s_0, which is initialized to zero, is needed to compute the first hidden layer. U is the weight value from the input to the hidden layer, W is the weight value contributed by the hidden layer at the previous time to the current hidden layer, and V is the weight value from the hidden layer to the output layer.

In RNNs, the value of s_t depends not only on x_t but also on s_{t-1}. The calculation formula is as follows, where $f()$ and $g()$ represent activation functions.

$$s_t = f(U^* x_t + W^* s_{t-1}) \tag{6.9}$$

$$o_t = g(V^* s_t) \tag{6.10}$$

Furthermore, based on the important characteristics of the shared parameters in each network layer, an RNN can be trained directly using the backpropagation over time algorithm (BPTT). The parameters of the model can then be properly adjusted and fitted by calculating the error from the model's output layer relative to its input

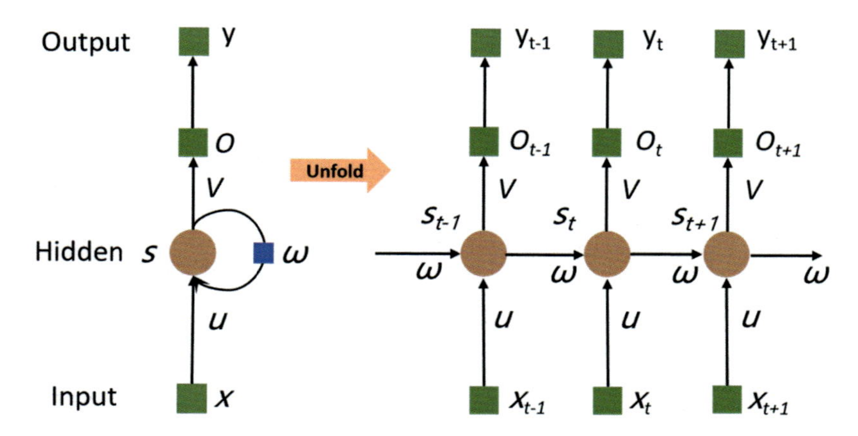

Figure 6.14 A typical RNN structure expanded in the temporal dimension.

layer [87]. However, unlike traditional backpropagation, BPTT sums the error at each time step, whereas feedforward networks do not.

As an efficient ML model, RNNs contain a hidden state distribution to store past information efficiently but also have nonlinear dynamics that allow them to perform complex updates of the hidden state. These features allow RNNs to save, remember, and process past complex signals for an extended period [88]. However, in practice, RNNs have the problem of short-term memory, in which the network fails to learn as the time since the input increases, making it difficult to effectively deal with long-term dependence. In addition, RNNs have the issues of long sequences and complex training, which result in serious gradient disappearance and gradient explosion problems [89]. Therefore, alternative approaches including the bidirectional recurrent neural network (BRNN), long short-term memory (LSTM), and gated recurrent unit (GRU) techniques have been proposed to optimize RNNs [90].

6.4.7 K-means

The k-means algorithm is a typical distance-based clustering algorithm that takes distance as the standard for measuring the similarity between data objects [91]. Its core goal is to divide a given dataset into k clusters where the distance between samples within the same cluster is small (i.e., high similarity), whereas samples in different clusters have large differences in similarity (Fig. 6.15).

The calculation steps in the k-means clustering approach are as follows:

1. Specify the number of clusters (k).
2. Initialize the cluster centroids. After randomly permuting the dataset, k points are randomly selected from the sample as initial centroids.

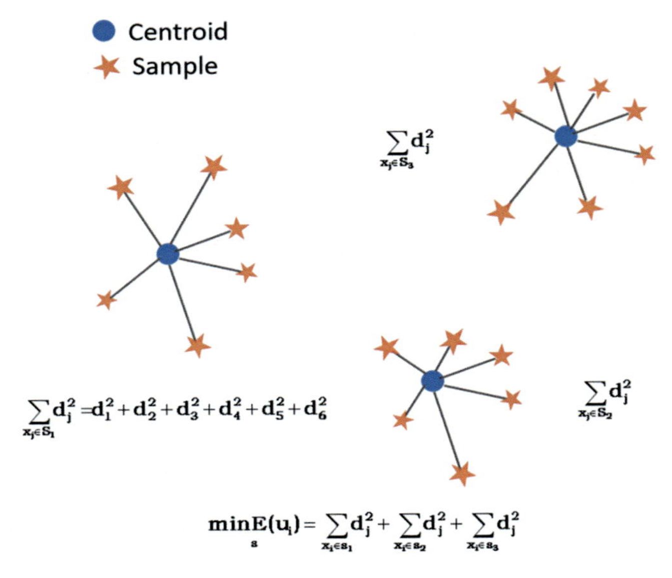

Figure 6.15 The core computing principle of k-means clustering.

3. Calculate the distance from each sample to each centroid (i.e., Euclidean distance) and divide the samples into clusters corresponding to the nearest centroid.
4. Calculate the mean of all samples in each cluster and update the cluster centroid using this mean [92].
5. Repeat steps 2 and 3 until some stop condition is reached. Typically, this condition is met when either the change in centroid position is below a specified threshold or the maximum number of iterations is reached.

As the simplest and most popular unsupervised clustering algorithm, the k-means approach is not only easy to understand but also has strong interpretability and rapid convergence [93]. Currently, this method is widely used in fields such as image segmentation, gene segment classification, article tag classification, species population classification, and abnormal data detection. However, the k-means approach also has some limitations. For example, due to the influence of initial values and outliers, the clustering results may not be globally optimal but instead locally optimal. In addition, e empty clusters can be easily produced during the clustering process.

Furthermore, a key challenge when applying the k-means method is how to determine the appropriate value of k. In practice, when using massive, high-dimensional datasets, it is difficult to estimate the appropriate k value, which may lead to poor results [94]. Given these difficulties, estimation methods such as the Elbow Method and ISODATA have been proposed. By using the Elbow Method to draw the relationship between the k-means cost function and the number of clusters k, the k value situated at the inflection point of the line can be chosen as the optimal number of clustering centroids [95]. The core concepts of the ISODATA technique are to (1) remove a cluster when the number of samples belonging to it is too small and (2) divide the cluster into two subclusters when the number of samples belonging to a cluster is too large and the degree of dispersion is large [96]. However, through comparative analysis of different techniques, a preferable approach is to review practical examples, manually specify a reasonable k value, and determine the most satisfactory result by randomly initializing the cluster centroid many times.

6.5 Implementation of machine learning

As one of the fastest-growing areas of technology, ML is developing at an exponential rate. To realize the use of ML to solve practical problems, it is essential to choose the right algorithm and strictly follow the model training process. However, selecting an appropriate programming language is also crucial to effectively implementing the ML process.

6.5.1 Programming language for machine learning

In the field of ML, one of the core skills required is programming, which involves a wide range of languages. At present, there are thousands of programming languages but, for ML, not all of them are equally important to learn and master, and there is no single "best" programming language. Each ML language has its own areas of specialism, and selecting the most appropriate programming language for a specific problem can be twice as effective.

According to statistics, the top 10 programming languages by popularity are Python, Java, JavaScript, C#, C++, PHP, R, TypeScript, Go, and Swift, as shown in Table 6.2. The change trends of the current top five programming languages over the past 20 years are shown in Fig. 6.16.

Python is an interpreted high-level programming language used for general-purpose programming, created by Guido van Rossum and first released in 1991 [97]. Python has become the most popular programming language in the field of ML due to its simplicity, flexibility, and ease of extension [98]. It is widely used in computer vision, natural language processing, general ML, data analysis/data visualization, and Kaggle competition source code. Python has a wide range of visual packages and core libraries, including NumPy, pandas, and Matplotlib, which can be used to build various ML frameworks [99]. However, pure Python code runs

Table 6.2 August 2022 ranking of programming language popularity compared with a year ago.

Ranking	Programming language	Ratings	Trend
1	Python	28.29%	− 1.8%
2	Java	17.31%	− 0.7%
3	JavaScript	9.44%	− 0.1%
4	C#	7.04%	− 0.1%
5	C/C++	6.27%	− 0.4%
6	PHP	5.34%	− 1.0%
7	R	4.18%	+ 0.3%
8	TypeScript	3.05%	+ 1.5%
9	Go	2.16%	+ 0.6%
10	Swift	2.11%	+ 0.5%

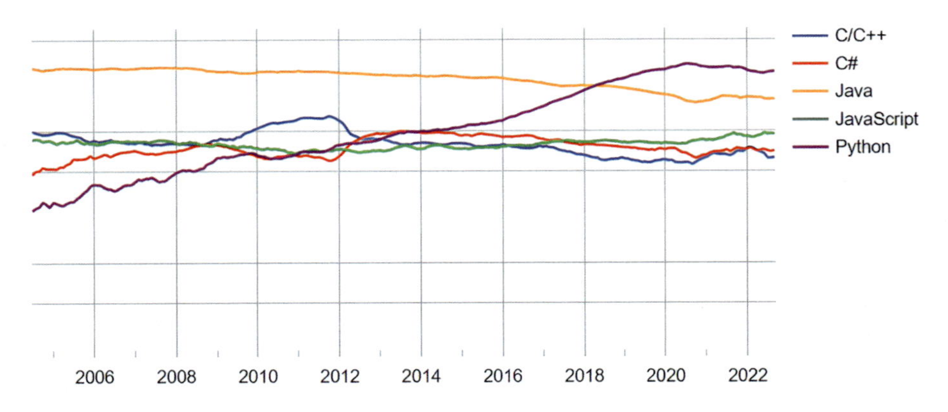

Figure 6.16 Trends in the popularity of five programming languages.

slowly, the code cannot be encrypted, and some packages are incompatible, leading to inevitable drawbacks.

Java is primarily an object-oriented language but also a multiparadigm language; its syntax is more complex than that of Python, but its code execution speed is quite high [100]. Java is fast, reliable, and "ancient," meaning that multiple infrastructures, applications, and data science frameworks such as Hadoop are built and written in Java. Java is thus not only used in data science processes such as data cleaning and statistical analysis but also widely used for developing ML applications. Java is equipped with various ML tools and IDE libraries such as Weka, Knime, RapidMiner, and Elka [101]. In addition, Java virtual machines enable developers to write identical code across multiple platforms and create custom tools at a faster speed, thus making Java a popular platform for ML development.

C and C++ permit interaction with low-level hardware, allow real-time performance, and are ideal languages for writing low-level software. C++ is an extension of the C language; of the two, C++ contains more functions and is more widely used [102]. C/C++ are not as user-friendly as Python; however, they are close to "standard languages" defined by roboticists. They can be used not only to complement existing ML projects but also to develop ML and DL libraries. Many ML frameworks such as TensorFlow, Torch, and wabbit are implemented in C++.

The R programming language was created by statisticians specifically for solving numerical/statistical problems [103]. R is an open-source system, completely free, and can be used across multiple platforms and systems. It allows for rapid prototyping and can effectively use datasets to build ML models, thus outperforming pure Python for data analysis and visualization. R also incorporates numerous ML libraries and advanced data analysis packages that contain many powerful functions. For example, ggplot2 is built based on graph syntax: this library can be used instead of traditional plotting methods to achieve complex and attractive graphs and data visualizations [104]. In addition, Shiny and flexdashboard can be used to directly construct interactive Web applications and make stand-alone visual apps [105]. However, there are also limitations—the R language is difficult to learn, its code is challenging to write, and the syntax for solving some problems is not clear, so it is best suited for exploratory work.

6.5.2 Python community for machine learning

As the most frequently used and popular ML language, Python, in addition to its simplicity and consistency, contains many useful tools and plugins that help ML engineers quickly solve common tasks and develop products faster, as shown in Fig. 6.17.

6.5.2.1 NumPy

NumPy is a basic software package for high-performance scientific computing and data analysis, as well as a third-party library of Python language, which supports numerous dimensional array and matrix operations [106]. As shown in

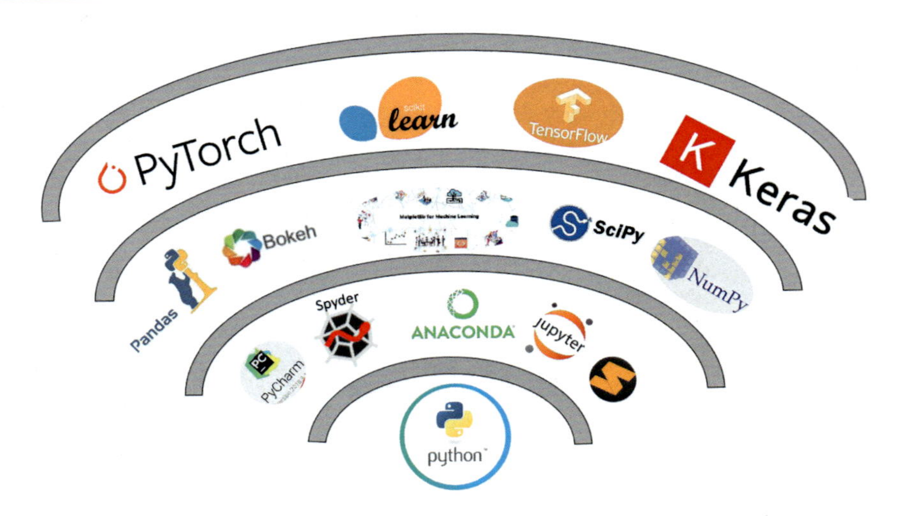

Figure 6.17 Effective tools and plugins in Python.

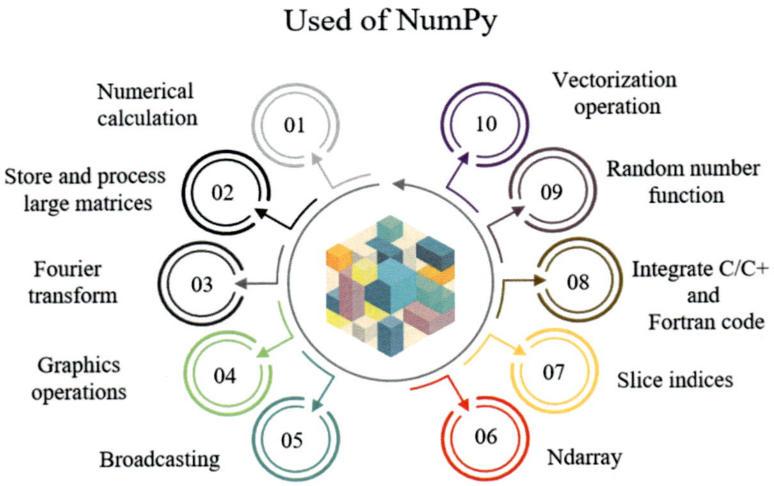

Used of NumPy

01 Numerical calculation
02 Store and process large matrices
03 Fourier transform
04 Graphics operations
05 Broadcasting
06 Ndarray
07 Slice indices
08 Integrate C/C+ and Fortran code
09 Random number function
10 Vectorization operation

Figure 6.18 Various functions and uses of NumPy.

Fig. 6.18, NumPy has a wide range of powerful functionality, and most of the basic numerical calculations in ML algorithms are performed by calling NumPy libraries.

The core of the NumPy package is its N-dimensional array object, ndarray, which is a zero-indexed collection of elements with the same data type [107]. Unlike native Python arrays, NumPy arrays are explicitly allocated with a fixed size. With some exceptions, the elements in a NumPy array must have the same data type. The contents of an ndarray are accessed and modified by indexing or slicing, using similar list slicing syntax to base Python. New arrays can be cut from the original array using the inbuilt slice functionality by setting the start, stop, and step parameters [108]. The ndarray object contributes to performing advanced mathematical operations on large amounts of data, thus making model operations more efficient.

In addition to the powerful ndarray object, NumPy includes comprehensive linear algebra, Fourier transform, and random number capabilities, supports a wide range of hardware and computing platforms, and works well with distributed computing, sparse array libraries, and GPUs that depend on NumPy library. To achieve its high calculation performance, the bottom layer of NumPy is written in C language. The "vectorization" and "broadcasting" operations allow NumPy to effectively handle arrays with different shapes, and the library's operation speed is not limited by the Python interpreter, thus providing high efficiency and flexibility [109].

6.5.2.2 Pandas

Pandas, derived from the terms "Panel Data" and "Python Data Analysis," is a powerful set of tools for analyzing structured data based on NumPy [110]. Pandas was developed by AQR Capital Management in April 2008 and went open source in late 2009 [111]. Following the continued development of this technology, a Pandas development sprint was implemented in 2019 (Fig. 6.19).

Pandas not only contains several libraries and standard data models but also provides a range of functions and methods for rapid data handling, providing a powerful and efficient data analysis environment within Python [112]. The critical code for Pandas is written in Cython or C for highly optimized performance. Pandas can read and write data from a variety of file formats such as CSV, JSON, SQL, and Microsoft Excel [113]. It also provides functionality to handle missing values in

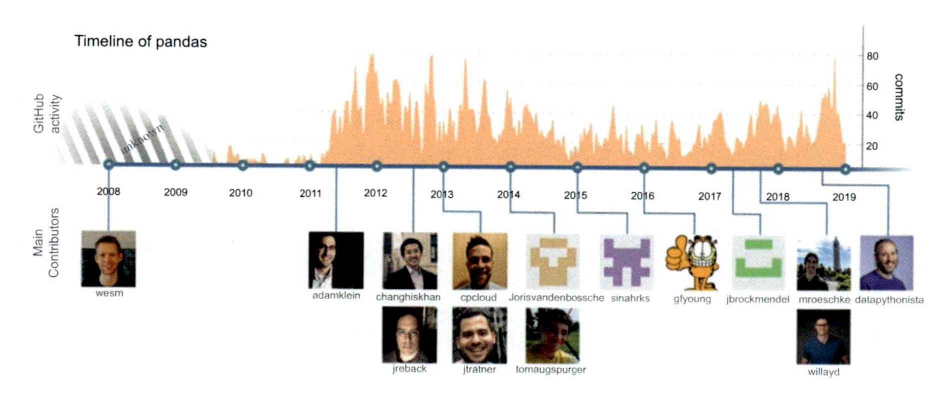

Figure 6.19 Timeline of the history of Pandas showing major GitHub contributors over time.

floating-point and non-floating-point data and can perform operations such as data merging, selecting, cleaning, and processing. In addition, Pandas provides flexible remodeling and rotation of datasets, intuitive merging and joining, and split-application-combination operations for datasets.

DataFrame and series objects are representative data structures of pandas that can effectively handle data for most typical use cases in areas such as finance, statistics, social sciences, and engineering [114]. The series object is similar to a one-dimensional array, consisting of a set of data and the labels (indexes) associated with it. This object type is also similar to an array in that each array can only store one data type for efficient memory use and computing efficiency. Series can be created from arrays or dictionaries, and their indexes can be added at the same time as the series is initialized or afterward.

A DataFrame is a two-dimensional tabular data structure containing both row and column labels, which can be viewed as a dictionary of series. A DataFrame is also known as a heterogeneous data table because it contains an ordered set of columns, each of which can contain a different data type, such as integers and strings [115] There are five main approaches for creating DataFrame objects: list creation, dictionary nested list creation, list nested dictionary creation, Series creation, and initializing an empty DataFrame object. Column indexes are mostly used for selecting, adding, and deleting data objects in DataFrames. The related Pandas Panel object is a 3D array that can be considered as a container for DataFrames.

6.5.2.3 SciPy

SciPy, an advanced scientific computing library distributed under the BSD license, is an open-source Python toolkit for math, science, and engineering. Most functions and routines in SciPy are compiled in low-level languages such as Fortran, C, and C++, allowing highly optimized performance while also benefiting from Python's flexibility and speed of compiling code [116].

SciPy relies on NumPy arrays as an underlying data structure for its scientific calculations and statistical analysis [117]. In addition, SciPy extends NumPy by providing specialized data structures such as sparse matrices and k-dimensional arrays as well as additional tools for array computation. SciPy contains task-specific submodules that provide algorithms for integration, interpolation, optimization, eigenvalue problems, algebraic equations, differential equations, and statistics, in addition to many other classes of problems [116]. Furthermore, SciPy makes interactive Python a data processing and system prototyping environment that rivals MATLAB, IOctave, and SciLab by providing users with high-level commands and classes to manipulate and visualize data [118].

SciPy is simple to use and provides algorithms and data structures that apply to a wide range of domains. In addition, the integration of SciPy within Python provides users access to a powerful programming language for developing complex and specialized applications, from parallel programming to Web applications, database subroutines, and classes.

6.5.2.4 Matplotlib

Matplotlib is a key, low-level, customizable, graph-rich, easy-to-use 2D plotting library for Python script [119], Python shell, Jupyter Notebook, Web, and other servers. Matplotlib can be used to create publication-quality diagrams in addition to interactive graphics that can be scaled, translated, and updated. This module is commonly used in combination with NumPy and Pandas [120].

Matplotlib plotting is mainly performed using its PyPlot module, which can programmatically generate a variety of charts [121]. Simple functions are used to customize charts independently and add text, points, lines, colors, images, and other elements. In addition to creating line plots, bar diagrams, scatter diagrams, and other 2D images, Matplotlib also provides a 3D plotting interface.

Matplotlib has two different drawing methods: one is a functional drawing method (similar to MATLAB), and the other is object-oriented drawing, in which an object-oriented drawing API is used to cooperate with WxPython or other Python GUI toolkits to easily embed diagrams in applications [122].

Matplotlib is one of the most important tools in Python-based data analysis and visualization. It is powerful with rich extensions and lays the foundation for the development of advanced visualization modules such as Seaborn [123].

6.5.2.5 Scikit-learn

Originally developed as a project by data scientist David Cournapeau in 2007 [124], scikit-learn has become one of the most popular ML-specific open-source frameworks in Python.

Scikit-learn, building on libraries such as Numpy and Matplotlib, forms an extension of SciPy. Scikit-learn offers dozens of built-in ML algorithms and models to perform both supervised and unsupervised ML, in addition to a variety of tools for aspects such as model selection, data preprocessing, and model evaluation [125]. Statistically, the most commonly used packages and tools from scikit-learn are shown in Fig. 6.20.

The functionality of scikit-learn can be summarized into six main packages: classification, regression, clustering, preprocessing, model selection and evaluation, and dimensionality reduction [126]. Among these, the preprocessing package includes modules such as "Feature_extraction" and "Transformer," which are used to transform, standardize and encode datasets. Model selection contains modules such as "cross_validation," "grid_search," "learning_curve," "classification," and "regression," which represent metrics for tuning hyperparameters and evaluating estimator performance. Furthermore, scikit-learn has a built-in "check" module for evaluating model assumptions and biases, designing better models, and diagnosing model performance problems, in addition to embedding some datasets and their loading and extraction procedures, etc.

The scikit-learn package is powerful and easy to learn. It's clear and consistent code style ensures that ML code is easy to understand and reproduce, which significantly lowers the entry barrier for coding ML models. However, scikit-learn also

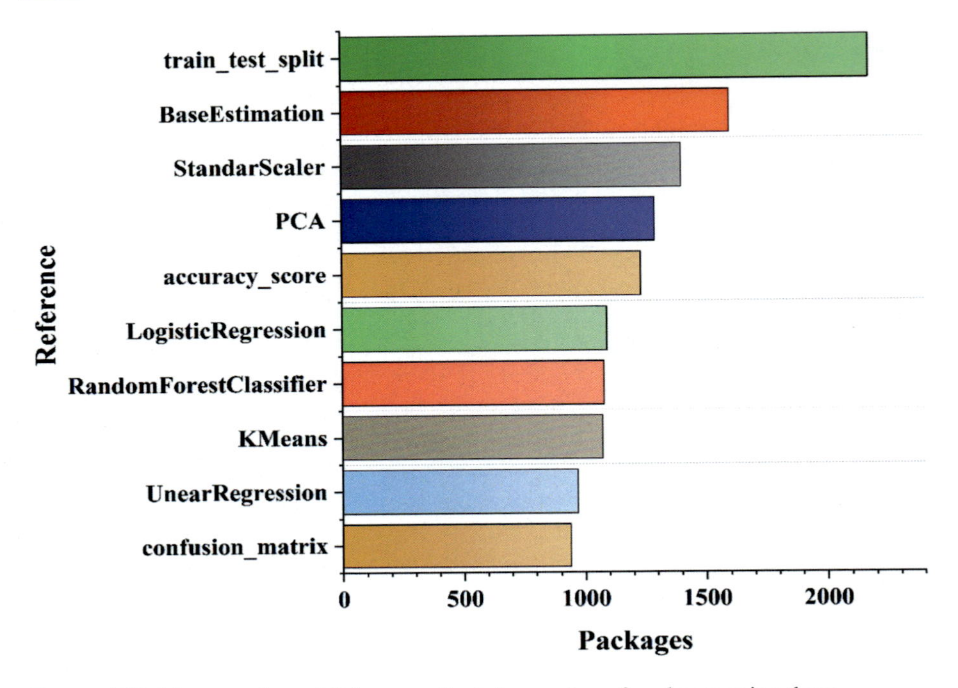

Figure 6.20 Most popular scikit-learn methods by number of packages using them.

has some disadvantages, for example, it does not support graph models, sequence prediction, PyPy, or GPU acceleration. Nonetheless, scikit-learn is still considered one of the best options for many ML projects.

6.5.2.6 TensorFlow

TensorFlow is a comprehensive and flexible open-source ML platform containing a variety of tools, libraries, and community resources. It was originally developed by engineers at Google to study ML and deep neural networks. As the era of big data has progressed and evolved, there have been regular new releases of TensorFlow that have introduced new functionality (Fig. 6.21), and this package is now widely used in other computing fields [127].

The term "Tensor" relates to n-dimensional arrays [128], and "Flow" means computation based on the data flow graph. Thus TensorFlow describes the flow of tensors from one end of the graph to the other, a numerical computation process based on the data flow graph's logic.

In TensorFlow, a calculation diagram must be initially built, then a session started that owns and manages all of the program's resources at runtime according to the calculation diagram. Finally, complete variable assignment and calculation are performed in the session to obtain the results. The entire computational framework is divided into two parts: (1) a construction phase, in which the graph is

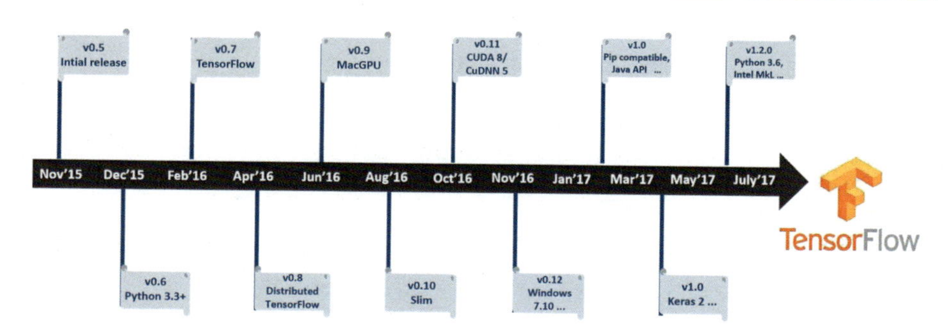

Figure 6.21 The timeline of TensorFlow version upgrades.

created for neural network training [127]; and (2) an execution phase, in which a session is initialized to perform the computation of nodes in the graph, including a series of training operations that may need to be performed repeatedly. When the calculation is complete, the session must be closed to allow the system to reclaim resources [129].

The principle of TensorFlow can be divided into single-machine implementation and distributed implementation. In single-machine mode, the graph is executed in order of dependencies between programs. In contrast, in distributed implementation, the client, master, worker process, and device must be managed. In addition, TensorFlow has various important extensions such as automatic derivation, subgraph execution, and compute graph control flow [130].

TensorFlow is not only simple and readable but is also highly efficient in terms of code compilation. Its automatic derivation of functions and distributed computation greatly save time during model training. In addition, TensorFlow supports distributed computing across heterogeneous devices and can automatically run models on various platforms and multi-GPU systems [131] as well as generate visualizations showing network topology and performance. Thus this package is highly important for promoting the development of advanced ML technologies, as well as building and deploying ML applications.

6.5.2.7 Keras

Keras is a high-level neural network API written in Python, comprising an open-source DL framework running TensorFlow or Theano as its back end [132]. However, there are key differences between Keras and TensorFlow, as shown in Fig. 6.22. Keras allows simple and fast prototyping design but also supports CNN and RNN, as well as a combination of the two [133]. In addition, Keras can run seamlessly on multiple platforms and backends as well as CPUs and GPUs. However, Keras focuses more on the network's hierarchy; thus it is less well-suited than TensorFlow for designing new networks and performing higher-order experiments.

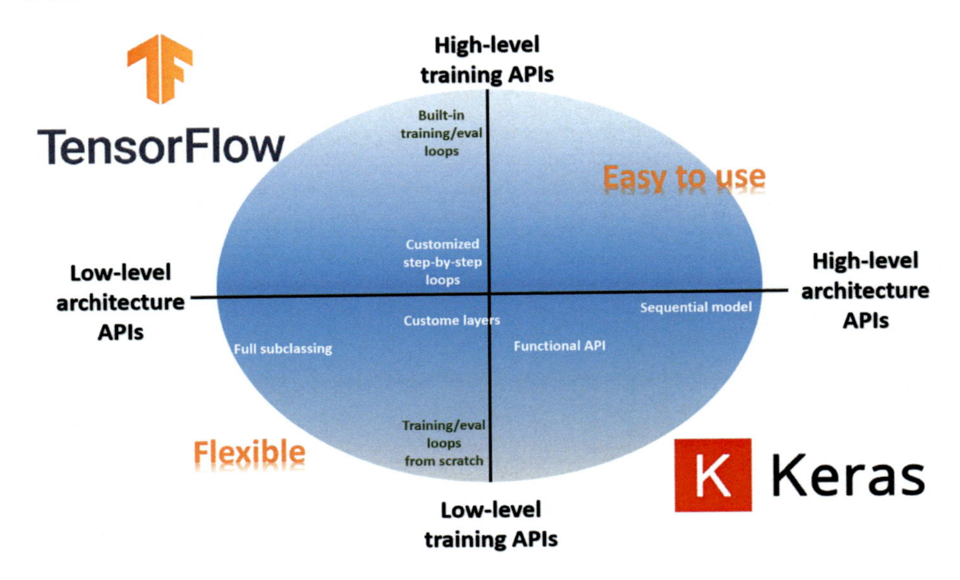

Figure 6.22 Comparison analysis of Keras and TensorFlow.

Keras focuses on the design principles of user-friendliness, modularity, and easy scalability, aiming to achieve fast definition of deep neural networks based on the smallest structure. Keras incorporates many advanced neural network modules such as Dense and Conv2D, making it possible to rapidly build usable neural network models [134].

The module structure of Keras comprises five components: back end, model, data preprocessing, network configuration, and network layer. The module's core is the Model class, a way of organizing network layers, the simplest being the Sequential Model in which multiple network layers are stacked linearly [135]. Furthermore, using Keras to build a neural network requires only the following five steps: (1) selecting the model, (2) building the network layer, (3) compiling, (4) training the model, and (5) predicting the problem.

6.5.2.8 *PyTorch*

PyTorch is an open-source DL framework developed by Facebook based on Torch [136], which was converted from the unpopular Lua language to Python language, thus providing a more flexible and fast scientific computing package. PyTorch includes powerful GPU-accelerated tensor computation and deep neural networks with automatic derivation systems [137], which have been widely used in natural language processing and other fields.

Due to the smooth integration of PyTorch within the Python data science stack, this framework can take advantage of all the services and functionality provided by the Python environment. The PyTorch interface is simple, making it easy to debug and understand the code. Rather than using a network structure that is defined and

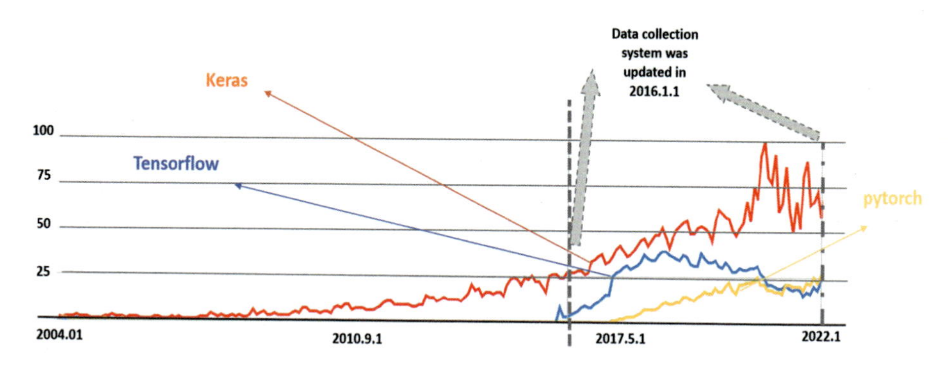

Figure 6.23 The popularity of the three frameworks over time.

then executed, PyTorch instead provides a dynamic computation graph that can be changed in real time depending on the computation [138]. Compared with Keras and TensorFlow, PyTorch is powerful, flexible, and fast. A popularity comparison between Tensorflow, Keras, and PyTorch over the past 20 years is shown in Fig. 6.23.

PyTorch has three abstraction levels: tensors, variables, and modules. The tensor is the core data structure in the neural network framework. It contains data and metadata, which can be understood as N-dimensional array container objects [136]. However, in contrast to objects such as NumPy arrays, PyTorch tensors can be directly used by GPUs to accelerate processing [139]. Tensors are connected by operations to form a calculation diagram. The variable is the node that stores data and gradients in the calculation graph, forming a key part of the autograd library. The autograd library provides automatic differentiation for all computing operations on Tensors [140]. The "autograd.variable" method wraps a tensor and supports almost all operations defined on it. Once the operation is complete, ".backward" can then be called to automatically compute all gradients. Furthermore, modules provide the basic constructing blocks of the neural network layer and create custom network models by inheriting Module classes [141].

PyTorch is relatively new compared with other competing technologies. However, the API of PyTorch is low, has automatic differentiation on DL networks, and supports high-power GPU applications with NN modules, optimize modules, and autograd modules [142]. It has better tuning capabilities than Keras and TensorFlow and is suitable for a wide range of applications.

References

[1] P. Langley, The changing science of machine learning, Machine Learning 82 (3) (2011) 275–279.

[2] A.L. Boulesteix, M. Schmid, Machine learning versus statistical modeling, Biometrical Journal 56 (4) (2014) 588–593.

[3] X. Teng, Y. Gong, Research on application of machine learning in data mining, IOP Conference Series: Materials Science and Engineering, IOP Publishing, 2018.
[4] J. McCarthy, E. Feigenbaum, In memoriam—Arthur Samuel (1901–1990), AI Mag. 11 (3) (1990) 10–11.
[5] T. Bayes, N. Price, An essay towards solving a problem in the doctrine of chances. by the late Rev. Mr. Bayes, F. R. S. communicated by Mr. price, in a letter to john canton, a. m. f. r. sLII Philosophical Transactions of the Royal Society of London 53 (1763) 370–418.
[6] A.M. Legendre, Nouvelles méthodes pour la détermination des orbites des comètes; par AM Legendre, 1806: chez Firmin Didot, libraire pour lew mathematiques, la marine, l.
[7] W.S. McCulloch, W. Pitts, A logical calculus of the ideas immanent in nervous activity, The Bulletin of Mathematical Biophysics 5 (4) (1943) 115–133.
[8] A.M. Turing, I.—computing machinery and intelligence, Mind; A Quarterly Review of Psychology and Philosophy 236 (1950) 433–460. LIX.
[9] M.M. Poulton, A brief history, Handbook of Geophysical Exploration: Seismic Exploration, Elsevier, 2001, pp. 3–18.
[10] F. Rosenblatt, The perceptron, A Perceiving and Recognizing Automaton Project Para, Cornell Aeronautical Laboratory, 1957.
[11] T. Cover, P. Hart, Nearest neighbor pattern classification, IEEE Transactions on Information Theory 13 (1) (1967) 21–27.
[12] J. Schmidhuber, Deep learning in neural networks: an overview, Neural Networks 61 (2015) 85–117.
[13] A. Chiamenti, P. Terna, S. Margarita, Portfolio optimization through genetic algorithms in an artificial stock market, 2015.
[14] J. Alzubi, A. Nayyar, A. Kumar, Machine learning from theory to algorithms: an overviewin Journal of Physics: Conference Series (2018). IOP Publishing.
[15] M. Eisenstein, Artificial intelligence powers protein-folding predictions, Nature 599 (7886) (2021) 706–708.
[16] P. Cunningham, M. Cord, S.J. Delany, Supervised learning, Machine Learning Techniques for Multimedia, Springer, 2008, pp. 21–49.
[17] B. Liu, Supervised learning, Web Data Mining, Springer, 2011, pp. 63–132.
[18] T. Hastie, R. Tibshirani, J. Friedman, Overview of supervised learning, in: T. Hastie, R. Tibshirani, J. Friedman (Eds.), The Elements of Statistical Learning: Data Mining, Inference, and Prediction, Springer New York, New York, NY, 2009, pp. 9–41.
[19] V. Nasteski, An overview of the supervised machine learning methods, Horizons. B 4 (2017) 51–62.
[20] S.B. Kotsiantis, I. Zaharakis, P. Pintelas, Supervised machine learning: a review of classification techniques, Emerging Artificial Intelligence Applications in Computer Engineering 160 (1) (2007) 3–24.
[21] R. Kumari, S.K. Srivastava, Machine learning: a review on binary classification, International Journal of Computer Applications 160 (7) (2017).
[22] M. Grandini, E. Bagli, G. Visani, Metrics for multi-class classification: an overview. arXiv Preprint arXiv:2008.05756, 2020.
[23] J.-H. Hong, S.-B. Cho, A probabilistic multi-class strategy of one-vs.-rest support vector machines for cancer classification, Neurocomputing 71 (16) (2008) 3275–3281.
[24] M. Galar, et al., An overview of ensemble methods for binary classifiers in multi-class problems: experimental study on one-vs-one and one-vs-all schemes, Pattern Recognition 44 (8) (2011) 1761–1776.

[25] G. Tsoumakas, I. Katakis, Multi-label classification: an overview, International Journal of Data Warehousing and Mining (IJDWM) 3 (3) (2007) 1–13.
[26] J. Li, Regression and classification in supervised learning, in Proceedings of the 2nd International Conference on Computing and Big Data, Association for Computing Machinery: Taichung, Taiwan, 2019, p. 99–104.
[27] Z. Ghahramani, Unsupervised learning, Summer School on Machine Learning, Springer, 2003.
[28] T. Hastie, R. Tibshirani, J. Friedman, Unsupervised learning, The Elements of Statistical Learning, Springer, 2009, pp. 485–585.
[29] M.E. Celebi, K. Aydin, Unsupervised Learning Algorithms, Springer, 2016.
[30] M.G. Omran, A.P. Engelbrecht, A. Salman, An overview of clustering methods, Intelligent Data Analysis 11 (6) (2007) 583–605.
[31] S. Huang, et al., Robust deep k-means: an effective and simple method for data clustering, Pattern Recognition 117 (2021) 107996.
[32] A. Malik, B. Bhushan, A. Kumar, Association rule-based routing protocol for opportunistic network, Innovations in Electronics and Communication Engineering, Springer, 2022, pp. 391–399.
[33] N. Jayawickreme, et al., Association rule learning is an easy and efficient method for identifying profiles of traumas and stressors that predict psychopathology in disaster survivors: the example of Sri Lanka, International Journal of Environmental Research and Public Health 17 (8) (2020) 2850.
[34] R. Zebari, et al., A comprehensive review of dimensionality reduction techniques for feature selection and feature extraction, Journal of Applied Science and Technology Trends 1 (2) (2020) 56–70.
[35] C.L. Chowdhary, D. Acharjya, Singular value decomposition–principal component analysis-based object recognition approach, Bio-Inspired Computing for Image and Video Processing, Chapman and Hall/CRC, 2018, pp. 323–341.
[36] J.E. van Engelen, H.H. Hoos, A survey on semi-supervised learning, Machine Learning 109 (2) (2020) 373–440.
[37] L. Xu, C. Hu, K. Mei, Semi-supervised regression with manifold: a bayesian deep kernel learning approach, Neurocomputing 497 (2022) 76–85.
[38] R. Nian, J. Liu, B. Huang, A review on reinforcement learning: introduction and applications in industrial process control, Computers & Chemical Engineering 139 (2020) 106886.
[39] A. Wachi, Y. Sui, Safe reinforcement learning in constrained markov decision processes, in International Conference on Machine Learning, PMLR, 2020.
[40] A.C. Mater, M.L. Coote, Deep learning in chemistry, Journal of Chemical Information and Modeling 59 (6) (2019) 2545–2559.
[41] C. Janiesch, P. Zschech, K. Heinrich, Machine learning and deep learning, Electronic Markets 31 (3) (2021) 685–695.
[42] T. Bouwmans, et al., Deep neural network concepts for background subtraction: a systematic review and comparative evaluation, Neural Networks 117 (2019) 8–66.
[43] G. Xie, J. Lai, An interpretation of forward-propagation and back-propagation of DNN, in Chinese Conference on Pattern Recognition and Computer Vision (PRCV), Springer, 2018.
[44] K. Dutta, et al., Improving CNN-RNN hybrid networks for handwriting recognition, in 2018 16th International Conference on Frontiers in Handwriting Recognition (ICFHR), IEEE, 2018.

[45] Y.-Y. Song, L. Ying, Decision tree methods: applications for classification and prediction, Shanghai Archives of Psychiatry 27 (2) (2015) 130.
[46] H.H. Patel, P. Prajapati, Study and analysis of decision tree based classification algorithms, International Journal of Computer Sciences and Engineering 6 (10) (2018) 74–78.
[47] S. Suthaharan, Decision tree learning, Machine Learning Models and Algorithms for Big Data Classification, Springer, 2016, pp. 237–269.
[48] M. Du, S.M. Wang, G. Gong, Research on decision tree algorithm based on information entropy, Advanced Materials Research (2011). Trans Tech Publ..
[49] S. Nowozin, Improved information gain estimates for decision tree induction, arXiv Preprint arXiv:1206.4620, 2012.
[50] M. Brijain, et al., A survey on decision tree algorithm for classification, 2014.
[51] M. Batra, R. Agrawal, Comparative analysis of decision tree algorithms, Nature Inspired Computing, Springer, 2018, pp. 31–36.
[52] R.J. Lewis, An introduction to classification and regression tree (CART) analysis, in Annual Meeting of the Society for Academic Emergency Medicine in San Francisco, California, Citeseer, 2000.
[53] G. Biau, E. Scornet, A random forest guided tour, Test 25 (2) (2016) 197–227.
[54] C. Qi, et al., Rapid identification of reactivity for the efficient recycling of coal fly ash: hybrid machine learning modeling and interpretation, Journal of Cleaner Production 343 (2022) 130958.
[55] J. Pi, D. Jiang, Q. Liu, Random forest algorithm for power system load situation awareness technology, in International Conference on Application of Intelligent Systems in Multi-modal Information Analytics, Springer, 2021.
[56] H. Pang, et al., Pathway analysis using random forests classification and regression, Bioinformatics (Oxford, England) 22 (16) (2006) 2028–2036.
[57] Y. Liu, Y. Wang, J. Zhang. New machine learning algorithm: random forest, in International Conference on Information Computing and Applications, Springer, 2012.
[58] Y. Qi, Random forest for bioinformatics, Ensemble Machine Learning, Springer, 2012, pp. 307–323.
[59] I. Babajide Mustapha, F. Saeed, Bioactive molecule prediction using extreme gradient boosting, Molecules (Basel, Switzerland) 21 (8) (2016) 983.
[60] H. Rao, et al., Feature selection based on artificial bee colony and gradient boosting decision tree, Applied Soft Computing 74 (2019) 634–642.
[61] Z. Peng, Q. Huang, Y. Han, Model research on forecast of second-hand house price in Chengdu based on xgboost algorithm, in 2019 IEEE 11th International Conference on Advanced Infocomm Technology (ICAIT), IEEE, 2019.
[62] W. XingFen, Y. Xiangbin, M. Yangchun, Research on user consumption behavior prediction based on improved xgboost algorithm, in 2018 IEEE International Conference on Big Data (Big Data), 2018.
[63] X. Ren, et al., A novel image classification method with CNN-XGBoost model, International Workshop on Digital Watermarking, Springer, 2017.
[64] L. Chao, Z. Wen-hui, L. Ji-ming, Study of star/galaxy classification based on the xgboost algorithm, Chinese Astronomy and Astrophysics 43 (4) (2019) 539–548.
[65] C. Zhang, et al. Interpretable learning algorithm based on xgboost for fault prediction in optical network, in 2020 Optical Fiber Communications Conference and Exhibition (OFC), IEEE, 2020.
[66] S. Suthaharan, Support vector machine, Machine Learning Models and Algorithms for Big Data Classification, Springer, 2016, pp. 207–235.

[67] H. Han, et al., Least squares support vector machine (LS-SVM)-based chiller fault diagnosis using fault indicative features, Applied Thermal Engineering 154 (2019) 540–547.
[68] K. Yamano, et al. Self-localization of mobile robots with RFID system by using support vector machine, in 2004 IEEE/RSJ International Conference on Intelligent Robots and Systems (IROS) (IEEE Cat. No. 04CH37566), IEEE, 2004.
[69] Y. Tian, et al., Nonparallel Support Vector Machines for Pattern Classification, IEEE Transactions on Cybernetics 44 (7) (2014) 1067–1079.
[70] M. Awad, R. Khanna, Support vector regression, Efficient Learning Machines, Springer, 2015, pp. 67–80.
[71] F. Zhang, L.J. O'Donnell, Support vector regression, Machine Learning, Elsevier, 2020, pp. 123–140.
[72] A. Yang, et al., Application of svm and its improved model in image segmentation, Mobile Networks and Applications (2021) 1–11.
[73] K. O'Shea, R. Nash, An introduction to convolutional neural networks. arXiv Preprint arXiv:1511.08458, 2015.
[74] S. Albawi, T.A. Mohammed, S. Al-Zawi, Understanding of a convolutional neural network, in 2017 International Conference on Engineering and Technology (ICET), 2017.
[75] R. Yamashita, et al., Convolutional neural networks: an overview and application in radiology, Insights into Imaging 9 (4) (2018) 611–629.
[76] Z. Li, et al., A survey of convolutional neural networks: analysis, applications, and prospects, IEEE Transactions on Neural Networks and Learning Systems (2021) 1–21.
[77] P. Kim, Convolutional neural network, MATLAB Deep Learning, Springer, 2017, pp. 121–147.
[78] J. Pamina, B. Raja, Survey on deep learning algorithms, International Journal of Emerging Technology and Innovative Engineering 5 (1) (2019).
[79] A.F. Agarap, Deep learning using rectified linear units (relu), arXiv Preprint arXiv:1803.08375, 2018.
[80] K. Fukushima, S. Miyake, Neocognitron: a self-organizing neural network model for a mechanism of visual pattern recognition, Competition and cooperation in neural nets, Springer, 1982, pp. 267–285.
[81] D. Yu, et al. Mixed pooling for convolutional neural networks. in International Conference on Rough Sets and Knowledge Technology, Springer, 2014.
[82] N. Ketkar, J. Moolayil, Convolutional neural networks, Deep Learning with Python, Springer, 2021, pp. 197–242.
[83] S. Sathasivam, W.A.T.W. Abdullah, Logic learning in hopfield networks. arXiv Preprint arXiv:0804.4075, 2008.
[84] L. Medsker, L.C. Jain, Recurrent neural networks: design and applications, CRC press, 1999.
[85] I. Banerjee, et al., Comparative effectiveness of convolutional neural network (CNN) and recurrent neural network (RNN) architectures for radiology text report classification, Artificial Intelligence in Medicine 97 (2019) 79–88.
[86] M. Kaur, A. Mohta, A review of deep learning with recurrent neural network, in 2019 International Conference on Smart Systems and Inventive Technology (ICSSIT), 2019.
[87] A. Kag, V. Saligrama, Training recurrent neural networks via forward propagation through time, in International Conference on Machine Learning, PMLR, 2021.
[88] G. Kanagachidambaresan, et al., Recurrent neural network, Programming with TensorFlow, Springer, 2021, pp. 53–61.
[89] R. Pascanu, T. Mikolov, Y. Bengio, On the difficulty of training recurrent neural networks, in International Conference on Machine Learning, PMLR, 2013.

[90] Y. Bai, et al., Regression modeling for enterprise electricity consumption: a comparison of recurrent neural network and its variants, International Journal of Electrical Power & Energy Systems 126 (2021) 106612.
[91] J. Wu, Advances in K-means clustering: a data mining thinking, Springer Science & Business Media, 2012.
[92] D. Steinley, K-means clustering: a half-century synthesis, British Journal of Mathematical and Statistical Psychology 59 (1) (2006) 1–34.
[93] M. Yedla, S.R. Pathakota, T. Srinivasa, Enhancing K-means clustering algorithm with improved initial center, International Journal of Computer Science and Information Technologies 1 (2) (2010) 121–125.
[94] G. Hamerly, C. Elkan, Learning the k in k-means, Advances in Neural Information Processing Systems (2003) 16.
[95] M. Cui, Introduction to the k-means clustering algorithm based on the elbow method, Accounting, Auditing and Finance 1 (1) (2020) 5–8.
[96] A.W. Abbas, et al., K-Means and ISODATA clustering algorithms for landcover classification using remote sensing, Sindh University Research Journal-SURJ (Science Series) 48 (2) (2016).
[97] M. Lutz, Programming Python, O'Reilly Media, Inc, 2001.
[98] G. Van Rossum, F.L. Drake Jr, Python Tutorial, 620, Centrum voor Wiskunde en Informatica Amsterdam, The Netherlands, 1995.
[99] S. Raschka, J. Patterson, C. Nolet, Machine learning in python: main developments and technology trends in data science, machine learning, and artificial intelligence, Information 11 (4) (2020) 193.
[100] K. Arnold, J. Gosling, D. Holmes, The Java programming language, Addison Wesley Professional, 2005.
[101] J. Gosling, et al., The Java language specification, Addison-Wesley Professional, 2000.
[102] J.C.M. Santos, Y. Fei, HATI: hardware assisted thread isolation for concurrent c/c++ programs, in 2014 IEEE International Parallel & Distributed Processing Symposium Workshops, IEEE, 2014.
[103] M. Gardener, Beginning R: The Statistical Programming Language, John Wiley & Sons, 2012.
[104] K. Ito, D. Murphy, Application of ggplot2 to pharmacometric graphics, CPT: Pharmacometrics & Systems Pharmacology 2 (10) (2013) 1–16.
[105] S. Haymond, Create laboratory business intelligence dashboards for free using R: a tutorial using the flexdashboard package, Journal of Mass Spectrometry and Advances in the Clinical Lab (2021).
[106] T.E. Oliphant, A Guide to NumPy. Vol. 1, Trelgol Publishing USA, 2006.
[107] W. McKinney, Python for Data Analysis: Data Wrangling with Pandas, NumPy, and IPython, O'Reilly Media, Inc, 2012.
[108] E. Bisong, NumPy, in Building Machine Learning and Deep Learning Models on Google Cloud Platform, Springer, 2019, pp. 91–113.
[109] M.R. Kristensen, et al., Battling memory requirements of array programming through streaming, in International Conference on High Performance Computing, 2016. Springer.
[110] H. Stepanek, Thinking in Pandas: How to Use the Python Data Analysis Library the Right Way, Apress, 2020.
[111] W. McKinney, P. Team, Pandas-Powerful Python Data Analysis Toolkit, Pandas—Powerful Python Data Analysis Toolkit, 2015, p. 1625.

[112] A. Sapre, S. Vartak, Scientific computing and data analysis using numpy and pandas, 2020.
[113] G. Rajagopalan, Prepping your data with pandas, A Python Data Analyst's Toolkit, Springer, 2021, pp. 147−241.
[114] W. McKinney, Pandas: a foundational python library for data analysis and statistics, Python for High Performance and Scientific Computing 14 (9) (2011) 1−9.
[115] N. Bantilan, Pandera: statistical data validation of pandas dataframes. in Proceedings of the Python in Science Conference (SciPy), 2020.
[116] P. Virtanen, et al., SciPy 1.0: fundamental algorithms for scientific computing in Python, Nature Methods 17 (3) (2020) 261−272.
[117] S. Van Der Walt, S.C. Colbert, G. Varoquaux, The NumPy array: a structure for efficient numerical computation, Computing in Science & Engineering 13 (2) (2011) 22−30.
[118] J. Ranjani, A. Sheela, K.P. Meena, Combination of NumPy, SciPy and matplotlib/pylab-a good alternative methodology to MATLAB-a comparative analysis, in 2019 1st International Conference on Innovations in Information and Communication Technology (ICIICT), IEEE, 2019.
[119] C. Schäfer, Extensions for scientists: numpy, scipy, matplotlib, pandas, Quickstart Python, Springer, 2021, pp. 45−53.
[120] E. Bisong, Matplotlib and seaborn, Building Machine Learning and Deep Learning Models on Google Cloud Platform, Springer, 2019, pp. 151−165.
[121] G. Moruzzi, Plotting with matplotlib, Essential Python for the Physicist, Springer, 2020, pp. 53−69.
[122] Y. Guan, F. Zhou, J. Zhou, Research and practice of image processing based on python, Journal of Physics: Conference Series (2019). IOP Publishing.
[123] A.H. Sial, S.Y.S. Rashdi, A.H. Khan, Comparative analysis of data visualization libraries matplotlib and seaborn in python, International Journal 10 (1) (2021).
[124] F. Nelli, Machine learning with scikit-learn, Python Data Analytics, Springer, 2018, pp. 313−347.
[125] E. Bisong, Introduction to scikit-learn, Building Machine Learning and Deep Learning Models on Google Cloud Platform, Springer, 2019, pp. 215−229.
[126] K. Jolly, Machine Learning with Scikit-Learn Quick Start Guide: Classification, Regression, and Clustering Techniques In Python, Packt Publishing Ltd., 2018.
[127] B. Pang, E. Nijkamp, Y.N. Wu, Deep learning with tensorflow: a review, Journal of Educational and Behavioral Statistics 45 (2) (2020) 227−248.
[128] P. Singh, A. Manure, Introduction to TensorFlow 2.0, in: P. Singh, A. Manure (Eds.), Learn TensorFlow 2.0: Implement Machine Learning and Deep Learning Models with Python, Apress: Berkeley, CA, 2020, pp. 1−24.
[129] H.B. Braiek, F. Khomh, TFCheck: a tensorflow library for detecting training issues in neural network programs, in 2019 IEEE 19th International Conference on Software Quality, Reliability and Security (QRS), IEEE, 2019.
[130] D.J. Gunn, et al., Touch-based active cloud authentication using traditional machine learning and LSTM on a distributed tensorflow framework, International Journal of Computational Intelligence and Applications 18 (04) (2019) 1950022.
[131] A. Sergeev, M. Del Balso, Horovod: fast and easy distributed deep learning in tensorflow. arXiv Preprint arXiv:1802.05799, 2018.
[132] N.K. Manaswi, Understanding and working with Keras, Deep Learning with Applications Using Python, Springer, 2018, pp. 31−43.
[133] J. Moolayil, An introduction to deep learning and keras, Learn Keras for Deep Neural Networks, Springer, 2019, pp. 1−16.

[134] J. Moolayil, J. Moolayil, S. John, Learn Keras for Deep Neural Networks, Springer, 2019.

[135] F.J.J. Joseph, S. Nonsiri, A. Monsakul, Keras and tensorflow: a hands-on experience, Advanced Deep Learning for Engineers and Scientists, Springer, 2021, pp. 85–111.

[136] P. Mishra, Introduction to pytorch, tensors, and tensor operations, in, PyTorch Recipes, Springer, 2019, pp. 1–27.

[137] Paszke, A., et al., Automatic Differentiation in Pytorch. 2017.

[138] G.W. Ding, L. Wang, X. Jin, AdverTorch v0. 1: an adversarial robustness toolbox based on pytorch. arXiv Preprint arXiv:1902.07623, 2019.

[139] H. Liu, et al., G3: when graph neural networks meet parallel graph processing systems on GPUs. Proceedings of the VLDB Endowment, 2020. 13(12): p. 2813–2816.

[140] T.L. Patti, et al., Tensorly-quantum: quantum machine learning with tensor methods. arXiv Preprint arXiv:2112.10239, 2021.

[141] T. Deleu, et al., Torchmeta: a meta-learning library for pytorch. arXiv Preprint arXiv:1909.06576, 2019.

[142] E. Stevens, L. Antiga, T. Viehmann, Deep learning with PyTorch, Manning Publications, 2020.

Machine learning modeling methodology for industrial solid ash

7

Abstract

This chapter presents a detailed explanation of the applications of machine learning (ML) in industrial solid ash management. Specifically, the ML modeling methodology, comprising data collection, data preprocessing, model construction, and parameter tuning, in addition to model evaluation, is explained in detail. Furthermore, due to the black box nature of ML, several model explanation methods are introduced to provide deeper insights into the decision-making process underlying ML-based models. Following the ML modeling methodology presented in this chapter, the readers can have a clear knowledge on the implementation of ML in industrial solid ash management. Moreover, this chapter also serves as a valuable reference for future application of ML in similar topics.

7.1 Introduction

Continued promotion of reform, opening up, and full implementation of the Five-Year Plan have encouraged sustained and healthy economic and social development in China. Industrial sectors including the metallurgical industry, electric power generation, and the chemical industry have been comprehensively developed and have maintained steady growth. These changes have improved standards of living and provided livelihoods for many; however, economic growth also brings potential threats to the environment. Solid ash is produced by the burning of coal, sewage sludge, domestic garbage, etc., in incinerators [1], which can cause serious harm to soil, air, water, and human health if not properly treated [2]. The management and recycling of industrial solid ashes have attracted considerable attention from both academia and industry. To date, efficient decision-making processes in industrial solid ash management and recycling have been hindered by their reliance on experience and expert knowledge. However, with the increasing availability of data and the development of ML techniques, ML-aided management and recycling of industrial solid ash have become important topics in sustainable development.

As shown in Fig. 7.1, constructing an efficient ML model typically requires the following steps. Firstly, a certain amount of data must be collected and preprocessed based on the problem being investigated [3]. For example, the physical and chemical properties of solid ash are required when investigating ash reactivity, while the ash generation rate from each plant must be collected for studies of ash generation prediction. Secondly, an appropriate ML algorithm must be selected for model construction. After the model is constructed, appropriate statistical indicators are then selected to evaluate the modeling performance. If the model is not sufficiently accurate,

Machine Learning Applications in Industrial Solid Ash. DOI: https://doi.org/10.1016/B978-0-443-15524-6.00013-3

© 2024 Elsevier Inc. All rights reserved.

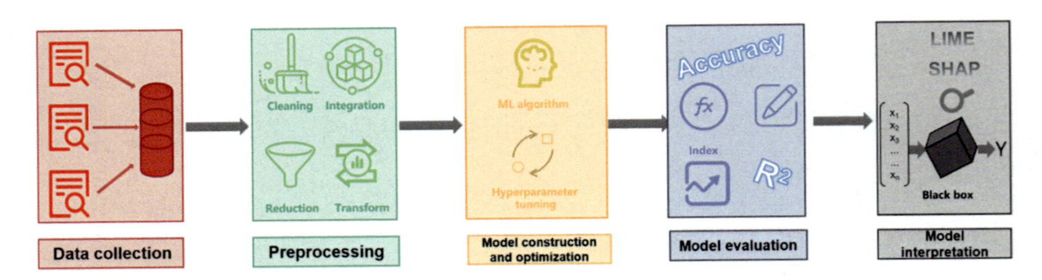

Figure 7.1 Flowchart illustrating the ML model construction process.

hyperparameter tuning can be used to improve the model's accuracy and ensure its feasibility [4]. Finally, the optimal model is applied to unknown data to predict and solve practical engineering problems. Furthermore, analyzing the constructed black box model, obtaining deep insights into the feature function, and explaining model behavior also form key parts of the modeling workflow.

7.2 Dataset collection and preprocessing

As ML becomes an increasingly widely used technology, considerable attention has been given to preparing datasets, including data collection, data preprocessing, feature engineering, and other aspects. A sufficient amount of data is a prerequisite for creating an effective model, and high-quality data make all the subsequent modeling process steps easier, which tends to lead to higher model accuracy [5].

7.2.1 Data collection

Data collection is not only an active area of research but also an important bottleneck in ML as there are few or no training data available for new applications, and DL-based approaches have increasingly large training data demands [6]. Accurate and scalable data collection methods are becoming essential in the big data era. Collecting datasets mainly refers to discovering or generating data that can be used to train ML models through a range of approaches, including observation, interviews, field investigations, lab experiments, and Internet-based collection (Fig. 7.2).

In recent years, qualitative data collection methods such as interviews and observations have been included in "ethnographic methods" [7]. Among these approaches, observation, in the simplest sense, refers to the use of the eyes or other senses to record data relating to objects or phenomena; however, other equipment and agents can be effectively used to track and record observation results. For example, Gonzalez et al. [8] observed the driving behavior of 12 human test subjects using a driving simulator to collect the data required to construct an ML model. An interview represents oral questioning of an interviewee (either an individual or a group) with some degree of flexibility [9]. For example, Cohen, Wright-Berryman [10] aimed to identify suicidal individuals by modeling and training

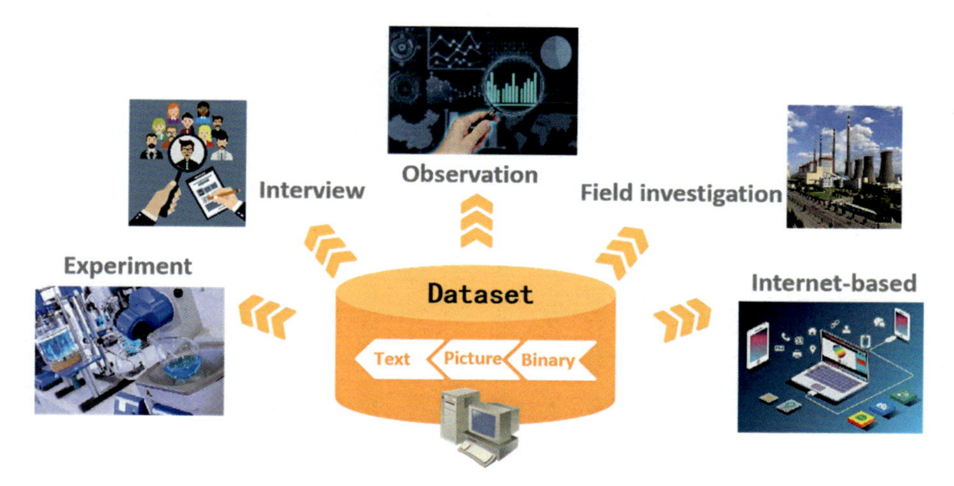

Figure 7.2 Various methods of data collection.

language samples with ML using speech samples collected from 60 students interviewed by 10 therapists. In addition, for specific problems, field or continuous investigation is an effective data collection approach. For example, Tziridis, Kalampokas [11] investigated a group of flights to the same destination (from Thessaloniki to Stuttgart) between December and July to explore the question of airfare forecasting.

For specific industries or problems such as industrial, biological, and environmental pollution, experiments are among the most common data collection methods. Experiments often require a robust study design involving causal inference, which allows researchers to manipulate one or more independent variables as designed and observe their effect on the dependent variable [12]. Experiments can produce informative and accurate data samples that can be used to advance the understanding of a research field [13]. In an ML context, de Farias, de Almeida [14] used two different steel workpieces to conduct machining experiments to verify machine tool wear and obtained a dataset to construct a linear associative memory neural network. The modeling features used by Liu, Lu [15] were obtained by experimental determination of 160 fly ash samples, which were collected from four incinerators. Nguyen, Nguyen [16] also conducted experimental work with 335 mixed ratios to obtain input and output variables for the model training and validation process.

With the advent of the era of big data, the Internet has become a popular medium for data collection (Fig. 7.3), providing researchers with a variety of options for data collection feedback [17]. Compared with approaches such as observations, experiments, and paper surveys, Internet-based data collection can shorten response times, reduce costs, and simplify data input [18]. Internet-based data collection mainly comprises three methods: website platform collection, direct usage of representative public datasets, and literature retrieval.

Figure 7.3 Internet-based data collection.

At present, there are a variety of platforms and websites that provide massive public datasets for free. Researchers can not only easily and quickly search for high-quality datasets they want but also preview sample data and labels online. Kaggle, Graviti Open Dataset [19], CEIC, and the microblogging service Twitter [20] provide researchers with rich sources of data; however, these data may require preprocessing. In addition, there are numerous classic public datasets on the web, such as the Iris dataset (provided by the UCI Machine Learning Library) [21], wine classification dataset, and Boston housing dataset, that can be used directly for model training. For example, Sharma et al. used the Wisconsin breast cancer dataset to construct KNN and other ML models [22]. Literature retrieval, a common method adopted by ML researchers, refers to reviewing a large number of published studies related to the research field of interest and extracting available data from them. Qi, Wu [23] and Qi, Wu [24] collected solid ash datasets using literature retrieval; they then used these data to trace the solid ash's origin, study its reactivity, etc. The specific operation steps of literature retrieval-based data collection in an ML context are as follows:

1. Analyze the subject, clarify the literature requirements, and translate the study's research questions into key topic terms. For example, for solid ash research, the terms "solid ash" and "machine learning" (ML) form the basis of the search. These can be extended by adding terms such as "physico-chemical properties" and "coal fly ash".
2. Select the retrieval tool. Access a database navigation system via an institutional library, select databases such as Web of Science, Elsevier, or Scopus, or directly use Google Scholar for article retrieval [25].
3. Determine the retrieval approach. The most popular approaches at present involve using "and," "or," and other Boolean operators to construct the search form and then entering multiple keywords to perform an advanced search [26]. In addition to specific queries, literature can also be queried by searching for the title of a book, journal, or article or the name of its author.

4. Preliminary determination of literature. Repeat steps 1, 2, and 3 for various queries to complete a preliminary search and obtain numerous studies.
5. Further screening of the literature. The titles and abstracts of the obtained studies are reviewed; those that do not conform to the research theme or those that are repeated can be excluded at this point [27]. The remaining studies can then be downloaded to local storage.
6. Mark key information. Obtain an overview of the literature to determine whether there is relevant information (e.g., physical and chemical property information such as SiO_2, CaO, and other oxide content, D_{10}, D_{90}, and other particle sizes for solid ash studies) and mark them. Delete studies lacking sufficient information.
7. Extract data from the literature. The remaining literature is then carefully reviewed, and the available data are extracted to obtain the final dataset.
8. If there is a complete dataset provided with the literature, this can be used directly or combined with other existing datasets. In this instance, steps 6 and 7 can be omitted.

In many cases, the collected datasets may not be complete, and additional information is needed to fill and enrich them. Deriving latent semantics from data, entity augmentation, and the data integration approach noted in step 8 earlier are increasingly used in ML. Alternatively, data can be obtained through crowdsourcing to manually generate data or using automated techniques to synthesize data [28].

7.2.2 Data preprocessing

An acceptable high-quality dataset meets the following key criteria: accuracy, completeness, consistency, and interpretability. Datasets such as those included in sklearn are of high quality with no missing values and no outliers, thus making them ideal for modeling. However, there are many problems in real datasets, such as different data types of data, poor data quality, nonuniform dimensions, and skewed data. Therefore, when the acquired new data are nonoptimal, they must be preprocessed to obtain standard, clean, continuous data [29].

As shown in Fig. 7.4, data preprocessing comprises four main tasks: data cleaning, data integration, data reduction, and data transformation. These processes are described in detail below.

7.2.2.1 Data cleaning

Data cleaning refers to removing anomalies, correcting errors, and accounting for missing values by discarding, filling, replacing, and rescaling the original data [30]. In practice, data are often dirty, with numerous missing values, outliers, and other noise types, or data duplication and information redundancy (Fig. 7.5), which are not conducive to successfully training ML models. Therefore, data cleaning is the most important step in the data preprocessing pipeline.

In the data cleaning process, the first necessary step is to conduct a descriptive statistical analysis of the dataset to understand its distribution [32]. Plotting data can indicate its quality by revealing the presence of any abnormal (outlier) points or noise. Unique attributes that do not characterize the overall distribution laws of the sample can also be removed. In addition, the integrity of datasets is essential for

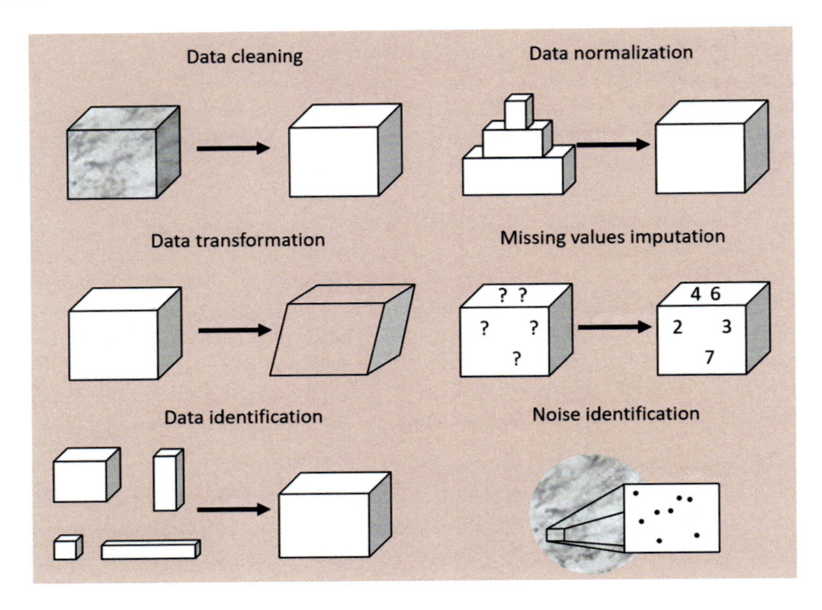

Figure 7.4 The main data preprocessing tasks.

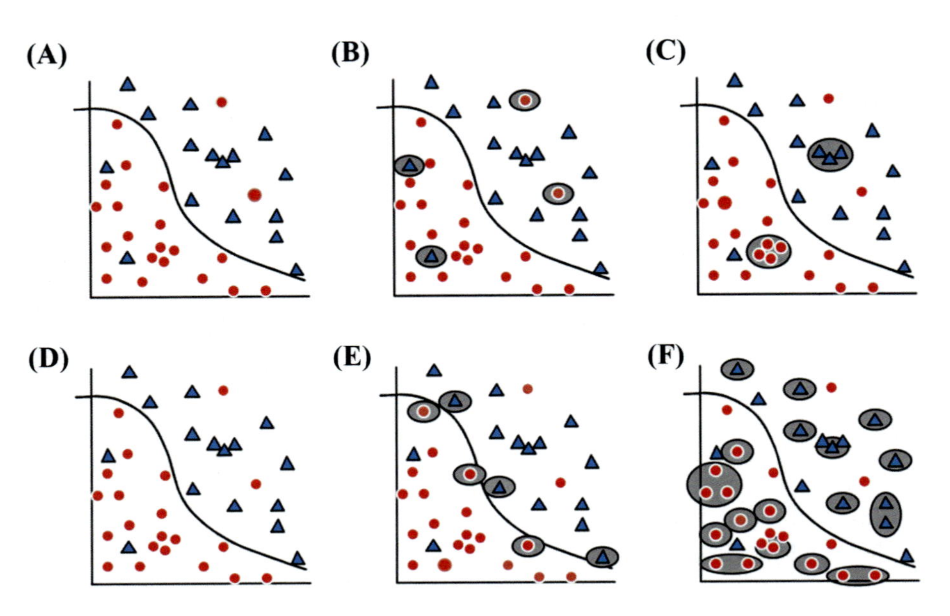

Figure 7.5 Examples of dirty data in practice: (A) missing data, (B) incorrect labeling, (C) data duplication, (D) outliers (E) borderlines, and (F) safe cases [31].

data mining. The presence of missing values can cause serious difficulties for researchers; thus, missing data points must be handled correctly. The treatment method should be determined based on the distribution proportion of the features with missing values in the whole sample and their degree of importance to the prediction results.

To avoid introducing new noise into the data, features containing missing values may not be preprocessed; however, this can lead to the failure of model training, and this approach is thus generally not adopted [33]. Furthermore, if the missing rate of the feature data is high and its importance is low, the feature can be directly deleted. However, if the missing rate is low and the feature has an obvious influence on the prediction results, one of the methods in Table 7.1 should be used to fill its missing values [34].

Data that are outside a specific distribution or range are usually defined as anomalies. Where there are few outliers and they only convey little information, they can be directly deleted, or the average and median values can be used instead. For noise caused by random error and variance, binning is usually performed. Specifically, bins of equal frequency or width are defined, and the average, median, or boundary

Table 7.1 Specific filling methods for missing values.

	Method		Explanation
1	Constant value fill		A constant value (usually 0) is used for filling.
2	Statistics	Average	This approach is suitable for numerical data and satisfies the normal distribution
		Median	This approach is suitable for numerical data and satisfies skewed distributions or those with discrete points.
		Mode	For nonnumerical data, "mode" is used to complete the missing attribute value.
3	Interpolate	Hot deck imputation	Identify a feature within the complete data that is most like the feature containing the missing value and fill it with the value of this similar feature.
		Multiple imputation	Based on repeated simulations, a complete dataset is generated from a dataset containing missing values.
4	High-dimensional mapping		The attribute value containing K discrete value range is extended to K + 1 attribute values. If the attribute value is missing, the extended K + 1 attribute value is set to 1.
5	Model-based fill		A supervised ML method is used to make predictions about the missing values and fill them.
6	Dummy variable fill		If the variable is discrete and has few values, it can be converted into a dummy variable.

value of each bin is used to replace all the numbers within the bin to smooth the data [35]. Constructing a regression model and smoothing data with function fitting is another feasible noise-removal approach [36].

7.2.2.2 Data integration

The collection of datasets is not a simple replication process. In addition to the possibility of "dirty" data arising from different data sources, there are issues such as fields sharing a name but having different meanings, disunity of units and structures, etc. To combine and place data from multiple data sources in a consistent database, data integration is required and plays an important role in the preprocessing workflows (Fig. 7.6).

Data integration mainly aims to solve four key problems: entity identification, data redundancy, data duplication, and data value conflict [35]. Entity identification is the primary problem in data integration, where multiple entities from different sources need to be matched [37]. The metadata of each attribute, including its name, meaning, data type, and allowed value range of its attributes, can be used to assist the matching process. In addition, rules can be applied to handle null or zero values.

Data redundancy generally arises from inconsistent feature naming or when a property can be derived from one attribute or another set of attributes, which can be detected by correlation analysis [38]. Specifically, for scalar data, the Chi-square test is applied using Pearson and χ^2 indices. For numerical features, covariance and the correlation coefficient R are used to evaluate the correlation between features. In addition to checking for feature redundancy, duplicates of a given unique data entity are detected and filtered.

Identifying and resolving data value conflicts are the last important stage of data integration. In many cases, due to differences in scale, scoring systems, or coding, the feature values of unified entities from different sources may vary. Whether or

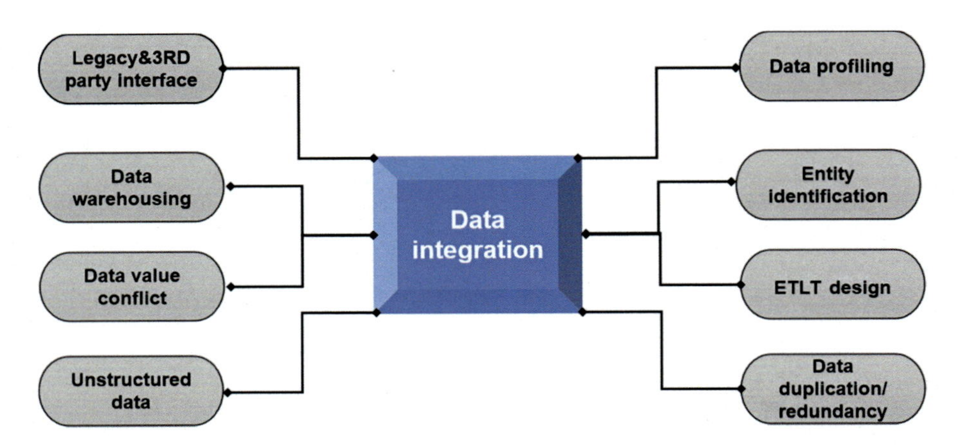

Figure 7.6 Various methods for data integration.

not conflicting data should be retained and how to select or omit data should be evaluated comprehensively in the data integration process [39].

7.2.2.3 Data reduction

Data reduction is a technique for reducing the scale of data, which can be used to obtain a reduced dataset from an original large dataset while also preserving the integrity of the original data information [40]. This approach not only shortens the ML modeling time but can also reduce the impact of invalid data on model accuracy.

The main data reduction methods include numerosity reduction, data compression, and dimension reduction [29]. Among these, numerosity reduction refers to replacing the original data with smaller data representations, which can be divided into two types: nonparametric and parametric. Nonparametric data reduction includes histograms, clustering, sampling, and data cube aggregation [41], while parametric data reduction can be implemented using regression or log–linear models. Data compression refers to the use of a transformation to achieve a one-time reduction of the original data either by lossy or lossless compression. Lossless compressed data can be reconstructed without loss of information, which is suitable for string compression [42], whereas lossy compression approximates the original data reconstruction, which is more suitable for audio or video data.

Dimensionality reduction is the process of mapping data from a high-dimensional feature space to a low-dimensional feature space and is the most common data reduction method in ML [43]. This approach removes unimportant features based on ensuring the integrity of data information, alleviates the curse of dimensionality, and reduces the learning task, which is convenient for calculation and visualization. Fig. 7.7 further illustrates the need for dimension reduction.

Common dimension reduction approaches fall into two categories: feature selection and mapping. Filtering, wrapping, and embedding methods are typical of feature selection and reduce the amount of data by removing irrelevant or redundant features [44]. Mapping approaches such as PCA, which was introduced in detail in Chapter 6, and the wavelet transform project or map the original data to a smaller space and form new features through the computational combination of features.

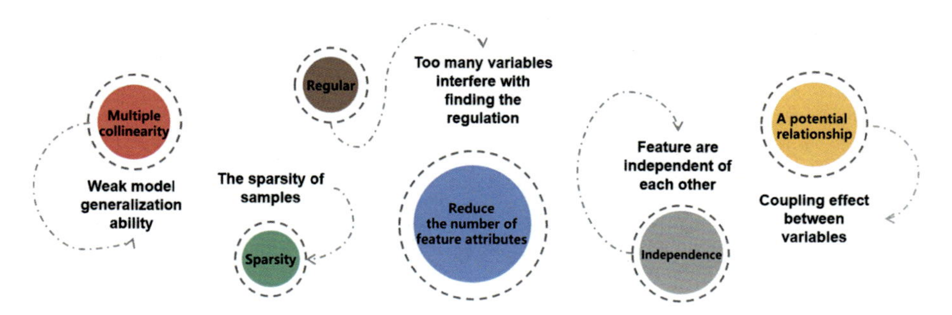

Figure 7.7 Reasons for data dimensionality reduction.

7.2.2.4 Data transformation

Data transformation refers to the standardization, discretization, sparsification, and generalization of data [45]. This process aims to unify the data into a form suitable for data mining and improve the degree of model fitting, as shown in Fig. 7.8.

The premise of data transformation involves reviewing the data type of the dataset and casting it. Data types can be broadly divided into numerical types that contain continuous and discrete values and nonnumerical types that contain categories and noncategories [46]. In this process, nonnumerical features must be transformed into numerical ones. For example, the male and female gender features in a dataset can be numerically encoded as 0 and 1 for further processing.

The dimensions of the transformed numerical data may be inconsistent, and different features' numerical values will vary, which can affect model performance [47]. To prevent small data from being "swallowed" by big data and make different indexes comparable, data normalization is necessary. The standard normalization process involves scaling feature data to fit a specific region and mainly comprises the following five methods:

1. Max−min standardization (normalization): scale the data to [0,1] [48].

$$x_{new} = \frac{x - x_{\min}}{x_{\max} - x_{\min}} \tag{7.1}$$

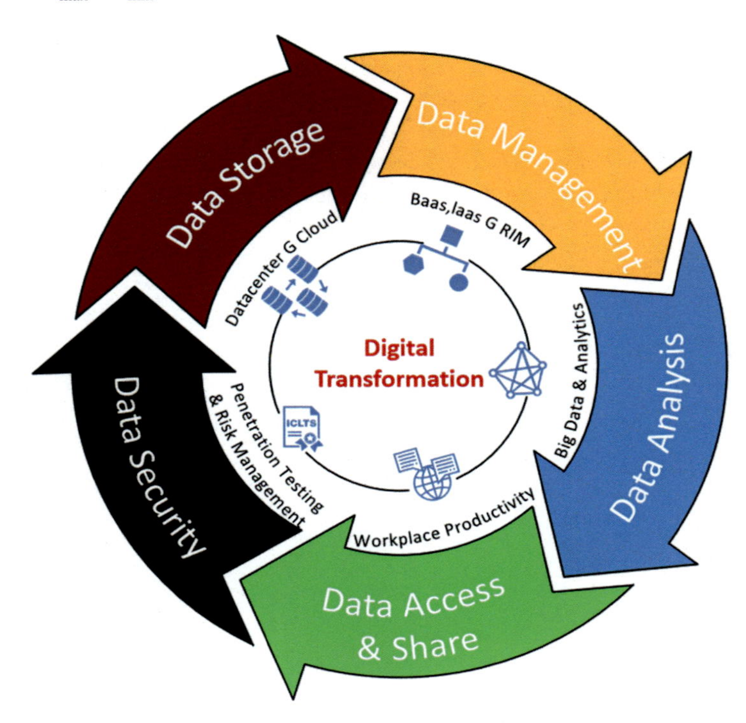

Figure 7.8 The specific steps and methods of data transformation.

Where x represents the sample data of each feature, $x_{\max}$ and $x_{\min}$ are the maximum and minimum values of the feature, respectively, and x_{new} is the feature value after scaling.

2. Z-score standardization: the data are converted to a standard normal distribution with a mean of 0 and variance of 1 using Formula (2).

$$x_{new} = \frac{x - \mu(x)}{\delta(x)} \tag{7.2}$$

Where $\mu(x)$ and $\delta(x)$ are the mean and variance of the feature, respectively.

3. Zero-centered: the feature minus its mean value.
4. Decimal scale standardization: the feature samples are divided by the nth power of 10 to move the decimal point position of the variable as a whole such that it falls within [-1,1] [49].
5. Log transformation: in time series data, the Log function transformation is usually applied to the variable with the largest difference in data magnitude.

Because many models run on discrete data, effective discretization can shorten the running time of the model and improve the model's sample classification ability. Therefore, in terms of specific problems, continuous data can be segmented into discrete intervals based on the principles of equal distance or equal frequency [37]. For some discrete nominal variables, sparsifying is usually performed to convert the categorical variables into dummy variables to make the model converge more quickly when an ordered label encoder cannot be used [50]. Data generalization refers to the conceptual layering of features, where high-level concepts are used to replace low-level or raw data, and is widely used to transform nominal data [51].

7.3 Machine learning modeling

After preprocessing the original data, such as by deleting outliers and inserting missing values, the next step is to use the data for model construction. This involves the following main issues: selection of the learning algorithm, segmentation of the dataset, cross-validation, and hyperparameter tuning, which are described in detail below.

7.3.1 Selection of the learning algorithm

Understanding the specific problem being investigated is a prerequisite for ML. Different ML problems require different strategies [52], and no single algorithm can handle achieve the best results in all scenarios [53]. Therefore, it is essential to choose different models according to engineering requirements. For example, multiple classification models should be selected to trace the source of solid ash, whereas regression models should be selected to predict the reactivity of coal fly ash.

In addition, the scale of the dataset should be considered in algorithm selection. When the amount of data is relatively small, commonly used ML algorithms are typically suitable. When there is a large amount of data but few features, bagging algorithms (RF) and deep learning (neural networks) are widely used. However, whether a selected initial model can be used as the final model largely depends on selecting appropriate indexes for objective model evaluation.

7.3.2 Dataset splitting

The essential concept of ML involves passing data into an algorithm to train a model and then using that model to make predictions about unknown data. Directly using all of the available data for model training and prediction will tend to yield better results on the known data; however, the model's performance on unknown data cannot be guaranteed. Given this limitation, the dataset is usually split in supervised ML strategies to obtain models with the best generalization ability.

There are two methods of dataset splitting depending on whether a validation set is required. One approach is to randomly split the dataset into two parts according to a given ratio, as shown in Fig. 7.9; common splitting ratios include 8:2 and 7:3 [54,55]. The larger data subset is used as the training set, while the smaller subset is used as the testing set [53]. The other approach is to split the data into three parts: training set, validation set, and testing set [56], which is often used in neural network models.

As shown in Fig. 7.10, the training set is used to train the model directly and obtain the model parameters [57]. The validation set is then used to test the model derived from the training data to prevent overfitting. If the model effect does not meet the required specifications, it is necessary to retrain the model, adjust the fitting times, and tune the hyperparameters. This process is repeated iteratively until a satisfactory final model is obtained. The testing set acts as new and unknown data and does not participate in any model construction or preparation. The performance of the ML model on the testing set data is used as the final verification.

The amount of data in the training set should generally account for the largest proportion, typically 70%–80%, while the fractions in the validation and testing sets should each represent around 10% of the total. However, the specific segmentation ratio of any dataset is not absolute, and the different proportions used for each step can be optimized to meet the needs of the modeling process [58,59].

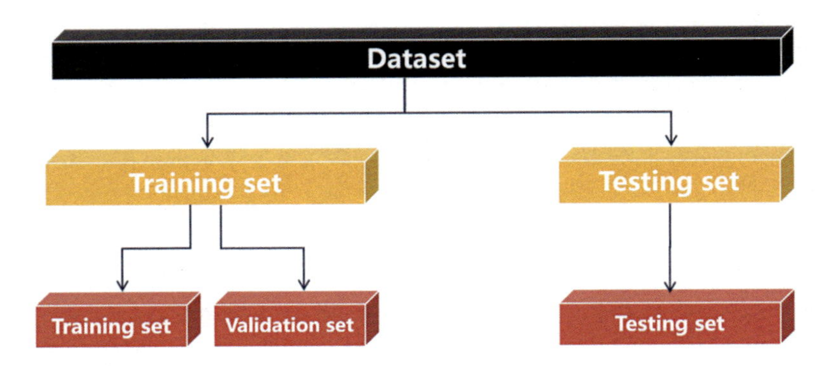

Figure 7.9 Workflow of dataset splitting.

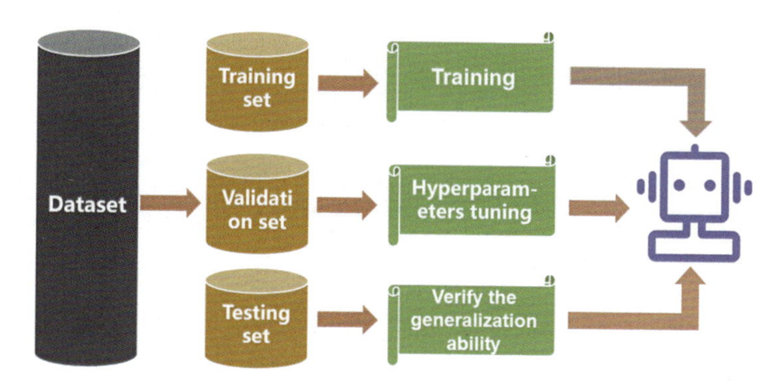

Figure 7.10 The role of three data subsets in model construction.

Because the dataset is usually split randomly, the result from a single random splitting does not necessarily represent the true performance of the ML model. When the dataset is only split once, the model is sensitive to the points in the training, validation, and testing sets and whether the distribution of the segmented data is the same as that of the original dataset. To make the model evaluation more convincing, the dataset should be randomly split several times, and the average of the results should be taken as the final performance of the model [60].

7.3.3 Cross-validation

The theory of cross-validation originated from Seymour Geisser [61]. This approach represents a practical method for cutting data samples into smaller subsets in statistics. It is mainly used to evaluate the predictive performance of a model and can also reduce overfitting to some extent [62]. Common cross-validation approaches include leave-one-out cross-validation (LOOCV) and k-fold cross-validation.

As shown in Fig. 7.11, assuming that there are n samples in the dataset, LOOCV iteratively removes one sample each time as the validation set, and the remaining n-1 samples are used as the training set for model training and hyperparameter tuning [63]. Finally, the average accuracy of the n classifiers or the average Mean Squared Error (MSE) of n regressors is taken as the model's performance index. The LOOCV method is not affected by the partition ratio of the validation and training sets, and n-1 data points are used to train the model, which reduces the bias. However, the calculation costs are high, and this method is slow to compute.

Given the computational disadvantages of LOOCV, a compromise method (k-fold cross-validation) is proposed (Fig. 7.12). Unlike LOOCV, the validation set consists of multiple data points, and the exact amount of data used in each set is determined by the selection of k.

The basic concepts and operational steps of k-fold cross-validation are as follows:

1. The original training data are randomly divided into k parts by nonrepeated sampling.
2. One part is selected as the validation set, and the remaining k-1 parts are used to form the model's training set [64].

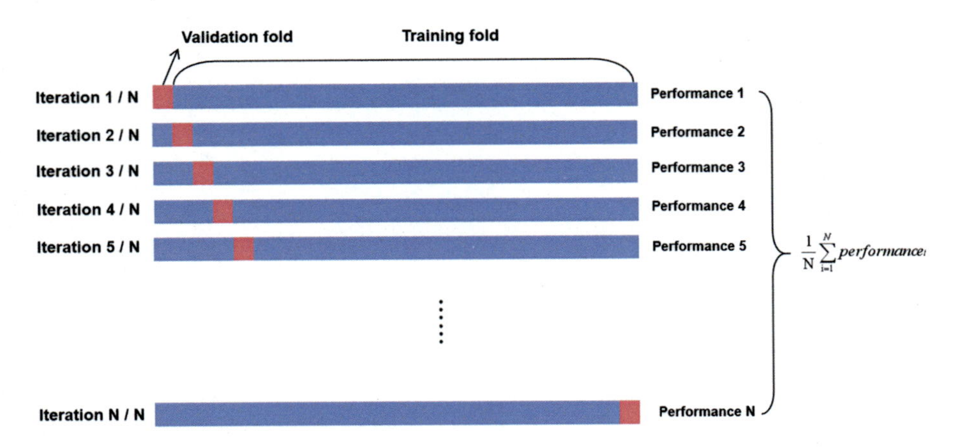

Figure 7.11 Principle of the LOOCV algorithm.

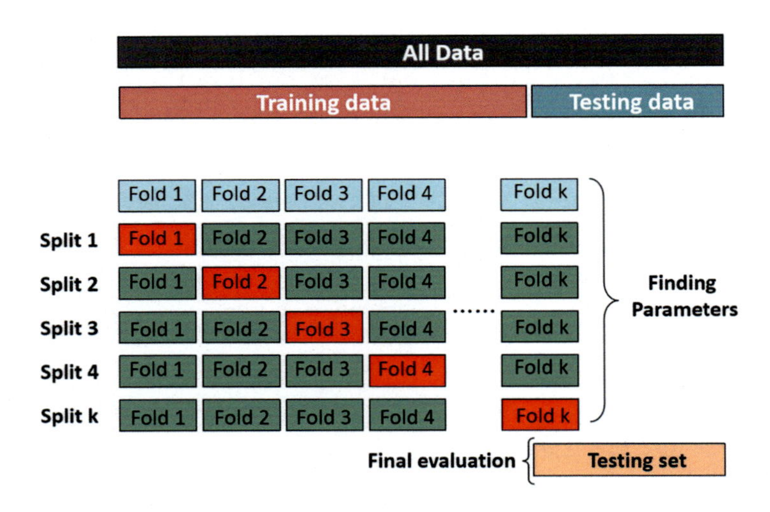

Figure 7.12 The algorithm principle of k-fold cross-validation.

3. Repeat step 2 k times to obtain k models, test them on the corresponding validation sets, and then calculate and save the model's evaluation result.
4. Take the average value of the k validation results as the performance of the model under k-fold cross-validation [65].

The k-fold cross-validation method helps to minimize model overlearning and underlearning and makes the model results more convincing. In this method, the selection of the k value is an important aspect: generally, k is set as 5 or 10 [66]. However, when the amount of data is small, k can be set as a larger value to make the training set account for a larger proportion of the whole. However, when the amount of data is large, k is set to a small value.

7.3.4 Hyperparameter tuning

There are two parameter types in ML models. One parameter type (model parameters) must be learned and estimated from data and represents the configuration variables inside the model [67]. The other type is the tuning parameters in ML algorithms, also known as hyperparameters; these must be manually set before model training and cannot be obtained through data learning [68]. One of the most difficult parts of the ML workflow is identifying the optimal hyperparameters for a model, which can minimize the predefined loss function and improve the model's performance [69].

Hyperparameter tuning comprises two main strategies: manual tuning and automatic optimization. Manually tuning usually involves the use of trial and error to search for a range of values based on default suggested values as well as intuition and experience [70]. However, this method is tedious, time-consuming, and unsuitable for models containing a wide range of hyperparameters [68]. Automatic hyperparameter optimization refers to the use of algorithms to search for the best value. This approach not only has good performance but is also simple and time-saving and can effectively be used to solve practical problems [71]. Several typical hyperparameter tuning approaches are described in detail as follows.

7.3.4.1 Grid search

The grid search approach is an exhaustive search method that optimizes model performance by traversing a given combination of parameters, also known as a brute force search [72], which is often combined with cross-validation. In this method, the possible values of each parameter are arranged and grouped then used for training, and cross-validation is then used to evaluate the model's performance. After the fitting function has tried all the parameter combinations, it returns a model whose parameters have automatically been adjusted to their optimal values [73].

The parameter tuning process used in the sklearn package's GridsearchCV method is shown in Fig. 7.13, and the parameters and common methods and attributes used in this approach are shown in Table 7.2.

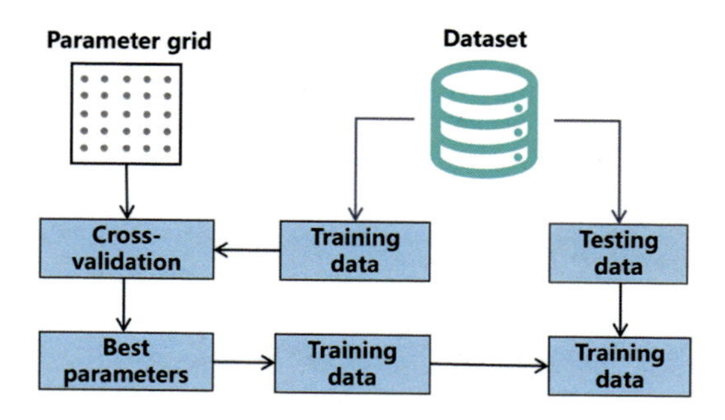

Figure 7.13 Overview of the process of parameter selection and model evaluation with GridSearchCV.

Table 7.2 Common parameters as well as methods and properties of GridSearchCV.

Parameters	Explanation	Method/ property	Implication
Estimator	An interface that must be supplied with a function or passed.	best_ params _	Describes the combination of hyperparameters that achieved the best result
param_grid	A parameter setting list with parameter names and corresponding values.	best_score_	Provides the best score observed during optimization
scoring	A strategy for evaluating the performance of a cross-validation model.	cv_results_	A dict object containing the results of hyperparameter optimization, with keys as column headers and values as column entries.
n_jobs	The number of jobs to be run in parallel.	best_index_	The index that corresponds to the best candidate hyperparameter setting.
refit	A bool option to use the best hyperparameter results to fit all datasets again at the end of the search.	grid.fit ()	Method to run a grid search.
cv	Strategies for cross-validation.	grid.score ()	Method to return the evaluation score after running a grid search.

The grid search approach can ensure to the most accurate hyperparameter values are identified within the specified parameter range, which is suitable for smaller datasets and those containing fewer hyperparameters [45]. However, as the amount of data and hyperparameters increases, the computational complexity will also increase exponentially, which is very time-consuming: in this case, the random approach search is more suitable.

7.3.4.2 Random search

The random search approach is a hyperparameter tuning algorithm proposed by James Bergstra and Yoshua Bengio in 2012 [74]. In contrast to the grid search, this method does not test all parameter combinations but randomly samples the hyperparameters of a given probability distribution and then combines the sample values (Fig. 7.14) [74].

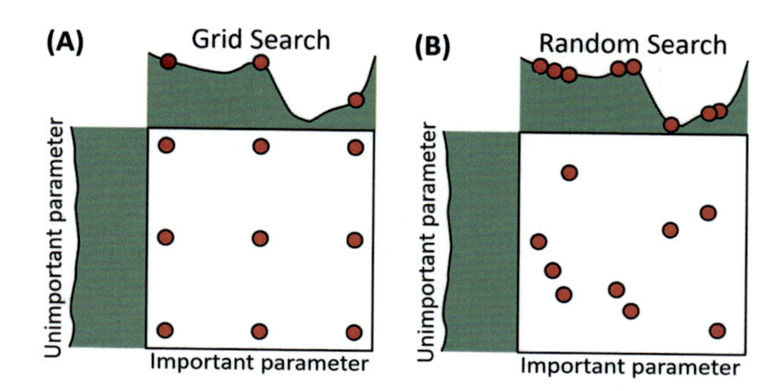

Figure 7.14 Difference between (A) grid search and (B) random search.

The principle of the random search is to generate random combinations in a certain interval and then calculate the value of the constraint function and objective function [75]. For the combinations that meet the constraint conditions, the objective function values are compared one by one; the bad combinations are then discarded and the good combinations retained and, finally, the optimal solution is approximated.

In scikit-learn, the RandomizedSearchCV strategy comprises two main steps. First, the parameters are sampled by the n_ iter _ method [76]. For hyperparameters whose search ranges are distributions, samples are randomly taken within the given distribution range, whereas for hyperparameters whose search ranges are lists, samples with medium probability in the given list are chosen. If all the hyperparameters are lists, this approach behaves similarly to a grid search. After sampling, the search and parallel method is used to solve the model's hyperparameters [77]. The main algorithm steps involved in the random search are as follows:

1. Define an algorithm
2. Define the parameter distribution.
3. Define a RandomizedSearchCV object and then pass the algorithm, parameter distribution, random seed, and other parameters
4. Model training and parameter tuning.

RandomizedSearchCV overcomes the disadvantages of GridSearchCV and returns comparable results in less time [67]. In addition, it ensures that the resulting model is not biased toward an arbitrarily chosen set of values. However, the results of the random search method may not represent the optimal combination of hyperparameters, which should be considered when selecting the optimization method.

7.3.4.3 Bayesian optimization

Bayesian Optimization (BO) is a principled technique based on Bayes' theorem, which is used to solve the global optimization problem of black-box functions [78]. This approach is now widely used in ML to tune the hyperparameters of a given well-performing model on a validation dataset [79]. As the objective function is unknown, the basic concept involves using a prior function to replace the target function, and then, based on the Gaussian modeling process, this function is used to estimate the posterior distribution of the alternative function, which is constantly updated by adding sample points [80]. The BO tuning process ends when the posterior distribution is essentially consistent with the real distribution and the globally optimal parameters are found. The complete BO workflow is shown in Fig. 7.15.

As the core function of BO, the acquisition function is mainly used to solve the balance between exploration and exploitation to achieve efficient sampling [81]. The general form of the acquisition function is a function related to x that indicates the probability that the target function value will exceed the current optimal value when evaluated at the next point. There are three main forms: upper confidence bound, probability of improvement, and expected improvement [82].

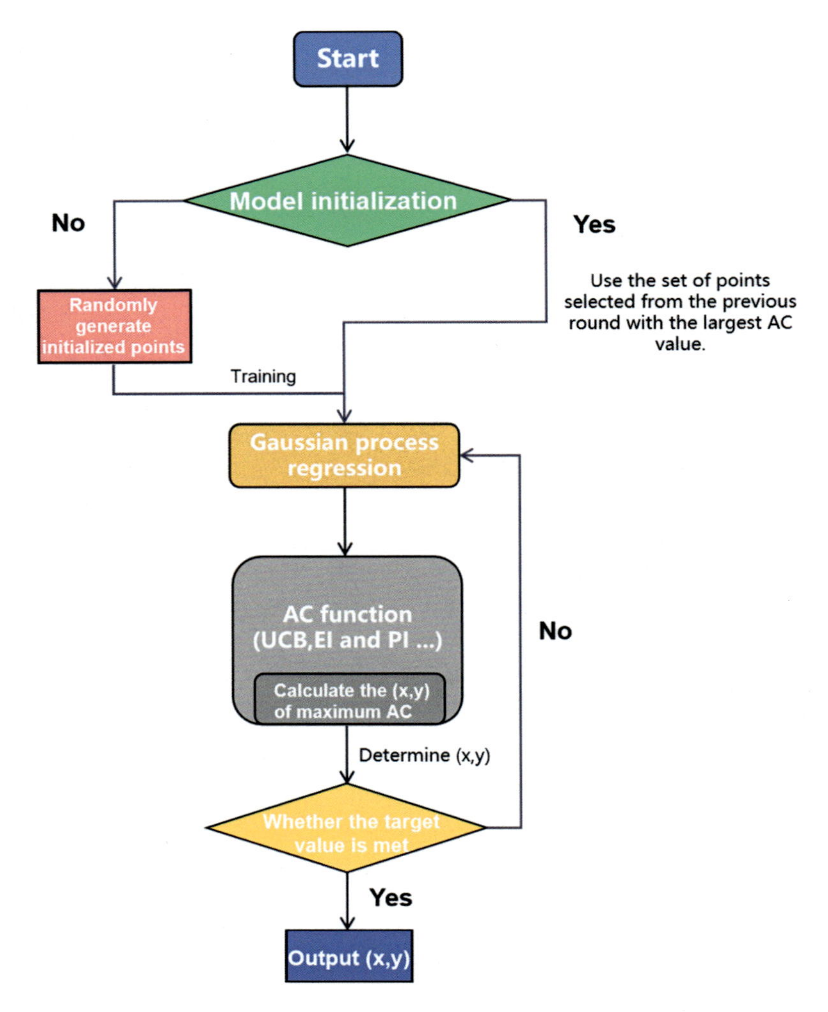

Figure 7.15 Flowchart of Bayesian optimization.

Compared with conventional grid search or random search methods, BO makes full use of the previous parameter information to tune the current parameters; it requires fewer iterations and is faster, while also being efficient and reliable for nonconvex problems. In practice, to avoid errors, improve speed, and take advantage of a wider range of configuration options, BO can be implemented through standards available in open-source libraries. Scikit-optimize and Hyperopt are two popular Bayesian optimization libraries in Python [83]. The scikit-optimize library optimizes the scikit-learn algorithm's hyperparameters by performing optimizations directly on the search space or using the BayesSearchCV class.

7.3.4.4 Particle swarm optimization

Particle swarm optimization (PSO) is a population-based evolutionary computing technology, which is derived from the study of bird predation behavior [84]. Each particle in the swarm represents a possible solution to a problem. PSO realizes intelligent problem-solving through the simple behaviors of individual particles and information interaction within the group.

PSO does not have the "crossover" and "mutation" operations of genetic algorithms and instead finds the global optimal based on the currently identified optimal value. The algorithm flow of PSO is shown in Fig. 7.16, and the specific steps in the process are explained as follows [85].

1. Set the maximum number of iterations, the number of independent variables of the objective function, the maximum velocity of the particles, and the location information as the whole search space and then set the particle swarm scale, initial position, and velocity within the search space.
2. Calculate the fitness of each particle according to the fitness function.
3. The current fitness of each particle is compared with the fitness corresponding to its individual historical best position (pbest). If the current fitness value is higher, the current position is used to update the individual particle's pbest value [86].
4. The current fitness of each particle is compared with the fitness corresponding to the global best position (gbest). If the current fitness is higher, the current position is used to update the historical optimal position (gbest) of the whole particle population.
5. Update particle velocity and position based on Eqs. (7.3) and (7.4). The corresponding parameters are described in Table 7.3.

$$V_{id}^{k+1} = \omega V_{id}^{k} + C_1 random(0,1)(P_{id,pbest}^{k} - X_{id}^{k}) + C_2 random(0,1)(P_{d,gbest}^{k} - X_{id}^{k}) \tag{7.3}$$

$$X_{id}^{k+1} = X_{id}^{k} + V_{id}^{k+1} \tag{7.4}$$

6. When the maximum iteration number Gk is reached or the optimal location searched meets the predetermined minimum adaptation threshold, the iteration process is terminated; otherwise, steps 2–5 are repeated.

PSO has been widely used in many fields, such as function optimization, image processing, and geodetic surveys because of its simple operation and high convergence speed [87]. PSO and other optimization algorithms (or strategies) can also be

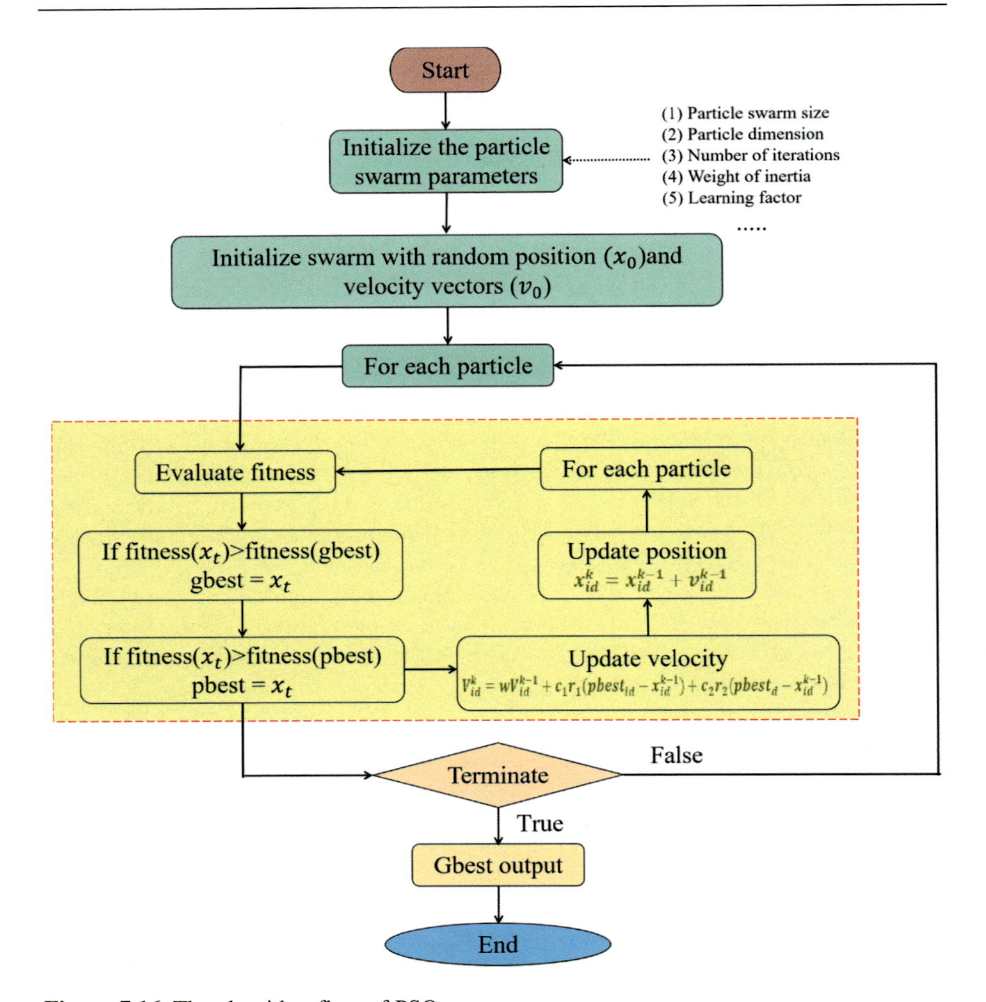

Figure 7.16 The algorithm flow of PSO.

combined to form a hybrid PSO algorithm, which is more effective for tuning the parameters of ML models. However, some problems may arise, such as premature convergence, dimension disaster, and falling into local extrema.

7.4 Modeling evaluation

Model evaluation refers to the measurement of model performance. Different tasks often require different evaluation indicators, and most of these indicators can only partially reflect a model's performance. To help effectively use evaluation indicators, identify model problems in time, and avoid drawing erroneous conclusions

Table 7.3 Description of parameters related to speed and position update.

Parameter	Definition	Parameter	Definition
I (i = 1,2,3. . .. N)	The serial number of the particle	N	The scale of the swarm
d (d = 1,2,3. . .. D)	The ordinal number of the particle's dimension	D	Dimension of particle
k	Number of iterations	ω	Weight of inertia
C_1	Learning factor of the individual	C_2	Learning factor of the group
V_{id}^k	The velocity vector of the *d-th* dimension of particle *i* in the *k-th* iteration	X_{id}^k	The position vector of the *d-th* dimension of particle *i* in the *k-th* iteration
$P_{id,pbest}^k$	The historical optimal position of the *d-th* dimension of particle *i* in the *k-th* iteration, that is, the individual optimal solution obtained by searching after *k* iterations.	$P_{d,gbest}^k$	The historical optimal position of the *d-th* dimension of the population in the *k-th* iteration, that is, the optimal solution within the whole particle population after the *k-th* iteration.
random(0,1)	Random numbers within the interval [0,1] to increase the randomness of the search		

from modeling, this provides a detailed introduction to performance evaluation metrics based on classification, regression, and clustering.

7.4.1 Classification evaluation metrics

As the most widely used ML model, classification model evaluation indicators have diversified as the field of ML has developed. Accuracy, recall, precision, F1-score, receiver operating characteristic (ROC), and area under the ROC curve (AUC) are the most commonly used classification model evaluation indicators and are calculated based on the confusion matrix.

A confusion matrix, also known as an error matrix or possibility table, presents a detailed and accurate visual representation of a model's classification results. As shown in Fig. 7.17, taking a binary classification problem as an example, the

Confusion Matrix			
Samples = P + N		Predicted condition	
		Positive (P)	Negative (N)
Actual condition	Positive (P)	True positive (TP)	False negative (FN)
	Negative (N)	False positive (FP)	True negative (TN)

Figure 7.17 Confusion matrix.

Table 7.4 Description of the evaluation indicators for binary classification.

Evaluation indicator	Implication	Formula
Accuracy	The number of all correctly classified samples divided by the total number of samples.	$\text{Accuracy} = \frac{TP+TN}{TP+TN+FP+FN}$
Precision	The proportion of the number of correctly classified positive samples to the total number of positive samples predicted by the classifier	$\text{Precision} = \frac{TP}{TP+FP}$
Recall (True positive rate)	The probability that a real positive sample is predicted to be positive by the model	$\text{Recall} = \frac{TP}{TP+FN}$
F1 Score	Harmonic average of precision and recall	$\text{F1-score} = \frac{2}{\frac{1}{\text{Recall}} + \frac{1}{\text{Precision}}} = \frac{2 \cdot \text{Recall} \cdot \text{Precision}}{\text{Recall} + \text{Precision}}$
False-positive rate	The probability that a negative sample is predicted to be positive by the model	$FPR = \frac{FP}{FP+TN}$

instances are divided into positive or negative classes [88]. Four result combinations are obtained by comparing the true classifications and model predictions.

True-positive (TP) results indicate that the model has correctly predicted a class as positive; similarly, true-negative (TN) values indicate correct negative classification. If an instance belongs to a negative class but is predicted to be positive, this is a false positive (FP). In the opposite case, this is designated as a false negative (FN) [89]. The definitions and formulas for evaluation indicators of binary classification based on the above are shown in Table 7.4 [90].

In addition, precision−recall (P-R) curves and ROC curves can also reflect model performance. A P-R curve describes changes in precision and recall, with recall plotted on the X-axis and precision on the Y-axis. Multiple P-R curves can be obtained using different classification thresholds. As shown in Fig. 7.18, the greater the precision, the lower the recall; conversely, the lower the precision, the greater the recall.

When P = R, there is a break-even point (BEP), which is used to measure the area between the model and the coordinate axes under the P-R curve. The closer the BEP is to the upper-right corner, that is, the larger the enclosed area, the better the performance of the model [91].

The ROC curve is plotted with the false-positive rate on the x-axis and the true-positive rate (recall) on the y-axis [92]. Similar to the P-R curve, an ROC curve can show the effect of the same model under different thresholds. The AUC represents the area below the ROC curve, which can quantitatively reflect the performance of the evaluated model [93]. In addition, the AUC approach considers the classification ability of the classifier for both positive and negative examples and can still achieve a reasonable evaluation in the case of unbalanced samples. As shown in Fig. 7.19, the closer the ROC curve is to the point (0,1), that is, the more it deviates from the 45-degree diagonal, the larger the AUC is and the better the performance of the model is [94]. The ROC and AUC methods are simple and intuitive and form a comprehensive representative of detection accuracy. However, in contrast to P-R curves, how to choose the curve varies depending on the practical problem being investigated.

When the classification task is changed from binary classification to multiclassification, the index metric is also changed [95]. To achieve a more comprehensive overall evaluation of a multiclassification model's performance, it is usually first converted into multiple binary classification models, and the indicators are then calculated and summarized based on various rules. There are three commonly used summary criteria [96]:

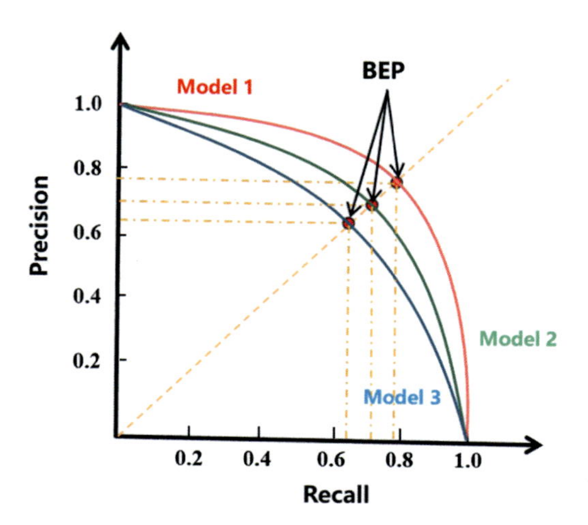

Figure 7.18 Schematic diagram showing P-R curves for different thresholds.

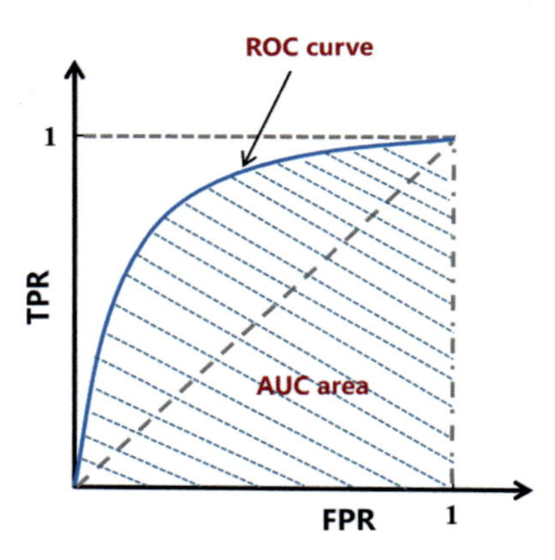

Figure 7.19 ROC curve and corresponding AUC under different thresholds.

(A)

Samples \ Class	Class A	Class B	Class C
1	P=0.7	P=0.2	P=0.1
2	P=0.3	P=0.55	P=0.15
3	P=0.25	P=0.1	P=0.65
4	P=0.4	P=0.25	P=0.35

(B)

Samples \ Class	Class A	Class B	Class C
1	0	0	1
2	1	0	0
3	0	1	0
4	1	0	0

Figure 7.20 Matrices associated with multiclassification problems: (A) probability matrix and (B) label matrix.

1. Macro-average. In this approach, the evaluation indicators of different classes (precision/recall/F1-score) are summed and averaged. All classes are given the same weight. This method is the simplest but is more affected by classes with small sample sizes.
2. Weighted average. In this method, different weights are assigned to each class (the weights are determined based on the proportion of the true distribution of the class) and then calculate the weighted average.
3. Micro-average. The TP, FP, and FN of each class are first added and then calculated according to the formula of binary classification.

In addition, there are two methods to draw ROC curves for multiclassification problems. As shown in Fig. 7.14, after the training is completed, the probability or confidence of each test sample for each class is calculated, and a matrix P with shape [m, n] is obtained. Each row represents the probability value of a test sample under each class, as sorted by class label [97] (Fig. 7.20A). Accordingly, the label of each test sample is transformed into a binary-like form. Each position is used to indicate whether a sample belongs to the corresponding class. Thus a label matrix L of [m, n] can also be obtained (Fig. 7.20B).

Based on the aforementioned assumptions, the first method involves calculating the FPR and TPR under each threshold according to each column corresponding to probability matrix P and label matrix L. N ROC curves are then drawn, and the final ROC curve is obtained by averaging them [98]. Another method is based on test samples, the core concept of which is as follows: (1) the labels are divided into two classes (0 and 1) corresponding to the binary classification step, where 0 represents other classes [99]. (2) If the classifier correctly classifies the test sample, the value of the corresponding position of sample label 1 in the probability matrix P is the probability value predicted by the model. Based on the aforementioned two points, label matrix L and probability matrix P are expanded in rows and transposed to form two columns that represent the binary classification; the ROC curve is then directly plotted based on this result [100], as shown in Fig. 7.21.

7.4.2 Regression evaluation metrics

Unlike the classification of predicted class labels, the regression task involves predicting specific values. Accordingly, specialized metrics must be used to evaluate the differences between the predicted and true values. For regression models, the most-used evaluation indicators include *R*, *R*-squared (R^2), explained variance score (EVS), MSE, mean absolute error (MAE), root mean squared error (RMSE), and mean absolute percentage error (MAPE).

The Pearson Correlation Coefficient, or *R*, is a statistic measuring the linear correlation between two variables, and its value is between [-1,1] [101]. *R* is used to describe the linear relationship between predicted and true values, as shown in Eq. (7.5). Broadly, when the *R* value is lower than 0.3, the two variables are interpreted as weakly correlated; when the *R* value is between [0.3,0.7], the two variables are interpreted as moderately correlated [102], and when the *R* value is above 0.7, the two variables are interpreted as strongly correlated. To date, *R* has been widely used as an index to evaluate regression model performance [103–105]. In an ML context, *R* values greater than 0.7 tend to indicate good model performance.

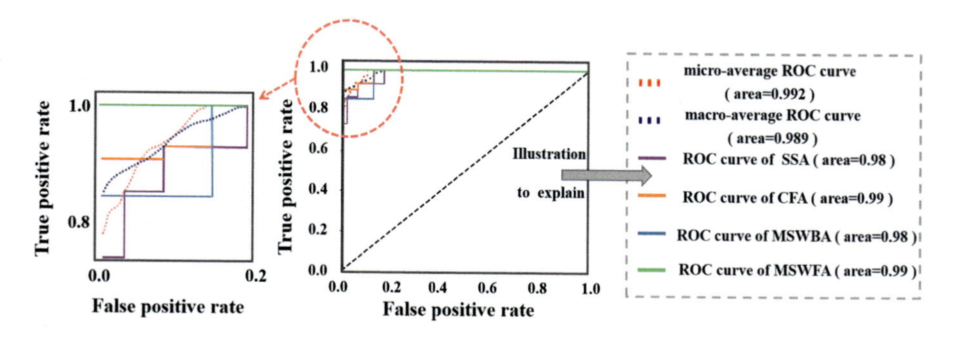

Figure 7.21 ROC curve of multiclass labels.

R^2 represents the square of the R value and can be used to measure the proportion of the variation in the dependent variable that can be explained by the independent variable, thus describing the explanatory power of the statistical model [106]. Compared with R, R^2 is more commonly used and a more convincing metric for model evaluation. The calculation of R^2 is shown in Formula (6) [107], whose value is between 0 and 1. The closer the value is to 1, the stronger the explanatory ability of the investigated feature is for the target variable. R^2 values closer to 0 indicate a worse-fitting model.

EVS measures how close the dispersion of the difference between all predicted sample values is to the dispersion of the sample itself [108]. As shown in Formula (7), the EVS value ranges from 0 to 1: the closer it is to 1, the more the independent variable can explain the variance change of the dependent variable. The MSE [109], RMSE [110], MAE [111], and MAPE [112] evaluation indicators are also summarized in Table 7.5.

$$R = \frac{\sum_{i=1}^{n}(y_i - \overline{y})(\hat{y}_i - \overline{\hat{y}})}{\sqrt{\sum_{i=1}^{n}(y_i - \overline{y})^2 \sum_{i=1}^{n}(\hat{y}_i - \overline{\hat{y}})^2}} \tag{7.5}$$

$$R^2 = 1 - \frac{\sum_i (\hat{y}_i - y_i)^2}{\sum_i (y_i - \overline{y})^2} \tag{7.6}$$

$$EVS = 1 - \frac{Var\{y_i - \hat{y}_i\}}{Var\{y_i\}} \tag{7.7}$$

Where $\hat{y}_i$ is the predicted value, y_i is the desired target response, $\overline{y}$ is the mean of the target response, $\overline{\hat{y}}$ is the mean of the predicted value, n is the number of samples, and Var is the variance.

Table 7.5 Description of four common evaluation indicators.

Evaluation indicator	Formula	Description information
MSE	$\frac{1}{n}\sum_{i=1}^{n}(y_i - \hat{y}_i)^2$	The square of the differences between the predicted values and true values of all samples are summed and then averaged
MAE	$\frac{1}{n}\sum_{i=1}^{n}\lvert(y_i - \hat{y}_i)\rvert$	The absolute values of the differences between the predicted and true values for all samples are summed and then averaged
RMSE	$\sqrt{\frac{1}{n}\sum_{i=1}^{n}(y_i - \hat{y}_i)^2}$	The square root of the MSE
MAPE	$\frac{1}{n}\sum_{i=1}^{n}\left\lvert\frac{\hat{y}_i - y_i}{y_i}\right\rvert$	For all samples, the deviation between the predicted value and the true value is divided by the true value; the average of the absolute deviation values is then calculated.

The range of the four evaluation indicators is [0, + ∞]. When the predicted values are exactly consistent with the real values (i.e., the error is 0), the model performs perfectly. However, the larger the magnitude of the error is, the larger the values of these indicators, and the worse the model performs. The MSE and MAE metrics are more applicable when the error is relatively large, while the RMSE is better suited to smaller errors. Compared with the MSE, the MAE can more effectively highlight outliers. Furthermore, multiple indicators can also be used together. For example, combining MAE and MAPE, which are based on $\bar{y}$ can estimate how well a model fits samples of different orders of magnitude.

Although the aforementioned indicators can effectively evaluate the performance of regression models, the impact of adding new predictor variables should also be considered. Based on the sensitivity of these indicators to model changes, more stable indicators, such as the adjusted *R*-squared approach, have been proposed to help identify optimal regression models [113]. The adjusted *R*-squared method offsets the effect of sample size on *R*-squared and demonstrates the ability of a set of predictors to explain variations in the response variables. This approach can be used to compare the fit of regression models containing different predictors and is calculated as follows:

$$R_{adj}^2 = 1 - \frac{(1 - R^2)(m - 1)}{(m - p - 1)} \tag{7.8}$$

Where m is the number of samples, and p is the number of features. In addition to this approach, the residual plot and result comparison plot are important visualization methods used for judging model quality. In the residual plot, the original values are plotted on the x-axis, and the residuals, that is, the differences between the true y-values and the model's predicted y-values, are plotted on the y-axis [111]. In the example shown in Fig. 7.22, the residuals all fall within a small range and fluctuate randomly around 0, indicating that the regression model has good performance.

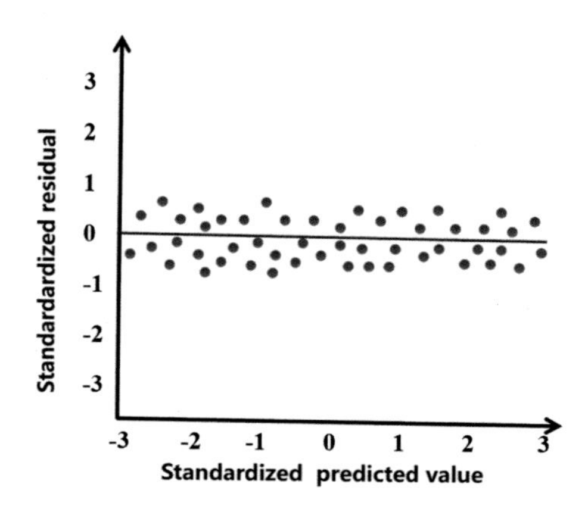

Figure 7.22 Residual plot based on coal fly ash reactivity.

However, when important variables are not considered, or the network structure is incorrect, the model is prone to underfitting and overfitting [114], as shown in Fig. 7.23.

In addition to the residual plot method, comparing the differences in the distributions of the true and predicted values is also a simple and intuitive way to visualize the model fit. As an example, Fig. 7.24 shows the differences between the actual and estimated generation of CFA. The data points were mostly concentrated around 0, indicating the ML model has good predictive performance.

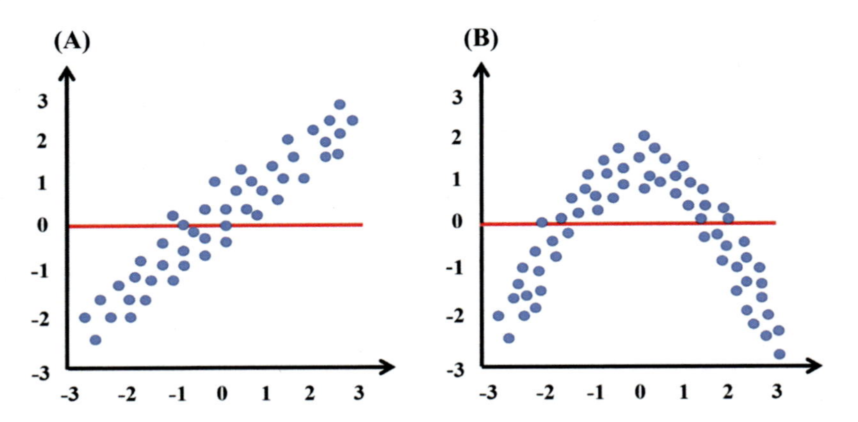

Figure 7.23 Residual plots illustrating two different biases: (A) model underfitting and (B) model overfitting.

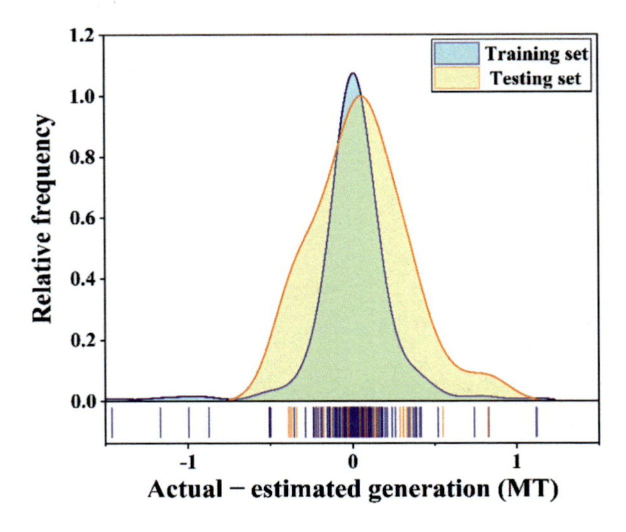

Figure 7.24 Comparison of results of RF model based on CFA reactivity.

7.4.3 Clustering evaluation metrics

After the clustering algorithm process is complete, performance evaluation metrics can be used to measure the clustering results and determine the selection of algorithm parameters. Clustering performance evaluation indicators can be divided into internal and external indicators [115]. Internal metrics are unsupervised and evaluate the performance based on the data clustering itself, without the need for external models [116]. Specifically, based on the dataset's structure information, the clustering effect is measured in terms of compactness, separation, connectivity, and overlap. In contrast, external indicators have reference standards, and the performance of the clustering algorithm is evaluated by comparing the clustering results with known labels. The two indicator types are described in detail as follows.

7.4.3.1 Internal evaluation

In practical engineering problems, it is challenging to directly obtain accurate data clustering labels. Therefore, when the actual class label is unknown, the distance between the sample points in the dataset and the cluster centroid is typically used to measure the quality of the clustering results. Common internal evaluation indicators include the silhouette coefficient, Calinski–Harabasz index (CHI), Davies–Bouldin index (DBI), and Dunn validity index (DVI).

The silhouette coefficient was first proposed by Peter J. Rousseeuw in 1986 [117]. This approach is based on using the original data in combination with the cohesion and separation degree to evaluate the impacts of different algorithms on clustering. The principle and steps of calculating this coefficient are as follows:

1. Calculate the average distance a_i between sample i and other samples in the same cluster, referred to as the within-cluster dissimilarity of sample i.
2. Calculate the average distance b_{ij} between sample i and all samples of some other cluster such as B_j, referred to as the dissimilarity between sample i and cluster B_j. The smallest average distance between sample i and the samples of all the other clusters is called the intercluster dissimilarity of sample i, that is, $b_i = \min\{b_{i1}, b_{i2}, b_{i3}\cdots, b_{ik}\}$
3. The formula of the silhouette coefficient is as follows [118]:

$$s(i) = \frac{b(i) - a(i)}{\max\{a(i), b(i)\}} = \begin{cases} 1 - \dfrac{a(i)}{b(i)}, & a(i) < b(i) \\ 0, & a(i) = b(i) \\ \dfrac{b(i)}{a(i)} - 1, & a(i) > b(i) \end{cases} \tag{7.9}$$

When the value of $s(i)$ is close to 1, the clustering effect is good; however, when $s(i)$ is close to -1, this indicates that sample i should belong to other clusters. When $s(i)$ is approximately 0, this implies that sample i is on the boundary of two clusters.

4. Calculate the average $s(i)$ value over all the samples to obtain the silhouette coefficient S of the overall clustering.

The value of S ranges from -1 to 1 [119]. The closer the distance between samples within the same cluster, the greater the distance between samples of different clusters—in this case, the value of S will be larger, indicating a better clustering effect. In contrast, negative values of S indicate a poor clustering effect.

The silhouette coefficient makes no a priori assumptions about the data distribution of data so it performs well on many datasets. However, the silhouette coefficient has a limitation in that it may produce falsely high values for convex clusters, making it unsuitable for the evaluation of density-based algorithms.

The CHI method is used to measure the closeness within a cluster by calculating the sum of the squares of the distances between each sample point in a cluster and the cluster's center. This approach also measures the separation degree of a dataset in terms of the between-cluster distance, that is, the sum of squares of the distances between each centroid and the center point of the dataset [120]. The calculation formula is as follows.

$$\mathrm{CHI} = \frac{tr(B_k)(N-K)}{tr(W_k)(K-1)} \tag{7.10}$$

Where N is the capacity of the dataset and K is the number of clusters, $B_K = \sum_{k=1}^{K} n_k(c_k - c_X)(c_k - c_X)^T$ is the between-cluster covariance matrix, $W_K = \sum_{k=1}^{K} \sum_{x \in C_k} (x - c_k)(x - c_k)^T$ is the within-cluster covariance matrix, C_k indicates a cluster centered on c_k, n_k is the number of samples in the cluster, c_X denotes the center of the dataset, and Tr is the trace of the matrix.

The smaller the intracluster covariance and the larger the between-cluster covariance, the higher the value of CHI and the better the clustering effect [121]. In comparison with the silhouette coefficient, the calculation of CHI is fast. However, similarly, this approach is not suitable for evaluating density-based clustering algorithms.

DBI was proposed by David L. Davis and Donald Bouldin. The essence of DBI is to sum the average within-cluster distance of any two clusters, divide it by the distance between the centroids of the two clusters, and then take the maximum value [122]. This index is calculated as follows.

$$DBI = \frac{1}{N}\sum_{i=1}^{N} \max_{j \neq i} \left(\frac{\overline{S}_i + \overline{S}_j}{\left\| w_i - w_j \right\|_2} \right) \tag{7.11}$$

Where N is the number of clusters, and $\overline{S}_i + \overline{S}_j$ represents the sum of the average distances of all samples in the two clusters to the centroids of the corresponding clusters. $\left\| w_i - w_j \right\|_2$ represents the Euclidean distance between the centroids of the two clusters.

Fig. 7.25 illustrates a schematic when the number of clusters $\mathrm{N} = 3. DBI = Max[(d(A) + d(B))/d(AB), (d(B) + d(C))/d(BC), (d(A) + d(C))/d(AC)]$.

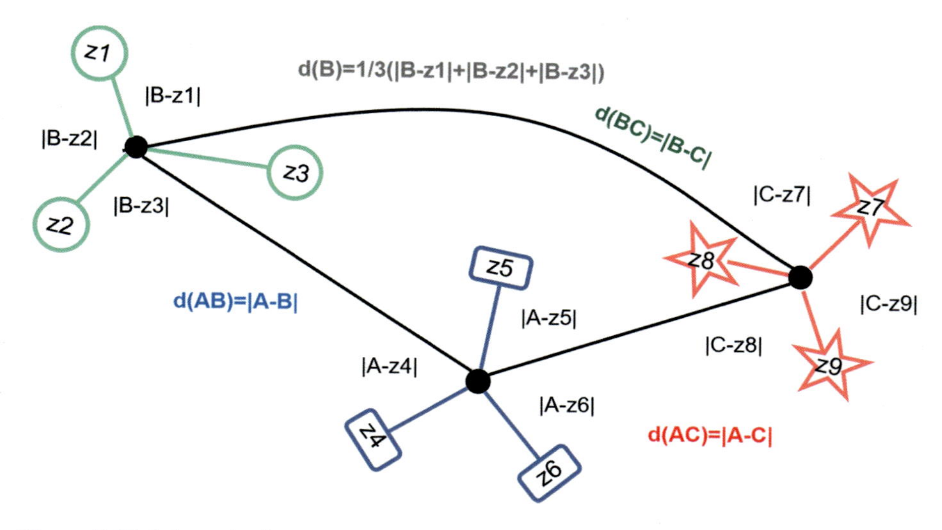

Figure 7.25 Schematic diagram of DBI calculation when the number of clusters is 3.

Smaller DBI values indicate closer samples within the same cluster and a greater distance between different clusters, indicating an overall better clustering effect.

The DVI approach was introduced by J. C. Dunn in 1974 as an internal evaluation index based on the clustered data itself [123]. Like the silhouette coefficient, CHI, and DBI, DVI aims to identify close cluster sets with small distance differences between the samples within the clusters and large between-class distances [124]. The DVI for a number of clusters q is defined as follows.

$$\mathrm{DVI} = \frac{\min\limits_{1 \le i < j \le q} \delta(C_i, C_j)}{\max\limits_{1 \le k \le q} \Delta_k} \tag{7.12}$$

Where C_i and C_j represent two clusters, $\delta(C_i, C_j)$ is the distance measure between the two clusters, and Δ_k refers to the distance between samples in any clusters, that is, the within-cluster distance.

For a given set of clusters, a higher DVI value indicates better clustering. However, as the number of clusters and data dimension increases, the computation cost of this index will be higher. In addition, this approach has good evaluation capability for discrete point clustering but has a poor evaluation effect for ring distributions.

7.4.3.2 External evaluation

External indicator evaluation refers to measuring the performance of different clustering algorithms by comparing the matching degree between cluster division and external criteria when the dataset's labels are available [125]. The principal external indicators are listed in Table 7.6.

Table 7.6 External evaluation index of clustering algorithm

Measure	Notation	Computation	Parameter specification
Purity	P	Purity $(\Omega, C) = \frac{1}{N}\sum_k \max\left|\omega_k \cap c_j\right|_j$	N is the total number of samples, $\Omega = \{\omega_1, \omega_2, \cdots \omega_k\}$ indicates the division of clusters, and $C = \{c_1, c_2, \cdots, c_j\}$ represents the division of the real classes
Mutual information	MI	$MI(U,V) = \sum_{i=1}^{\|U\|}\sum_{j=1}^{\|V\|} P(i,j)\log\left(\frac{P(i,j)}{P(i)P(j)}\right)$	U and V are labels allocated to N samples. $p(i)$ is the proportion of the data belonging to class i as a fraction of the total data and $p(j)$ is analogous. $P(i,j)$ indicates the joint probability that a sample belongs to both i and j.
Adjusted mutual information	AMI	$AMI(U,V) = \frac{MI(U,V) - E(MI)}{F(H(U),H(V)) - E(MI)}$	$E(MI)$ is the expectation of mutual information. $F(H(U),H(V))$ can be the maximum and minimum function or find the geometric average.
F-measure	F	$F = (1+\beta^2)\frac{Precision\ Recall}{\beta^2 \cdot Precision + Recall}$	β is a parameter, usually with a value of 1.
Rand Index	RI	$RI = \frac{a+b}{C_2^n}$	a denotes the logarithm of elements of the same class in both the real class and the cluster, b represents the logarithm of elements that are in different classes in both the real class and the cluster, and C_2^n is all the possible sample pairs.
Adjusted Rand Index	ARI	$ARI = \frac{RI - E(RI)}{\max(RI) - E(RI)}$	$E(RI)$ is the expected value and $\max(RI)$ is the maximum function.

In Table 7.6, the purity of clustering and accuracy of classification problems are similar. The basic concept of purity in this context is to assign the true label of the samples that appear most times in a cluster to that cluster. The sample numbers of the corresponding labels of all clusters are then summed, and the average is calculated [126]. As shown in the example in Fig. 7.26, $Purity = \frac{1}{17} \times (5 + 4 + 3) = 0.7$. This indicates that the clustering effect is good; however, this metric cannot be used to evaluate the relationship between clustering quality and number.

MI is used to measure the degree of coincidence between two data distributions, as shown in Fig. 7.27, MI can be calculated using a statistical table for clustering results and external labels. To better compare different clustering effects, the AMI concept has also been proposed. AMI corrects the disadvantage of higher MI values for clusters containing more samples by adjusting the probability of the clusters

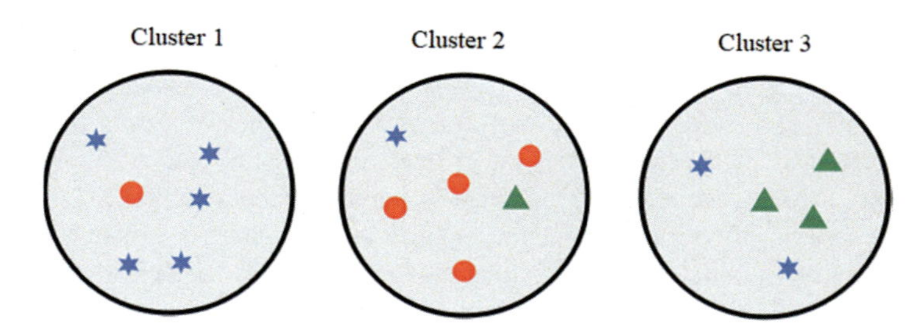

Figure 7.26 Purity as an external evaluation criterion for cluster quality.

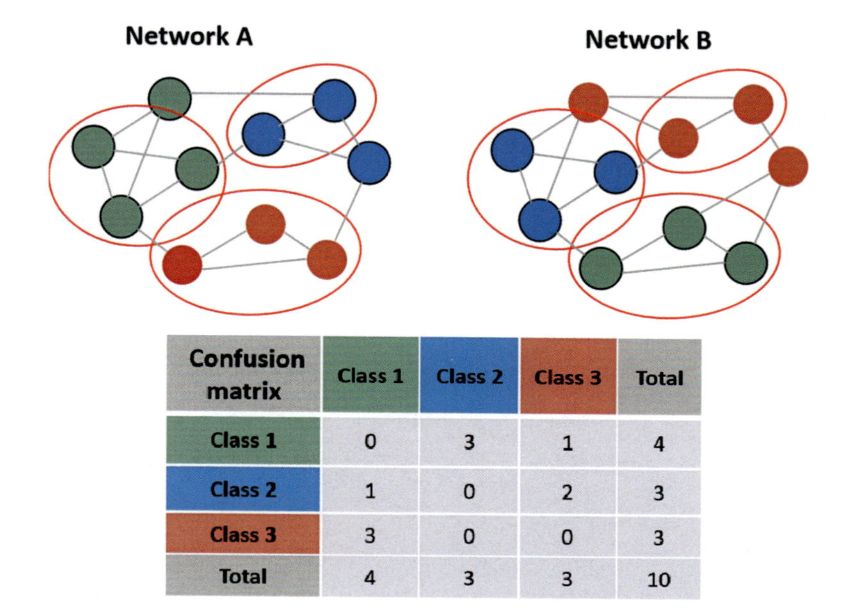

Confusion matrix	Class 1	Class 2	Class 3	Total
Class 1	0	3	1	4
Class 2	1	0	2	3
Class 3	3	0	0	3
Total	4	3	3	10

Figure 7.27 Comparison of two data distributions with three classes.

[127]. The MI ranges from 0 to 1, and the value range of AMI is [-1,1] [128]. The closer these metrics are to 1, the more consistent the clustering results are with the classification of the real data.

The RI is calculated based on a contingency table, reflecting the degree of overlap between the instance class and clustering division [129]. The value range of RI is [0,1]: the larger the value, the better the clustering effect. Given that the RI method does not necessarily produce a value close to 0 when labels are randomly assigned, the ARI method with a range of [-1,1] was proposed, which has a higher discrimination degree [130]. The F-measure index measures the accuracy of a clustering result, which is directly proportional to the precision and recall [131]; the greater the precision and recall, the higher the F-measure value.

7.5 Machine learning–based knowledge discovery

ML models are data-driven. They are often referred to as "black-box models" due to the difficulties involved in analyzing their underlying mechanisms, which thus limits their interpretability [132]. Understanding the decision-making process behind ML-based models is highly significant for guiding feature engineering, adjusting the models, improving the models' credibility, and solving practical problems. On this basis, a variety of model interpretability methods have been proposed.

Model interpretability methods are approaches used to visually explain a model's behavior. ML interpretability can be divided into intrinsic and ex-post interpretability [130], model-specific and model-agnostic interpretation, local interpretability [133], and global interpretability [134] based on when the interpretability method is used, the method's matching relationship with the model, and the scope of the method.

Intrinsic interpretability refers to limiting the architecture of a model so that its working principles and results can be easily understood (e.g., linear, parametric, or tree-based models), while ex-post interpretability refers to interpreting trained models through various statistics, visualization methods, causal reasoning, and other methods [135]. Model-specific interpretation methods are fully dependent on the functions and features of each model [135], while model-agnostic interpretation methods can be applied to any ML model and are independent of the model's internal structure. Global interpretability implies being able to understand the model's overall decision logic, whereas local interpretability focuses on the interpretation of individual predictions [136].

Model interpretable approaches are complementary, and there is no single best way to interpret a model. In this section, we introduce several commonly used approaches for reference.

7.5.1 Feature importance

Feature importance is a built-in interpretation of tree-based models by providing a score indicating the worth of each feature in the model's decision tree and then ranking and comparing the features on this basis [137].

For a single decision tree, the default performance measure is the Gini index, which calculates the increase in purity when using each variable to split a node [138]. There are two principal methods to measure the importance of features in RF models. In addition to using the OOB error, the Gini index is also widely used to measure the importance of features [139] by averaging the contribution of each feature on all trees.

Assuming that there are J features X_1, X_2.... X_j, I decision trees, and C classes, then the main calculation steps of the RF importance score are as follows:

1. Calculate the Gini index of node q of the i-th tree.

$$\mathrm{GI}_q^{(i)} = \sum_{c=1}^{|C|} \sum_{c' \neq c} p_{qc}^{(i)} p_{qc'}^{(i)} = 1 - \sum_{c=1}^{|C|} (p_{qc}^{(i)})^2 \tag{7.13}$$

   Where the Gini index is denoted by GI, and p_{qc} indicates the proportion of class c on node q.
2. The exponential change of the Gini index before and after node q branching is $IM_{jq}^{(Gini)(i)} = GI_q^{(i)} - GI_1^{(i)} - GI_r^{(i)}$ [140]. IM indicates the importance measure, and $GI_1^{(i)}$ and $GI_r^{(i)}$ represent the Gini index of the two new nodes after branching.
3. The importance of feature X_j in the i-th tree is $IM_j^{(Gini)(i)} = \sum_{q \in Q} IM_{jq}^{(Gini)(i)}$, and the set Q is the node in which X_j appears in decision tree i.
4. Normalize the importance score of I trees [141].

$$IM_j^{(Gini)} = \frac{\sum_{i=1}^{I} \mathrm{IM}_j^{(Gini)(i)}}{\sum_{j'=1}^{J} IM_{j'}^{(Gini)}} \tag{7.14}$$

In addition, other ensemble models such as XGBoost and GBDT quantify feature importance by recording the total number of split features or the total/average information gain [142]. Given the randomness of feature splitting, the model is generally run several times, and the important scores obtained from it are averaged before sorting.

7.5.2 Permutation importance

Permutation importance (PI) is a common type of feature importance that is calculated after the model has been fitted and represents an ex-post interpretability method that is unrelated to the model itself [143]. PI is suitable for tabular data, and its core concept is the permutation test. Specifically, in this process, the arrangement of a feature column in the data table is changed, the rest of the features are kept unchanged, and a change in model accuracy is observed [144] (Fig. 7.28). PI is implemented in the following simple steps:

1. Randomly permute the values of the i-th feature in the training set (i = 1,2,3... N, N is the total number of features in the dataset), leaving the other features unchanged.

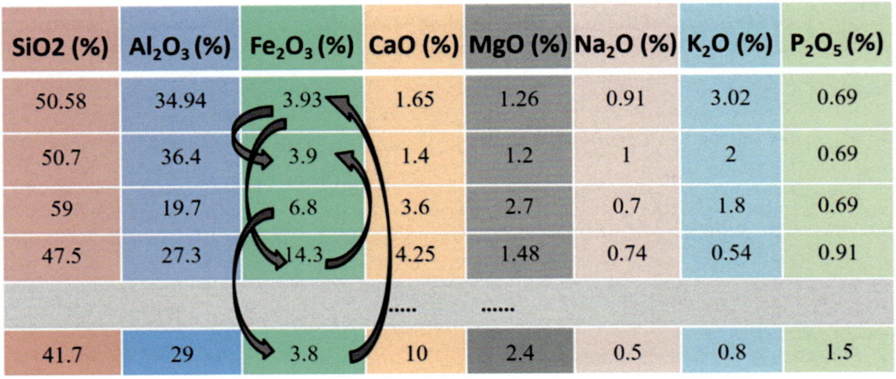

SiO_2 (%)	Al_2O_3 (%)	Fe_2O_3 (%)	CaO (%)	MgO (%)	Na_2O (%)	K_2O (%)	P_2O_5 (%)
50.58	34.94	3.93	1.65	1.26	0.91	3.02	0.69
50.7	36.4	3.9	1.4	1.2	1	2	0.69
59	19.7	6.8	3.6	2.7	0.7	1.8	0.69
47.5	27.3	14.3	4.25	1.48	0.74	0.54	0.91
			......	......			
41.7	29	3.8	10	2.4	0.5	0.8	1.5

Figure 7.28 The principles of PI.

2. The training dataset after permutation and change is used for prediction and score calculation.
3. The difference between the old and new scores is used to indicate the PI of the feature.
4. Restore the disrupted i-1-th column, and repeat steps (1)–(3) on the next column of data until the PI of all columns has been calculated.

To avoid the influence of uncertainty introduced by random permutation, the aforementioned steps are repeated several times and then averaged.

PI can be straightforwardly implemented in two ways in Python. One is to use the permutation _ importance function of the Inspection module in sklearn [145]; the other is to take advantage of the permutation _ importance module in ELI5 [146]. The ELI5 package provides an estimator permutation importance wrapper that is compatible with sklearn, whereas other estimators must be wrapped in sklearn-compatible objects. However, for deep learning models (such as DNN approaches), methods related to sklearn are not applicable. Given this limitation, the ELI5 package provides the eli5.permutation_importance.get_score_importances method to measure the importance of features for deep learning models.

7.5.3 Partial dependence plot

Feature importance and the IP score reflect the impact of features on model prediction accuracy, but they cannot show how these features affect the prediction itself. Therefore, a global interpretation method called the partial dependence plot (PDP) has been proposed and successfully applied [147]. Like IP, the PDP can only be calculated after the model is fitted.

The PDP is an intuitive and easy-to-understand visualization of the impact of input variables on the predicted output. Specifically, the PDP presents the linear, monotonic, or nonlinear relationships between one or two feature variables and the predicted response [148]. For the PDP of a single variable, the underlying premise that the target feature is independent of the other feature must be satisfied to show the marginal effect of the target feature on the prediction results of the previously fitted model [149]. In addition, two-way PDPs can be used to show the interaction between two features [150].

The Inspection module in sklearn provides a convenient function called plot_partial_dependence that can create both single-variable and bivariate PDPs, as shown in Fig. 7.29. In Fig. 7.29A, the *solid black line* shows the average effect of a single feature on the output response. When this curve is almost horizontal or jitters irregularly, this indicates invalidity of the feature, that is, it does not affect the target feature. However, a steep PDP curve implies that the contribution of the feature is large. Fig. 7.29B illustrates an example of the coupling effect of two oxides on the amorphous phase in coal fly ash.

The PDP approach is simple, intuitive, and has high interpretability [151]. However, due to visualizing limitations, this method only supports bivariate interaction at most and cannot be used to investigate higher-dimensional interactions. In addition, this approach cannot reflect the distribution of the feature variable itself and has major assumptions. Using the PDP as an interpretation method tends to highlight the overall change trends rather than the smaller local fluctuations caused by the algorithm.

7.5.4 Individual conditional expectation

Similar to the PDP, the individual conditional expectation (ICE) plot depicts the dependency between the target response and a single variable of interest, subject to the independence assumption. The ICE method also has the same limitation as the

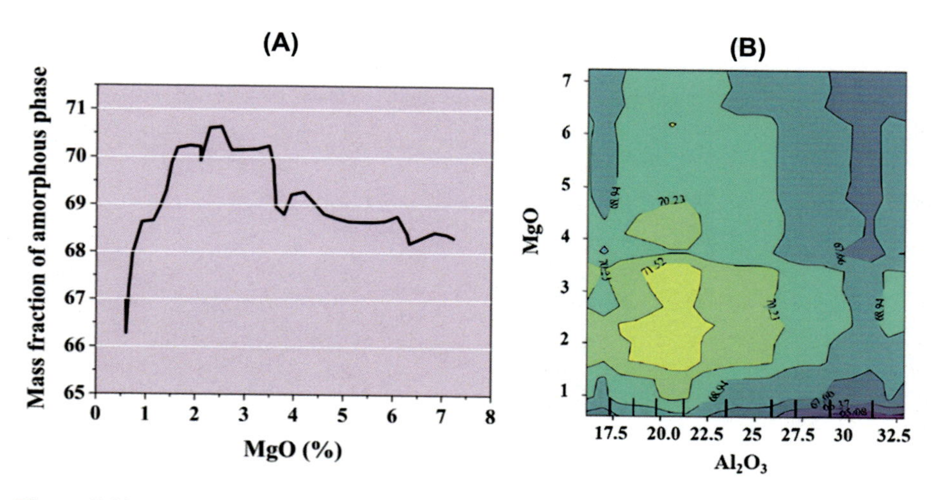

Figure 7.29 Implied relationship between oxide and coal fly ash reactivity explored by PDP: (A) single-variable PDP and (B) bivariate variable PDP. In (A), the horizontal axis represents the value of the feature, and the vertical axis represents the predicted value; however, in (B), the horizontal and vertical axes both represent features. The *yellow area* indicates higher predicted values, while the *green area* represents lower predicted values.

PDP approach of being unable to depict more than two features. However, unlike the average response of the PDP, the ICE plot shows each instance (one line per instance) to reveal heterogeneity among individual instances [152]. For a certain instance, the other features are kept unchanged, and the value of the selected feature variable is randomly replaced and put into the black box model to output prediction results. Finally, an ICE plot is constructed.

However, plotting multiple ICE curves can be visually confusing and may make interpreting the average difficult. Therefore, combining PDP and ICE approaches while limiting the number of curves is a good alternative approach (Fig. 7.30). As shown in Fig. 7.30A, the single marginal effect curves are consistent with the average marginal effect—both follow similar trends with only small changes. In contrast, different marginal effect curves do not show similar effects on the target when the considered feature interacts with other features [153]. As shown in Fig. 7.30B, the ICE curves become heterogeneous, and the average marginal effect cannot be readily interpreted from the individual features.

To eliminate differences in intercepts and highlight the heterogeneity caused by interactions, the central ICE plot and derivative ICE plot methods are also recommended [154]. The central ICE plot focuses the curve at a point in the feature and shows only the predicted differences from that point. The derivative ICE map helps to highlight the range of features that have changed and the regions in the feature curve where the derivatives are heterogeneous (i.e., where the target feature interacts with other features).

7.5.5 SHapley Additive exPlanations

SHapley Additive exPlanations (SHAP) is an additive feature attribution method inspired by the Shapley value, which originates from coalitional game theory [155]. SHAP is an ex-post model interpretation method that can be applied both globally and to individual samples. The key advantage of the SHAP approach is that it can

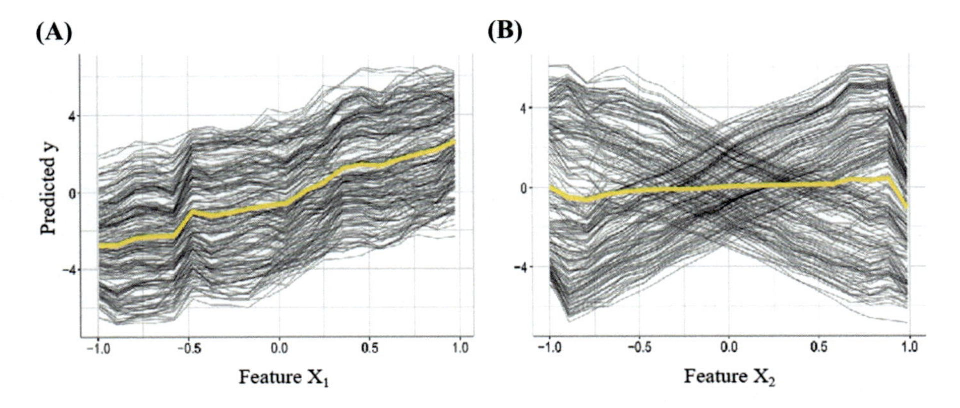

Figure 7.30 Graphs combining ICE (*black lines*) and PDP (*yellow lines*) curves. (A) Marginal effect of feature X_1 and (B) marginal effect of X_2 interacting with other features.

reflect both the positive and negative influence of features in each sample [156] (Fig. 7.31).

Let us assume that the *i-th* sample is x_i, the *j-th* feature of the *i-th* sample is x_{ij}, and the predicted value of the model for this sample is y_i. In addition, the baseline of the whole model, also known as the constant attribution value (i.e., the mean of the target variable of all samples), is given by y_{base}. In this case, SHAP obeys the basic concept of local accuracy, and the formula is as follows [157]:

$$y_i = y_{base} + f(x_{i1}) + f(x_{i2}) + \ldots + f(x_{ij}) + \ldots + f(x_{ik}) \tag{7.15}$$

Where there is a total of k features, and $f(x_{ik})$ represents the contribution (SHAP value) of the *j-th* feature in the *i-th* sample to the final predicted value y_i. When $f(x_{ij}) > 0$, feature j improves the predicted value and has a positive effect. In contrast, when $f(x_{ij}) < 0$, feature j makes the predicted value decrease and has a negative effect.

In addition, missingness and consistency are two other important attributes of SHAP [158]. When a feature is not observed in instances, the missingness property attributes its Shapley value as zero. Consistency means that if the model changes so that the marginal contribution of a feature increases or remains the same, the attribution value will also accordingly increase or remain the same [159].

SHAP provides a range of extremely powerful data visualization approaches. In addition to beeswarm, bar, and dependence scatter plots for global interpretation, there are waterfall and force plots for local interpretation. Furthermore, feature distribution heatmaps [160] and hierarchical clustering maps can also be plotted to explore the multicollinearity of features and the problem of interpretive similarity.

Taking different solid ash models as the research objects, three commonly used SHAP plots are used for model interpretation from overall and local perspectives. The summary plot presented in Fig. 7.32 explains all the samples and presents them in two forms [161]. In Fig. 7.32A, a standard bar plot is shown indicating the average absolute SHAP value for each feature. In Fig. 7.32B, all of the SHAP values of each feature are plotted in a scatter diagram; in this representation, the relationship

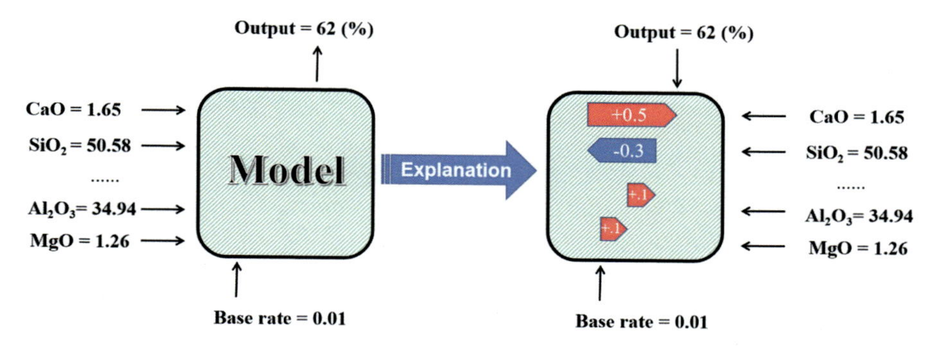

Figure 7.31 Interpretation of the black box model using SHAP.

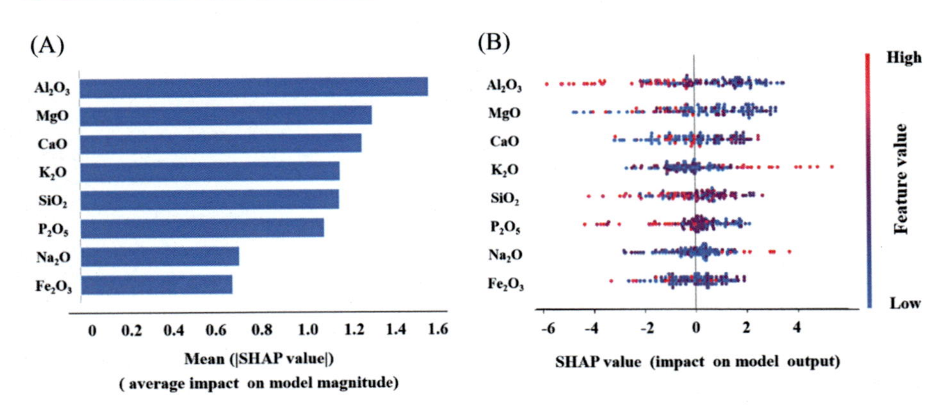

Figure 7.32 Global importance interpretation based on SHAP: (A) bar plot and (B) scatter plot.

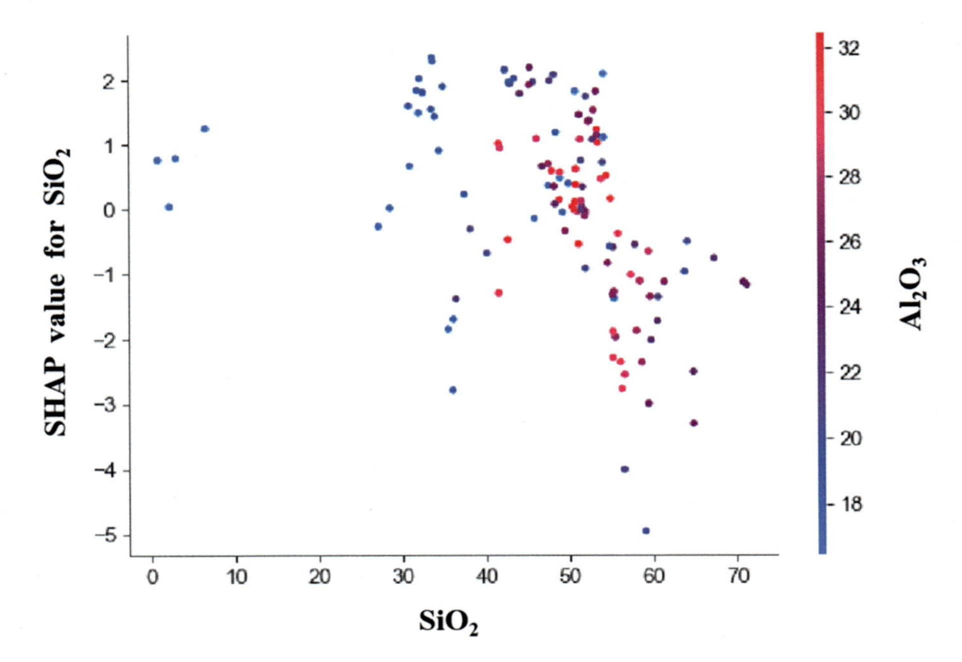

Figure 7.33 Dependency plot of the interaction of features.

between the size of the feature and its predicted impact is clearly shown, and the distribution of each feature is obvious. The feature importance values in both plots in Fig. 7.32 are ranked based on the mean absolute SHAP value, where each row represents a feature. In particular, in Fig. 7.32B, vertically taller areas of the plot indicate regions where many samples are clustered, and the gradient color indicates the original value of that feature (high values in red and low values in blue) [162]. Fig. 7.33 illustrates the synergistic effect among variables, which further shows

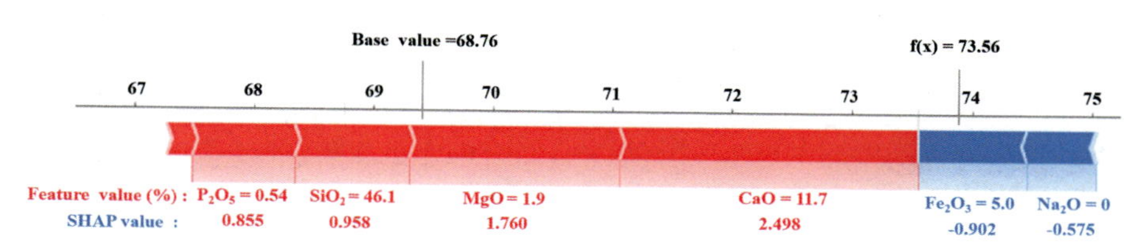

Figure 7.34 Local interpretation force plot for individual samples.

how one feature interacts with another feature to influence the model prediction results [163].

The force plot shown in Fig. 7.34 displays the prediction interpretation for a single sample. Each feature has its own contribution, and the SHAP values are visualized as arrows that push the prediction result of the model from the base value to its final output [164]. The red arrows indicate that the feature pushes the predicted value higher, whereas the blue values indicate that the feature contributes negatively to the model's prediction.

Creating an explainer forms the first step in the use of SHAP for model interpretation. The explainer types in SHAP include the tree explainer, deep explainer, gradient explainer, and kernel explainer [165], which should be selected based on the practical problem being investigated. Specifically, the tree explainer applies to tree-based integration algorithms [162], the linear explainer is suitable for linear models with independent and uncorrelated features, the deep explainer and gradient explainer are used to calculate the deep learning model, while the kernel explainer works with any model, but its performance is not necessarily optimal [166].

7.5.6 Local Interpretable Model-agnostic Explanations

The Local Interpretable Model-agnostic Explanations (LIME) method is an interpretation approach that provides local fidelity to agnostic models [167]. The core concept of LIME is to explain a single sample by simulating the local behavior of a complex model f through an interpretable simple model g. The framework of LIME is as follows [168]:

$$\xi(x) = \arg\min_{g \in G} L(f, g, \pi_x) + \Omega(g) \tag{7.16}$$

Where $\Omega(g)$ is defined to measure the complexity of explaining model g. $L(f, g, \pi_x)$ represents the unfaithful degree of the model g approximation f in domain π_x. The smaller the value of $L(f, g, \pi_x)$, the higher the approximation degree of g and f. To trade off fidelity and interpretability, it is necessary to minimize $L(f, g, \pi_x)$ while keeping the value of $\Omega(g)$ sufficiently low.

The LIME method generates perturbations around selected sample points to generate new sample points, which are then predicted using the initially trained complex model. At the end of the process, a simple interpretable model is trained using the new dataset to obtain a good local approximation of the black-box model [169]. As shown in Fig. 7.35, the nonlinear decision function of the original complex model is represented by a blue/pink background, and the explained sample X is represented by a thick, bright red cross [167]. After defining the similarity calculation method and selecting K features, perturbation and sampling are carried out around X, and the sample weight is given according to the distance from these sample points to X. The original model is then used for prediction. Finally, a linear model (dashed line) is trained to approximate the original model near X [170].

LIME can use a wide range of data types from tabular data to image or text data [171]. Additionally, this approach can be used to interpret any classification or regression model, including RF, GBDT, neural networks, etc. Although interpreting regression and classification modes is similar within the LIME approach, there are some differences. The classification mode can be considered as the regression of the probability of each classification situation, that is, the explanation of N classification models is equivalent to the explanation of N regression models [172]. As shown in Fig. 7.36, predicted values and prediction probabilities are the output prediction of the original ML model for the sample point. The positive/negative and XXX/not XXX plots are weight plots. Features that plot to the left play a negative role, whereas those that plot to the right have a positive role in the prediction.

The LIME algorithm has simple principles, strong universality, and good effect. In addition to the interpretation of image classification results, this approach can be applied to related tasks such as natural language processing. However, the LIME method is slow, and its neighborhood range must be specified: with different neighborhood ranges, the resulting local interpretability models obtained may also be very different. Furthermore, LIME sampling ignores the correlation between features, and its interpretation is also unstable [173], thus limiting the large-scale applicability of the LIME method.

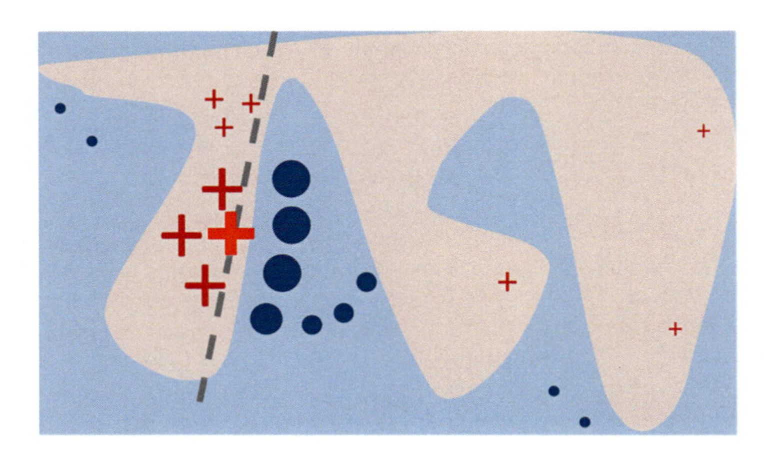

Figure 7.35 The principle of LIME specific to a single sample.

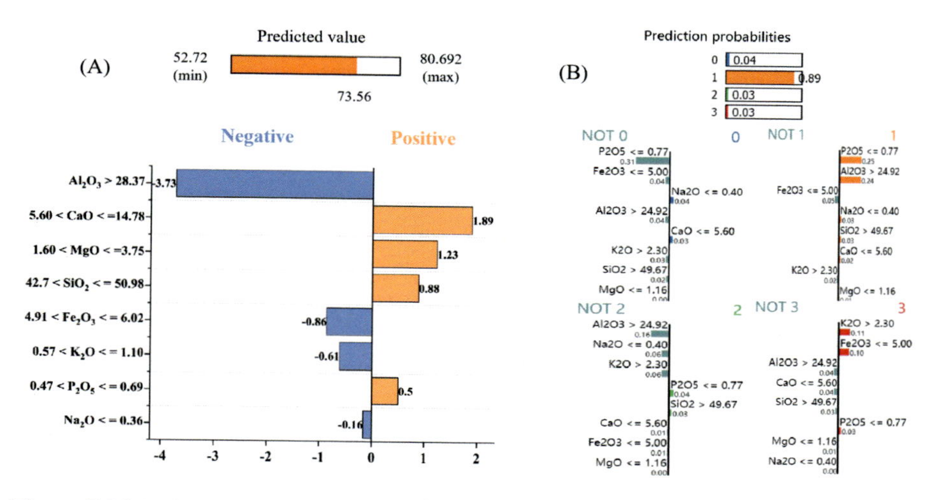

Figure 7.36 Weight chart and detailed information table for both modes. (A) Regression mode and (B) multiple classification mode.

References

[1] L.K. Sear, Properties and Use of Coal Fly Ash, Thomas Telford Limited, 2009.

[2] S. Donatello, C.R. Cheeseman, Recycling and recovery routes for incinerated sewage sludge ash (ISSA): a review, Waste Management 33 (11) (2013) 2328–2340.

[3] L.V. Zhang, A. Marani, M.L. Nehdi, Chemistry-informed machine learning prediction of compressive strength for alkali-activated materials, Construction and Building Materials 316 (2022) 126103.

[4] C. Birgen, et al., Machine learning based modelling for lower heating value prediction of municipal solid waste, Fuel 283 (2021) 118906.

[5] Z. Dauter, Data-collection strategies, Acta Crystallographica Section D: Biological Crystallography 55 (10) (1999) 1703–1717.

[6] Y. Roh, G. Heo, S.E. Whang, A survey on data collection for machine learning: a big data-ai integration perspective, IEEE Transactions on Knowledge and Data Engineering 33 (4) (2019) 1328–1347.

[7] B.B. Kawulich, Participant observation as a data collection method, Forum Qualitative Sozialforschung / Forum: Qualitative Social Research 6 (2) (2005).

[8] A.J. Gonzalez, et al., Detection of driver health condition by monitoring driving behavior through machine learning from observation, Expert Systems with Applications 199 (2022) 117167.

[9] K. Chaleunvong, Data collection techniques, Training Course in Reproductive Health Research Vientine, 2009.

[10] J. Cohen, et al., A feasibility study using a machine learning suicide risk prediction model based on open-ended interview language in adolescent therapy sessions, International Journal of Environmental Research and Public Health 17 (21) (2020) 8187.

[11] K. Tziridis, et al., Airfare prices prediction using machine learning techniques, in 2017 25th European Signal Processing Conference (EUSIPCO), 2017.

[12] J.J. Hox, H.R. Boeije, Data collection, Primary Versus Secondary, 2005.
[13] A. Radovic, et al., Machine learning at the energy and intensity frontiers of particle physics, Nature 560 (7716) (2018) 41–48.
[14] A. de Farias, et al., Simple machine learning allied with data-driven methods for monitoring tool wear in machining processes, The International Journal of Advanced Manufacturing Technology 109 (9) (2020) 2491–2501.
[15] Z. Liu, et al., Identification of heavy metal leaching patterns in municipal solid waste incineration fly ash based on an explainable machine learning approach, Journal of Environmental Management 317 (2022) 115387.
[16] K.T. Nguyen, et al., Analyzing the compressive strength of green fly ash based geopolymer concrete using experiment and machine learning approaches, Construction and Building Materials 247 (2020) 118581.
[17] S.J. Best, B.S. Krueger, Internet Data Collection, Sage, 2004.
[18] D.H. Granello, J.E. Wheaton, Online data collection: strategies for research, Journal of Counseling & Development 82 (4) (2004) 387–393.
[19] C. Shan, J. Ou, X. Chen, Matrix-product neural network based on sequence block matrix product, The Journal of Supercomputing 78 (6) (2022) 8467–8492.
[20] C. Allen, et al., Applying gis and machine learning methods to twitter data for multiscale surveillance of influenza, PLoS One 11 (7) (2016) e0157734.
[21] A. Tsymbal, S. Puuronen, Bagging and boosting with dynamic integration of classifiers, in European Conference on Principles of Data Mining and Knowledge Discovery, Springer, 2000.
[22] L. Liu, Research on logistic regression algorithm of breast cancer diagnose data by machine learning, in 2018 International Conference on Robots & Intelligent System (ICRIS), 2018.
[23] C. Qi, et al., Chemical signatures to identify the origin of solid ashes for efficient recycling using machine learning, Journal of Cleaner Production 368 (2022) 133020.
[24] C. Qi, et al., Rapid identification of reactivity for the efficient recycling of coal fly ash: hybrid machine learning modeling and interpretation, Journal of Cleaner Production 343 (2022) 130958.
[25] Y. Xiao, M. Watson, Guidance on conducting a systematic literature review, Journal of planning education and research 39 (1) (2019) 93–112.
[26] W. Abdelkader, et al., Machine learning approaches to retrieve high-quality, clinically relevant evidence from the biomedical literature: systematic review, JMIR Medical Informatics 9 (9) (2021) e30401.
[27] H. Kim, J.S. Sefcik, C. Bradway, Characteristics of qualitative descriptive studies: a systematic review, Research in Nursing & Health 40 (1) (2017) 23–42.
[28] D. Oleson, et al., Programmatic gold: targeted and scalable quality assurance in crowdsourcing, in Workshops at the Twenty-Fifth AAAI Conference on Artificial Intelligence, 2011.
[29] S. García, J. Luengo, F. Herrera, Data Preprocessing in Data Mining, 72, Springer, 2015.
[30] S. Xu, et al., Data cleaning in the process industries, Reviews in Chemical Engineering 31 (5) (2015) 453–490.
[31] G.L. Libralon, A.C.P.D.L.F. de Carvalho, A.C. Lorena, Pre-processing for noise detection in gene expression classification data, Journal of the Brazilian Computer Society 15 (1) (2009) 3–11.
[32] E. Rahm, H.H. Do, Data cleaning: problems and current approaches, IEEE Data Eng. Bull 23 (4) (2000) 3–13.

[33] S. García, et al., Big data preprocessing: methods and prospects, Big Data Analytics 1 (1) (2016) 9.
[34] S.K. Kwak, J.H. Kim, Statistical data preparation: management of missing values and outliers, Korean Journal of Anesthesiology 70 (4) (2017) 407–411.
[35] S.A. Alasadi, W.S. Bhaya, Review of data preprocessing techniques in data mining, Journal of Engineering and Applied Sciences 12 (16) (2017) 4102–4107.
[36] A. Karmaker, S. Kwek, An iterative refinement approach for data cleaning, Intelligent Data Analysis 11 (5) (2007) 547–560.
[37] H. Yang, Data Preprocessing, Citeseer, 2018.
[38] W. Zhang, T. Chen, Data preprocessing for web data mining, in advances in electronic commerce, Web Application and Communication, Springer, 2012, pp. 303–307.
[39] W. Fan, et al., Discovering and reconciling value conflicts for numerical data integration, Information Systems 26 (8) (2001) 635–656.
[40] S.H. Raju, M.N. Rao, Improvement of time complexity on external sorting using refined approach and data preprocessing, International Journal of Computer Sciences and Engineering 4 (11) (2016) 82–86.
[41] S. Christa, L. Madhuri, V. Suma, An effective data preprocessing technique for improved data management in a distributed environment, in International Conference on Advanced Computing and Communication Technologies for High Performance Applications, International Journal of Computer Applications, Cochin, Citeseer, 2012.
[42] J. Abel, W. Teahan, Universal text preprocessing for data compression, IEEE Transactions on Computers 54 (5) (2005) 497–507.
[43] M.A. Carreira-Perpinán, A review of dimension reduction techniques. Department of computer science, University of Sheffield. Technical Report 9 (1997) 1–69. CS-96-09.
[44] M.A. Hall, Correlation-Based Feature Selection for Machine Learning, The University of Waikato, 1999.
[45] I. Syarif, A. Prugel-Bennett, G. Wills, SVM parameter optimization using grid search and genetic algorithm to improve classification performance, TELKOMNIKA (Telecommunication Computing Electronics and Control) 14 (4) (2016) 1502–1509.
[46] J. Brownlee, Data preparation for machine learning: data cleaning, feature selection, and data transforms in Python, Machine Learning Mastery, 2020.
[47] T.A. Runkler, Data preprocessing, Data Analytics, Springer, 2016, pp. 23–36.
[48] L. Yongxiang, et al., A data-driven prognostics approach for RUL based on principle component and instance learning, in 2016 IEEE International Conference on Prognostics and Health Management (ICPHM), IEEE, 2016.
[49] S. Panigrahi, H. Behera, Effect of normalization techniques on univariate time series forecasting using evolutionary higher order neural network, International Journal of Engineering and Advanced Technology 3 (2) (2013) 280–285.
[50] S. Zhu, Investigation of vehicle-bicycle hit-and-run crashes, Traffic Injury Prevention 21 (7) (2020) 506–511.
[51] J. Wei, Research on data preprocessing in supermarket customers data mining, in 2010 2nd International Conference on Information Engineering and Computer Science, IEEE, 2010.
[52] S. Studer, et al., Towards CRISP-ML (Q): a machine learning process model with quality assurance methodology, Machine Learning and Knowledge Extraction 3 (2) (2021) 392–413.
[53] T. Hastie, et al., The Elements of Statistical Learning: Data Mining, Inference, and Prediction, 2, Springer, 2009.

[54] Z. Mirikharaji, et al., D-LEMA: Deep learning ensembles from multiple annotations-application to skin lesion segmentation, in Proceedings of the IEEE/CVF Conference on Computer Vision and Pattern Recognition, 2021.
[55] J. Goldberger, et al., Neighbourhood components analysis, Advances in Neural Information Processing Systems (2004) 17.
[56] V.R. Joseph, A. Vakayil, Split: an optimal method for data splitting, Technometrics 64 (2) (2022) 166–176.
[57] X. Zeng, T.R. Martinez, Distribution-balanced stratified cross-validation for accuracy estimation, Journal of Experimental & Theoretical Artificial Intelligence 12 (1) (2000) 1–12.
[58] D. Karaboga, C. Ozturk, Neural networks training by artificial bee colony algorithm on pattern classification, Neural Network World 19 (3) (2009) 279.
[59] V.C. Raykar, A. Saha, Data split strategiesfor evolving predictive models, in Joint European Conference on Machine Learning and Knowledge Discovery in Databases, Springer, 2015.
[60] E. Jain, et al., A diagnostic approach to assess the quality of data splitting in machine learning. arXiv Preprint arXiv:2206.11721, 2022.
[61] A. Beschorner, M. Voigt, K. Vogeler. Monte Carlo cross-validation for response surface benchmark, in 12th International Probabilistic Workshop. 2014.
[62] A.Y. Ng, "Preventing" overfitting of "cross-validation data", in ICML. Vol. 97, Citeseer, 1997.
[63] Z. Reitermanova, Data splitting, in WDS, Matfyzpress Prague, 2010.
[64] Y. Xu, R. Goodacre, On splitting training and validation set: a comparative study of cross-validation, bootstrap and systematic sampling for estimating the generalization performance of supervised learning, Journal of Analysis and Testing 2 (3) (2018) 249–262.
[65] T.-T. Wong, Performance evaluation of classification algorithms by k-fold and leave-one-out cross validation, Pattern Recognition 48 (9) (2015) 2839–2846.
[66] J.D. Rodriguez, A. Perez, J.A. Lozano, Sensitivity analysis of k-fold cross validation in prediction error estimation, IEEE Transactions on Pattern Analysis and Machine Intelligence 32 (3) (2009) 569–575.
[67] L. Yang, A. Shami, On hyperparameter optimization of machine learning algorithms: theory and practice, Neurocomputing 415 (2020) 295–316.
[68] M. Kuhn, K. Johnson, Applied Predictive Modeling, 26, Springer, 2013.
[69] G. Luo, A review of automatic selection methods for machine learning algorithms and hyper-parameter values, Network Modeling Analysis in Health Informatics and Bioinformatics 5 (1) (2016) 18.
[70] T. Koike-Akino, K. Kojima, Y. Wang, AutoML hyperparameter tuning of generative DNN architecture for nanophotonic device design, CLEO: QELS_Fundamental Science, Optica Publishing Group, 2022.
[71] R. Elshawi, M. Maher, S. Sakr, Automated machine learning: state-of-the-art and open challenges. arXiv Preprint arXiv:1906.02287, 2019.
[72] P. Liashchynskyi, P. Liashchynskyi, Grid search, random search, genetic algorithm: a big comparison for NAS. arXiv Preprint arXiv:1912.06059, 2019.
[73] K.B. Ensor, P.W. Glynn, Stochastic optimization via grid search, Lectures in Applied Mathematics-American Mathematical Society 33 (1997) 89–100.
[74] J. Bergstra, Y. Bengio, Random search for hyper-parameter optimization, Journal of Machine Learning Research 13 (2) (2012).

[75] W. Price, Global optimization by controlled random search, Journal of Optimization Theory and Applications 40 (3) (1983) 333–348.
[76] E. Bisong, More supervised machine learning techniques with scikit-learn, Building Machine Learning and Deep Learning Models on Google Cloud Platform, Springer, 2019, pp. 287–308.
[77] S. Brajesh, I. Ray, Ensemble approach for sensor-based human activity recognition, in Adjunct Proceedings of the 2020 ACM International Joint Conference on Pervasive and Ubiquitous Computing and Proceedings of the 2020 ACM International Symposium on Wearable Computers, 2020.
[78] P.I. Frazier, Bayesian optimization, in recent advances in optimization and modeling of contemporary problems, Informs (2018) 255–278.
[79] J. Snoek, H. Larochelle, R.P. Adams, Practical bayesian optimization of machine learning algorithms, Advances in Neural Information Processing Systems (2012) 25.
[80] M. Pelikan, Bayesian optimization algorithm, Hierarchical Bayesian Optimization Algorithm, Springer, 2005, pp. 31–48.
[81] H. Wang, et al., A new acquisition function for Bayesian optimization based on the moment-generating function, in 2017 IEEE International Conference on Systems, Man, and Cybernetics (SMC), IEEE, 2017.
[82] M. Hoffman, E. Brochu, N. De Freitas. Portfolio allocation for bayesian optimization, in UAI, Citeseer, 2011.
[83] M.T. Young, et al., Distributed Bayesian optimization of deep reinforcement learning algorithms, Journal of Parallel and Distributed Computing 139 (2020) 43–52.
[84] M. Clerc, Particle Swarm Optimization, 93, John Wiley & Sons, 2010.
[85] Y. Shi, Particle Swarm Optimization, 2, IEEE connections, 2004, pp. 8–13 (1).
[86] A. Lazinica, Particle Swarm Optimization, BoD–Books on Demand, 2009.
[87] D. Wang, D. Tan, L. Liu, Particle swarm optimization algorithm: an overview, Soft Computing 22 (2) (2018) 387–408.
[88] M. Makhtar, D.C. Neagu, M.J. Ridley, Binary classification models comparison: on the similarity of datasets and confusion matrix for predictive toxicology applications, in International Conference on Information Technology in Bio-and Medical Informatics, Springer, 2011.
[89] M. Hasnain, et al., Evaluating trust prediction and confusion matrix measures for web services ranking, IEEE Access 8 (2020) 90847–90861.
[90] E. Beauxis-Aussalet, L. Hardman, Visualization of confusion matrix for non-expert users, in IEEE Conference on Visual Analytics Science and Technology (VAST)-Poster Proceedings, 2014.
[91] A. Tharwat, Classification assessment methods, Applied Computing and Informatics (2020).
[92] P.A. Flach, ROC analysis, Encyclopedia of Machine Learning and Data Mining, Springer, 2016, pp. 1–8.
[93] S.B. Nellore, Various performance measures in binary classification-an overview of ROC study, IJISET-International Journal of Innovative Science, Engineering & Technology 2 (9) (2015) 596–605.
[94] S. Raschka, An overview of general performance metrics of binary classifier systems. arXiv Preprint arXiv:1410.5330, 2014.
[95] S. Dreiseitl, L. Ohno-Machado, M. Binder, Comparing three-class diagnostic tests by three-way ROC analysis, Medical Decision Making 20 (3) (2000) 323–331.
[96] M. Heydarian, T.E. Doyle, R. Samavi, MLCM: multi-label confusion matrix, IEEE Access 10 (2022) 19083–19095.

[97] Y. Yuan, et al., Occupancy estimation in buildings based on infrared array sensors detection, IEEE Sensors Journal 20 (2) (2019) 1043–1053.
[98] Y. Yuan, et al., Approach of personnel location in roadway environment based on multi-sensor fusion and activity classification, Computer Networks 148 (2019) 34–45.
[99] J. Yan, Multi-class ROC random forest for imbalanced classification, State University of New York at Stony Brook, 2017.
[100] S. Sharma, R. Mehra, Conventional machine learning and deep learning approach for multi-classification of breast cancer histopathology images—a comparative insight, Journal of Digital Imaging 33 (3) (2020) 632–654.
[101] P. Sedgwick, Pearson's correlation coefficient, BMJ (Clinical Research ed.) (2012) 345.
[102] B. Ratner, The correlation coefficient: Its values range between + 1/ − 1, or do they? Journal of Targeting, Measurement and Analysis for Marketing 17 (2) (2009) 139–142.
[103] F. Mekanik, et al., Multiple regression and artificial neural network for long-term rainfall forecasting using large scale climate modes, Journal of Hydrology 503 (2013) 11–21.
[104] A. Vlachogianni, et al., Evaluation of a multiple regression model for the forecasting of the concentrations of NOx and PM10 in athens and helsinki, Science of the Total Environment 409 (8) (2011) 1559–1571.
[105] C. Qi, A. Fourie, Q. Chen, Neural network and particle swarm optimization for predicting the unconfined compressive strength of cemented paste backfill, Construction and Building Materials 159 (2018) 473–478.
[106] I.S. Helland, On the interpretation and use of R2 in regression analysis, Biometrics (1987) 61–69.
[107] A.G. Asuero, A. Sayago, A. González, The correlation coefficient: an overview, Critical Reviews in Analytical Chemistry 36 (1) (2006) 41–59.
[108] N.V. Sailaja, et al., Hybrid regression model for medical insurance cost prediction and recommendation, in 2021 IEEE International Conference on Intelligent Systems, Smart and Green Technologies (ICISSGT), 2021.
[109] D. Chicco, M.J. Warrens, G. Jurman, The coefficient of determination R-squared is more informative than SMAPE, MAE, MAPE, MSE and RMSE in regression analysis evaluation, Peer J Computer Science 7 (2021) e623.
[110] M. Ćalasan, S.H.A. Aleem, A.F. Zobaa, On the root mean square error (RMSE) calculation for parameter estimation of photovoltaic models: a novel exact analytical solution based on lambert W function, Energy Conversion and Management 210 (2020) 112716.
[111] M.R. Daudpoto, et al., A residual analysis for the removal of biological oxygen demand through rotating biological contactor, Mehran University Research Journal of Engineering & Technology 40 (2) (2021) 459–464.
[112] S. Kohli, G.T. Godwin, S. Urolagin, Sales prediction using linear and KNN regression, Advances in Machine Learning and Computational Intelligence, Springer, 2021, pp. 321–329.
[113] D.N. Prata, W. Rodrigues, P.H. Bermejo, Temperature significantly changes COVID-19 transmission in (sub) tropical cities of Brazil, Science of the Total Environment 729 (2020) 138862.
[114] M.J. Sajid, S.A.R. Khan, E.D. Gonzalez, Identifying contributing factors to China's declining share of renewable energy consumption: no silver bullet to decarbonisation, Environmental Science and Pollution Research (2022) 1–16.
[115] E. Rendón, et al., Internal versus external cluster validation indexes, International Journal of Computers and Communications 5 (1) (2011) 27–34.

[116] M. Hassani, T. Seidl, Using internal evaluation measures to validate the quality of diverse stream clustering algorithms, Vietnam Journal of Computer Science 4 (3) (2017) 171–183.
[117] P.J. Rousseeuw, Silhouettes: a graphical aid to the interpretation and validation of cluster analysis, Journal of Computational and Applied Mathematics 20 (1987) 53–65.
[118] H. Řezanková, Different approaches to the silhouette coefficient calculation in cluster evaluation, in 21st International Scientific Conference AMSE Applications of Mathematics and Statistics in Economics, 2018.
[119] D. Dey, et al., Instance selection in text classification using the silhouette coefficient measure, Mexican International Conference on Artificial Intelligence, Springer, 2011.
[120] J. Baarsch, M.E. Celebi, Investigation of internal validity measures for k-means clustering, in Proceedings of the International Multiconference of Engineers and Computer Scientists, 2012.
[121] Y. Wang, Y. Xu, T. Gao, Evaluation method of wind turbine group classification based on Calinski Harabasz, in 2021 IEEE 5th Conference on Energy Internet and Energy System Integration (EI2), 2021.
[122] U. Maulik, S. Bandyopadhyay, Performance evaluation of some clustering algorithms and validity indices, IEEE Transactions on Pattern Analysis and Machine Intelligence 24 (12) (2002) 1650–1654.
[123] J.C. Dunn, Well-separated clusters and optimal fuzzy partitions, Journal of Cybernetics 4 (1) (1974) 95–104.
[124] T.C. Havens, et al., Dunn's cluster validity index as a contrast measure of VAT images, in 2008 19th International Conference on Pattern Recognition, IEEE, 2008.
[125] D. Xu, Y. Tian, A comprehensive survey of clustering algorithms, Annals of Data Science (New York, N.Y.) 2 (2) (2015) 165–193.
[126] S.C. Sripada, M.S. Rao, Comparison of purity and entropy of k-means clustering and fuzzy c means clustering, Indian Journal of Computer Science and Engineering 2 (2011) 03.
[127] A. Amelio, C. Pizzuti, Correction for closeness: adjusting normalized mutual information measure for clustering comparison, Computational Intelligence 33 (3) (2017) 579–601.
[128] Z.J. Wang, P.W.-H. Lee, M.J. McKeown, A novel segmentation, mutual information network framework for EEG analysis of motor tasks, Biomedical Engineering Online 8 (1) (2009) 1–19.
[129] L. Hubert, P. Arabie, Comparing partitions, Journal of Classification 2 (1) (1985) 193–218.
[130] P. Das, L.R. Varshney, Explaining artificial intelligence generation and creativity: human interpretability for novel ideas and artifacts, IEEE Signal Processing Magazine 39 (4) (2022) 85–95.
[131] J. Wu, H. Xiong, J. Chen, Towards understanding hierarchical clustering: a data distribution perspective, Neurocomputing 72 (10) (2009) 2319–2330.
[132] Z. Zhou, Statistical inference for machine learning: feature importance, Uncertainty Quantification and Interpretation Stability, Cornell University, 2021.
[133] R. Guidotti, et al., A survey of methods for explaining black box models, ACM Computing Surveys (CSUR) 51 (5) (2018) 1–42.
[134] M.P. Neto, F.V. Paulovich, Explainable matrix-visualization for global and local interpretability of random forest classification ensembles, IEEE Transactions on Visualization and Computer Graphics 27 (2) (2020) 1427–1437.

[135] M. Du, N. Liu, X. Hu, Techniques for interpretable machine learning, Communications of the ACM 63 (1) (2019) 68−77.
[136] L. Kopitar, et al., Local vs. global interpretability of machine learning models in type 2 diabetes mellitus screening, Artificial Intelligence in Medicine: Knowledge Representation and Transparent and Explainable Systems, Springer, 2019, pp. 108−119.
[137] M. Al-Sarem, et al., Feature selection and classification using catboost method for improving the performance of predicting Parkinson's disease, Advances on Smart and Soft Computing, 2021, pp. 189−199.
[138] S.S. Sundhari, A knowledge discovery using decision tree by Gini coefficient, in 2011 International Conference on Business, Engineering and Industrial Applications, 2011.
[139] A. Altmann, et al., Permutation importance: a corrected feature importance measure, Bioinformatics (Oxford, England) 26 (10) (2010) 1340−1347.
[140] D. Niu, et al., Short-term photovoltaic power generation forecasting based on random forest feature selection and CEEMD: a case study, Applied Soft Computing 93 (2020) 106389.
[141] H. Liu, et al., Overall grouting compactness detection of bridge prestressed bellows based on RF feature selection and the GA-SVM model, Construction and Building Materials 301 (2021) 124323.
[142] Li, Y., et al., Feature importance recap and stacking models for forex price prediction. arXiv Preprint arXiv:2107.14092, 2021.
[143] A. Fisher, C. Rudin, F. Dominici, Model class reliance: variable importance measures for any machine learning model class, from the "Rashomon" perspective. arXiv Preprint arXiv:1801.01489, 2018. 68.
[144] X. Mi, et al., Permutation-based identification of important biomarkers for complex diseases via machine learning models, Nature Communications 12 (1) (2021) 1−12.
[145] V.Q. Tran, Hybrid gradient boosting with meta-heuristic algorithms prediction of unconfined compressive strength of stabilized soil based on initial soil properties, mix design and effective compaction, Journal of Cleaner Production 355 (2022) 131683.
[146] S. Kalra, R. Lamba, M. Sharma, Machine learning based analysis for relation between global temperature and concentrations of greenhouse gases, Journal of Information and Optimization Sciences 41 (1) (2020) 73−84.
[147] J.H. Friedman, Greedy function approximation: a gradient boosting machine, Annals of Statistics (2001) 1189−1232.
[148] A. Inglis, A. Parnell, C.B. Hurley, Visualizing variable importance and variable interaction effects in machine learning models, Journal of Computational and Graphical Statistics (2022) 1−13.
[149] T.R. Cook, et al., Explaining machine learning by bootstrapping partial dependence functions and Shapley values. Federal Research Bank of Kansas City, 2021.
[150] C. Ardito, et al., Interacting with features: visual inspection of black-box fault type classification systems in electrical grids, in XAI. it@ AI* IA, 2020.
[151] H. Sarvaiya, et al., Explainable artificial intelligence (XAI): towards malicious SCADA communications, in ISUW 2020, 2022, Springer. p. 151−162.
[152] A. Yeh, A. Ngo, Bringing a ruler into the black box: uncovering feature impact from individual conditional expectation plots, Machine Learning and Principles and Practice of Knowledge Discovery in Databases, Springer International Publishing, Cham, 2021.
[153] C. Molnar, et al., General pitfalls of model-agnostic interpretation methods for machine learning models, International Workshop on Extending Explainable AI Beyond Deep Models and Classifiers, Springer, 2022.

[154] A. Goldstein, et al., Peeking inside the black box: visualizing statistical learning with plots of individual conditional expectation, Journal of Computational and Graphical Statistics 24 (1) (2015) 44–65.
[155] Y. Wu, Y. Zhou, Hybrid machine learning model and Shapley additive explanations for compressive strength of sustainable concrete, Construction and Building Materials 330 (2022) 127298.
[156] D.V. Urista, et al., Prediction of antimalarial drug-decorated nanoparticle delivery systems with random forest models, Biology 9 (8) (2020) 198.
[157] L. Zhong, et al., Soil properties: their prediction and feature extraction from the LUCAS spectral library using deep convolutional neural networks, Geoderma 402 (2021) 115366.
[158] A. Datta, S. Sen, Y. Zick. Algorithmic transparency via quantitative input influence: theory and experiments with learning systems, in 2016 IEEE Symposium on Security and Privacy (SP), IEEE, 2016.
[159] M. Sundararajan, A. Taly, Q. Yan, Axiomatic attribution for deep networks, in International Conference on Machine Learning, PMLR, 2017.
[160] E. Shakeri, et al., Exploring features contributing to the early prediction of sepsis using machine learning, in 2021 43rd Annual International Conference of the IEEE Engineering in Medicine & Biology Society (EMBC), 2021.
[161] M. Vega García, J.L. Aznarte, Shapley additive explanations for NO2 forecasting, Ecological Informatics 56 (2020) 101039.
[162] S.M. Lundberg, G.G. Erion, S.-I. Lee, Consistent individualized feature attribution for tree ensembles. arXiv Preprint arXiv:1802.03888, 2018.
[163] Y. Nohara, et al., Explanation of machine learning models using Shapley additive explanation and application for real data in hospital, Computer Methods and Programs in Biomedicine 214 (2022) 106584.
[164] Y. Kim, Y. Kim, Explainable heat-related mortality with random forest and SHapley additive explanations (SHAP) models, Sustainable Cities and Society 79 (2022) 103677.
[165] S. Mangalathu, S.-H. Hwang, J.-S. Jeon, Failure mode and effects analysis of RC members based on machine-learning-based SHapley additive explanations (SHAP) approach, Engineering Structures 219 (2020) 110927.
[166] M.T. Ribeiro, S. Singh, C. Guestrin, Why should i trust you? Explaining the predictions of any classifier, in Proceedings of the 22nd ACM SIGKDD International Conference on Knowledge Discovery and Data Mining, 2016.
[167] M.T. Ribeiro, S. Singh, C. Guestrin, Model-agnostic interpretability of machine learning. arXiv Preprint arXiv:1606.05386, 2016.
[168] A. Messalas, C. Aridas, Y. Kanellopoulos, Evaluating MASHAP as a faster alternative to LIME for model-agnostic machine learning interpretability, in 2020 IEEE International Conference on Big Data (Big Data), 2020.
[169] J.A. Recio-Garcia, B. Diaz-Agudo, V. Pino-Castilla, et al.: A case-based reasoning approach to provide specific local interpretable model-agnostic explanations, in Case-Based Reasoning Research and Development, ICCBR 2020, 2020, p. 179–194.
[170] S. Gupta, G. Sikka. Explaining HCV prediction using LIME model, in 2021 2nd International Conference on Secure Cyber Computing and Communications (ICSCCC), 2021.
[171] D. Garreau U. Luxburg, Explaining the explainer: a first theoretical analysis of LIME, in International Conference on Artificial Intelligence and Statistics, PMLR, 2020.

[172] M. Graziani, et al. Sharpening local interpretable model-agnostic explanations for histopathology: improved understandability and reliability, in International Conference on Medical Image Computing and Computer-Assisted Intervention, Springer, 2021.
[173] M.R. Zafar, N. Khan, Deterministic local interpretable model-agnostic explanations for stable explainability, Machine Learning And Knowledge Extraction 3 (3) (2021) 525–541.

The application of clustering algorithms for industrial solid ashes based on physicochemical properties

8

Abstract

This chapter presents how unsupervised clustering analysis can be applied to the industrial solid ash. As introduced in previous chapters, unsupervised clustering can be used to reveal statistical relationships and correlations within a dataset. In this chapter, we present oxide content data comprising four industrial solid ash types across 310 samples. Clustering analysis using k-means clustering, Gaussian mixture model and agglomerative nesting algorithms was performed on the prepared dataset, and solid ash source detection was then conducted. The results indicate that clustering analysis can identify the underlying correlations between oxide contents of different ash types and could serve as a potential technique for identifying solid ash sources.

8.1 Background

Continuous industrialization throughout the 20th and 21st centuries has greatly increased the production and usage of fossil fuels such as coal. As a by-product of coal combustion, approximately 800 million tonnes of coal fly ash (CFA) are produced each year [1]. Given the finite nature of fossil fuel resources, the use of biomass as an alternative source of clean energy has received increasing attention [2,3]. To avoid resource wastage and meet urban hygiene standards, large volumes of municipal waste and dehydrated sludge are burnt to generate electricity [4], resulting in an increase in biomass ash types, such as municipal solid waste fly ash (MSWFA), bottom ash (MSWBA), and sewage sludge fly ash (SSA). The dramatic increase in solid waste production has caused serious environmental damage and endangered human health [5], thus resulting in huge demand for the treatment and reuse of solid ash.

The disposal and reuse strategies for solid ash vary according to the material's properties and origin. MSWB is used as secondary raw material in construction due to its typically low concentrations of pollutants [6]. MSWFA has significant leaching potential and contains high concentrations of heavy minerals, dioxins, and furane, which are highly toxic [7]. It is generally regarded as a hazardous waste type, and its management requires special treatment such as cement reinforcement [8]. Due to the physical and chemical properties of SSA,

Machine Learning Applications in Industrial Solid Ash. DOI: https://doi.org/10.1016/B978-0-443-15524-6.00002-9
© 2024 Elsevier Inc. All rights reserved.

toxicological and activation analyses are normally required, and the material is then decontaminated and cured before it can be recycled for use in roadbed materials [9]. CFA is suitable for use as a supplementary material for cement due to its high reactivity [10]. Overall, given these markedly different management workflows, source identification is a crucial step to ensure that the solid waste is properly disposed of and effectively reclaimed. Traditional solid ash identification requires complex component detection, which costs a lot of time and money. Therefore, an efficient technology for source identification and analysis of solid ash is urgently needed.

With the advent of the big data era, clustering analysis has been widely applied in pattern recognition, machine learning, and other fields. Clustering algorithms, which are used to locate samples and distribute them into unified classes based on similarity among data objects [11], include hierarchical clustering, zonal clustering, and clustering based on density, grid, and model approaches [12]. In recent years, clustering analysis has been widely used in analyses of mining safety and environmental issues due to its strong methodological advantages. For example, using hierarchical clustering analysis, Muneeb et al. [13] showed that coal mining has caused different levels of environmental pollution in areas around mines. Chen et al. [14] proposed a network of underground wireless sensors using chain-based clustering topology that could be used to provide network connectivity and meet the transmission requirements of mine roads. Smolenski et al. [15] used hierarchical clustering analysis to examine the similarities and differences in coal ash samples and displayed their results in a tree chart to further examine the relationships of coal samples to their measured rare earth element concentrations.

On the other hand, solid ash has been widely reported in terms of both recovery and usage for filling mines. For example, Li et al. [16] used coal gangue and CFA as backfill materials, measured the compression deformation characteristics of samples under different ratios, and took the coal mine backfilling face as an example, mixed gangue and CFA at a ratio of 1:0.35 to backfill the goaf; Yin et al. [17] studied the effect of CFA as fine aggregate on the performance of cement backfill and found that the early strength of cemented waste rock-CFA backfill samples increased slowly with the increase of CFA content. Nonetheless, only a few studies to date have focused on solid ash origins. In addition, to our knowledge, there are no studies that have used clustering analysis for ash source detection. Therefore, there is an urgency to develop a method to readily detect the source of solid ash without requiring expert expertise.

In this chapter, we collected global solid ash data through a literature search and analyzed its chemical properties. Based on this, we adopted the k-means, the Gaussian mixture model (GMM), and agglomerative nesting algorithms (AGNES) and combined two different approaches to study the clustering of solid ash samples. We used four indicators to evaluate the consistency of clustering results with experimental data. The selected cluster method in this chapter can be used to identify ash sources without requiring labeled data to construct complex models, thereby facilitating efficient solid ash disposal and recycling.

8.2 Methodology

8.2.1 Data acquisition

Two main methods are widely adopted to obtain data: the first is to analyze and process collected solid ash samples, and the other is to use data from public sources. In this chapter, we searched Web of Science, Elsevier, Scopus, and other databases for references relating to solid ash. More than 1000 references were identified initially; however, after preliminary screening, only 114 articles with data on chemical oxides and ash source information were retained. Overall, a global dataset was generated that contains four different solid ash source types.

8.2.2 Clustering analysis

GMM, AGNES, and k-means approaches were adopted to study the solid ash origins. GMM is a generative model that can fit any data distribution type to a weighted combination of multiple Gaussian distribution models. The basic principle of GMM is to calculate a probabilistic distribution from a given dataset and identify the parameters of each distribution using an expectation-maximum (EM) algorithm to estimate the model likelihood [18,19]. AGNES is a hierarchical clustering algorithm that employs a bottom-up aggregation strategy. AGNES regards each sample in the dataset as an initial cluster, then combines the two clusters closest to each other; this process is then repeated until the preset number of clusters is reached. The core principle of the AGNES algorithms is to determine the distance between the clusters based on Eqs. (8.1), (8.2), and (8.3). Details of the principles of the k-means algorithm are provided in Section 6.3.7.

$$d_{\min}(C_i, C_j) = \min_{x \in C_i, z \in C_j} dist(x, z) \tag{8.1}$$

$$d_{\max}(C_i, C_j) = \min_{x \in C_i, z \in C_j} dist(x, z) \tag{8.2}$$

$$d_{avg}(C_i, C_j) = \frac{1}{|C_i||C_j|} \sum_{x \in C_i} \sum_{z \in C_j} dist(x, z) \tag{8.3}$$

where C_i and C_j refer to two sample sets, and x and z are sample points in two clusters. When the cluster distance is calculated by $d_{\min}$, $d_{\max}$, and d_{avg}, the AGNES algorithm is correspondingly called "single-link," "complete-link" or "uniform-link" [20].

8.2.3 Evaluation of clustering effect

The effect of the clustering algorithm can be measured using external and internal indicators [21]. Internal indicators of a sample are unsupervised, and only the distance between a sampling point in the dataset and a clustering center is used to measure the clustering effect. In comparison, the external index is a supervised measure that compares the clustering results with known (labeled) clustering to measure the performance of the clustering algorithm [22].

In this chapter, we use the Rand index (RI), adjusted Rand index (ARI), adjusted mutual information (AMI), and accuracy (ACC) approaches to evaluate the effects of six methods. Detailed explanations of these evaluation indicators can be found in Section 7.4.3.2.

8.3 Results and discussion

8.3.1 Statistical analysis of the dataset

The dataset comprises 310 samples, where SiO_2, Al_2O_3, Fe_2O_3, CaO, MgO, Na_2O, and K_2O are the input features and CFA, MSWFA, MSWBA, and SSA are target variables, represented by the labels 0, 1, 2, and 3, respectively. In this chapter, the chemical oxide content of solid ash was statistically analyzed both overall and from the perspective of different ash origins, as shown in Figs. 8.1 and 8.2.

In Fig. 8.1, the distributions of SiO_2 and CaO were relatively discrete, with mass fractions ranging from 0% to 72%. Al_2O_3 was evenly distributed, whereas Fe_2O_3, MgO, Na_2O, and K_2O showed relatively concentrated distributions. The mass fractions of MgO, Na_2O, and K_2O were mostly within the 0%–10% range. The average

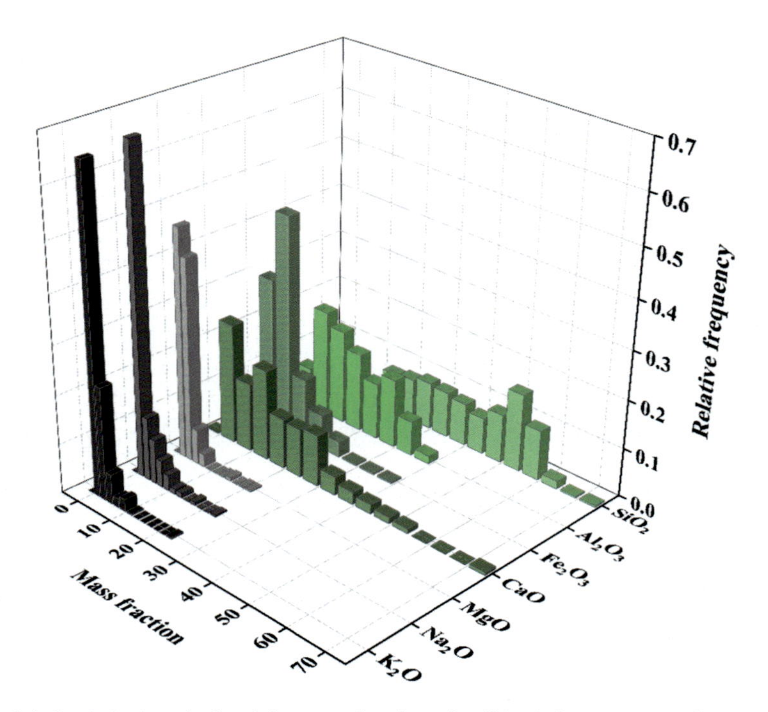

Figure 8.1 Statistical analysis of the mass fraction of solid ash from an overall perspective.

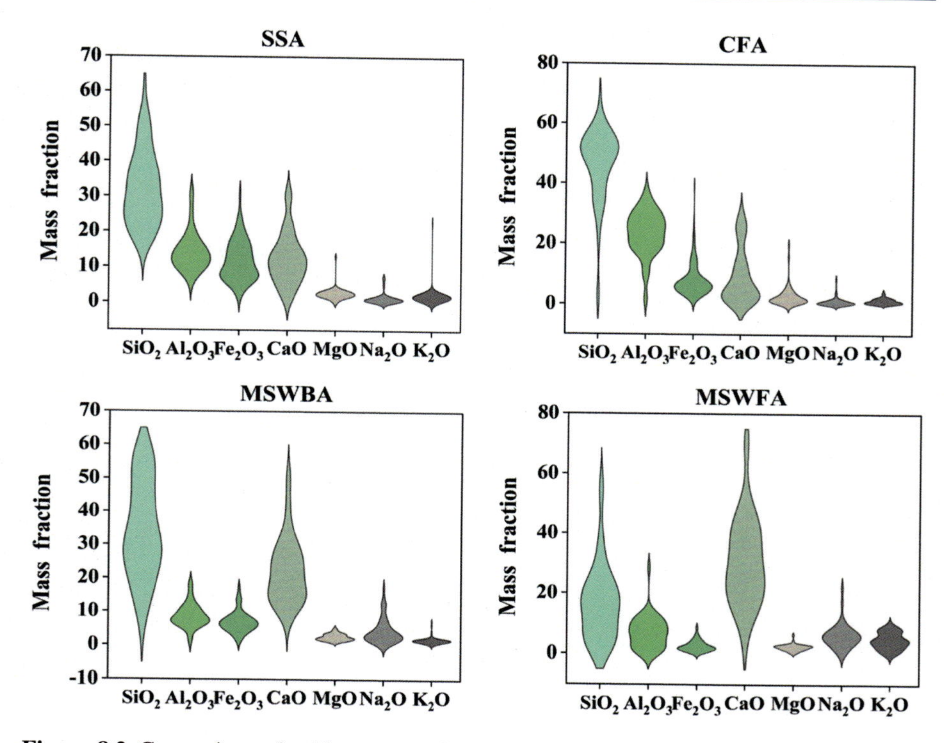

Figure 8.2 Comparison of oxide content of solid ash from different sources.

SiO_2 content in solid ash was the highest with a value of around 40%. Those of MgO, Na_2O, and K_2O were the lowest, with values of around 3%.

Solid ash from different origins is affected by many factors, and its oxide content and physical and chemical properties will differ slightly. In Fig. 8.2, we statistically analyzed the mass fractions of solid ash samples from different origins. In all four solid ash types, the mass fractions of SiO_2, Al_2O_3, CaO, and Fe_2O_3 accounted for more than 80% of the total chemical oxide content, which highlights the essential nature of the SiO_2-Al_2O_3-CaO-Fe_2O_3 quaternary system in solid ash research and recycling [23–25].

In CFA and SSA, the average SiO_2 content was the highest, implying that CFA and SSA can be used as supplementary cementitious materials [26]. In contrast, in SSA, the average Fe_2O_3 content was higher than those in CFA, which was mainly affected by flocculation in different physical deposition processes [27]. The composition of SSA was similar to that of clay, and this ash type can accordingly be used for firing bricks. For MSWBA and MSWFA, the average CaO content clustered around 20%, with a maximum content of 70%. Different incineration techniques can affect the CaO content of solid ash, consistent with the findings of Hu et al. [28]. Furthermore, the high calcium content of MSWFA and the high silicon content of MSWBA determined their respective high melting points; however, increasing Na_2O content lowers the melting temperature of the solid ash. This finding is consistent with the study of Tsvetkov et al.

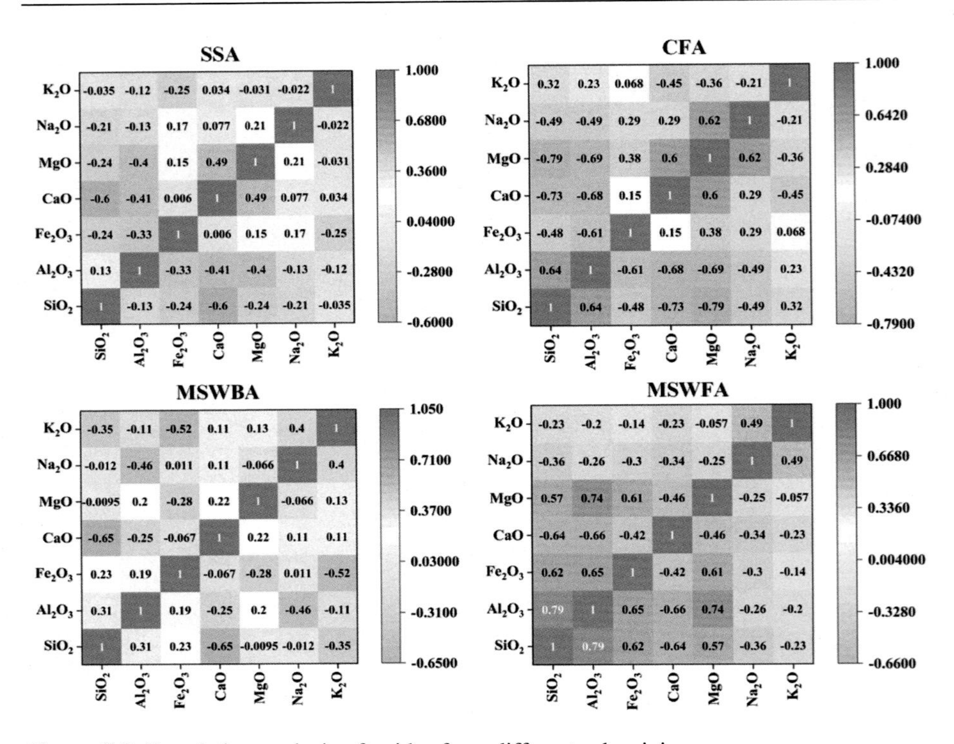

Figure 8.3 Correlation analysis of oxides from different ash origins.

[29], who performed a thermodynamic study on the chemical equilibrium of CaO-SiO_2-Al_2O_3-Na_2O systems containing 5%–25% Na_2O and found that the Na_2O content affects the melting point of solid waste ash.

8.3.2 Correlation between chemical oxides in solid ash

As shown in Fig. 8.3, we used the Pearson correlation coefficient matrix to describe the relationship between the oxides of solid ash samples from different sources. For CFA, the R value of Al_2O_3-SiO_2 was the largest positive value, reaching 0.64, whereas the R value of SiO_2 and MgO was the largest negative value, reaching −0.79, indicating a strong correlation of the Al_2O_3-SiO_2-MgO ternary system. In MSWFA, the R value between SiO_2 and Al_2O_3 was 0.79, the R value between CaO and SiO_2 was -0.64, and the negative correlation coefficient between CaO and Al_2O_3 was the largest, corresponding to an R value of −0.66. Except for the SiO_2-Al_2O_3-CaO ternary system, the R value between other oxides was small for this ash type. In SSA, SiO_2 had the largest negative correlation with CaO, corresponding to an R value of −0.6. The positive correlation coefficient between MgO and CaO was the largest, with an R value of 0.49. Similarly, the maximum negative R value in MSWBA was between CaO and SiO_2 (−0.65), while the maximum positive R value was between Na_2O and K_2O (0.4).

Table 8.1 Comparison of the evaluation results of the six studied methods.

Fitting method	Clustering algorithm	RI	ARI	AMI	ACC
All features	K-means	0.71	0.32	0.27	0.58
	GMM	0.74	0.37	0.37	0.46
	AGNES	0.68	0.26	0.23	0.25
Screened features (CaO, SiO_2 and Al_2O_3)	K-means	0.69	0.27	0.27	0.21
	GMM	0.64	0.16	0.24	0.33
	AGNES	0.69	0.29	0.26	0.21

For the same solid ash type, there were strong correlations between oxides. However, the correlation degree between oxides varied for different solid ash types, providing a theoretical basis for clustering analysis.

8.3.3 Clustering analysis

The correlation between features can often improve the clustering effect, the effect of different features may vary in different clustering and feature selection algorithms [30]. Based on the correlation analysis in Section 8.3.2, two different model fittings were performed for k-means, GMM, and AGNES clustering algorithms. One approach involved using all features for fitting, while the other involved screening SiO_2, CaO, and Al_2O_3 for fitting and prediction through feature selection. The evaluation results of the six methods are shown in Table 8.1.

As shown, the indicators of RI, ARI, and AMI obtained by the six methods were all within the range of [-1,1]. The RI value for the GMM clustering algorithm when used to fit all features was 0.74, close to 1, while the corresponding ARI and AMI values were both 0.37, above the average value of 0 for these indicators. The RI, ARI, and AMI values of the k-means clustering algorithm when fitting all features were 0.71, 0.32, and 0.27, respectively, indicating that the clustering effect was consistent with the data. In terms of accuracy, the k-means clustering algorithm, when using all features for fitting, had the highest ACC value of 0.58, followed by the GMM clustering algorithm with a value of 0.46. In summary, these results indicate that the k-means clustering algorithm, when fitting all features, was the most suitable for tracing the origin of solid ash.

In addition, the four evaluation indicators for the AGNES clustering algorithm when fitted using the screened features (SiO_2, CaO, and Al_2O_3) were all higher than those of the AGNES algorithm fitted with all the features. This finding is consistent with the clustering performance of similarity-based feature selection [31]. However, the k-means and GMM clustering algorithms showed different results from those of AGNES, which indirectly implies that feature selection plays a variable role in different clustering algorithms and also

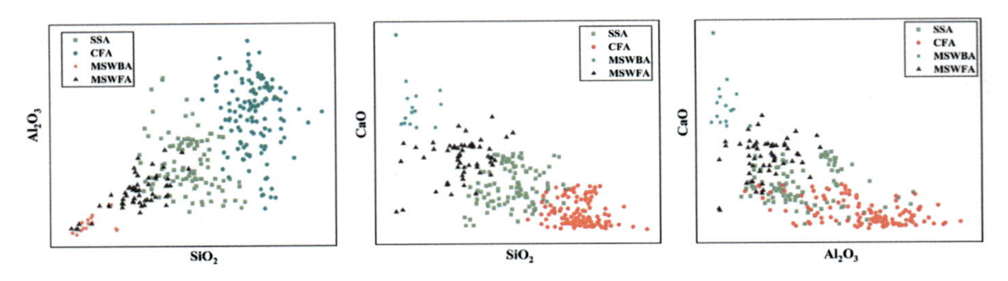

Figure 8.4 Clustering effect of the k-means algorithm fitted using all features. The red squares represent the SSA samples, the blue circles represent CFA, the orange stars represent MSWBA, and the green triangles represent MSWFA.

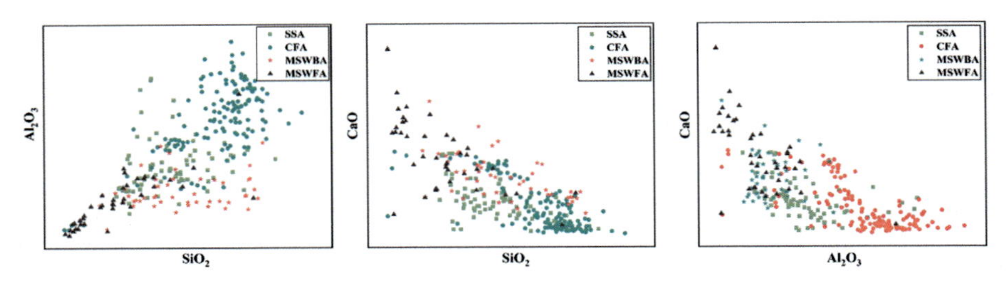

Figure 8.5 The actual distribution of oxide pairs for solid ash samples from different origins.

demonstrates the necessity of testing various fitting methods to identify the optimal clustering effect.

To more intuitively represent the clustering effect, we took the k-means clustering algorithm fitted using all features as an example and selected three combinations (Al_2O_3-SiO_2, CaO-SiO_2, and CaO-Al_2O_3) for analysis, as shown in Fig. 8.4. The three different representations all show that solid ash can be classified into four origins, which is approximately consistent with the real data distribution shown in Fig. 8.5.

8.4 Summary

This chapter analyzed the chemical properties of solid ash through statistical description and correlation analysis. Three clustering algorithms combined with two fitting methods were used for clustering analysis. The effects of the six clustering methods were compared using four indicators (RI, ARI, AMI, and ACC), and the most suitable method for detecting the origin of solid ash was identified. The key conclusions are as follows:

1. The physicochemical properties of solid ashes from different sources vary, as reflected by the oxide mass fractions.
2. The identified correlation between oxides in solid ash provided a theoretical basis for clustering analysis to some extent.

3. Feature selection based on correlation can improve the clustering effect; however, there may be differences between different clustering algorithms.
4. The RI and ACC indicators of the k-means clustering algorithm fitted using all features are 0.71 and 0.58, respectively, indicating that this algorithm was the most suitable for detecting the solid ash source.

References

[1] M. Ganesapillai, et al., Sustainable recovery of plant essential nitrogen and phosphorus from human urine using industrial coal fly ash, Environmental Technology & Innovation 24 (2021) 101985.

[2] O. Karlström, et al., Role of ash on the NO formation during char oxidation of biomass, Fuel 190 (2017) 274–280.

[3] H. Khodaei, et al., Air staging strategies in biomass combustion-gaseous and particulate emission reduction potentials, Fuel Processing Technology 157 (2017) 29–41.

[4] S. Huang, et al., Robust deep k-means: an effective and simple method for data clustering, Pattern Recognition 117 (2021) 107996.

[5] M. Nabozny, et al., Method of aggregate production from power plant ashes by sintering in the shaft furnace: economic aspects, Energy Sources 22 (3) (2000) 227–233.

[6] R.V. Silva, et al., Use of municipal solid waste incineration bottom ashes in alkali-activated materials, ceramics and granular applications: a review, Waste Management 68 (2017) 207–220.

[7] Y. Zhang, et al., Treatment of municipal solid waste incineration fly ash: state-of-the-art technologies and future perspectives, Journal of Hazardous Materials 411 (2021) 125132.

[8] F. Huber, D. Laner, J. Fellner, Comparative life cycle assessment of MSWI fly ash treatment and disposal, Waste Management 73 (2018) 392–403.

[9] C.J. Lynn, R.K. Dhir, G.S. Ghataora, Environmental impacts of sewage sludge ash in construction: leaching assessment, Resources, Conservation and Recycling 136 (2018) 306–314.

[10] Z. Li, G. Xu, X. Shi, Reactivity of coal fly ash used in cementitious binder systems: a state-of-the-art overview, Fuel (2021) 301.

[11] K.M.A. Patel, P. Thakral, IEEE, The best clustering algorithms in data mining, in 2016 International Conference on Communication and Signal Processing (ICCSP), Vol. 1. 2016. p. 2042–2046.

[12] Y. Lai, et al. Scalable clustering for large high-dimensional data based on data summarization. in 2007 IEEE Symposium on Computational Intelligence and Data Mining, 2007.

[13] A. Muneeb, et al., Status of soil and water pollution at the largest coal mining area of Punjab, Pakistan, Fresenius Environmental Bulletin 30 (1) (2021) 441–447.

[14] C. Wei, Y. Sun, X. Hui. Clustering chain-type topology for wireless underground sensor networks. in 2010 8th World Congress on Intelligent Control and Automation, 2010.

[15] A. Smolinski, M. Stempin, N. Howaniec, Determination of rare earth elements in combustion ashes from selected Polish coal mines by wavelength dispersive x-ray fluorescence spectrometry, Spectrochimica Acta Part B-Atomic Spectroscopy 116 (2016) 63–74.

[16] M. Li, et al., Reutilisation of coal gangue and fly ash as underground backfill materials for surface subsidence control, Journal of Cleaner Production 254 (2020) 120113.

[17] S. Yin, et al., Effect of fly-ash as fine aggregate on the workability and mechanical properties of cemented paste backfill, Case Studies in Construction Materials 16 (2022) e01039.

[18] F. Najar, et al., A comparison between different gaussian-based mixture models. in 2017 IEEE/ACS 14th International Conference on Computer Systems and Applications (AICCSA), 2017.
[19] J.L. Solé, Book review: pattern recognition and machine learning, Cristopher M. Bishop. Information Science and Statistics, 2006, Springer, 2007, p. 738.
[20] X. Geng, et al., A novel hybrid clustering algorithm for microblog topic detection. 2017.
[21] C. Tomasini, et al., A study on the relationship between internal and external validity indices applied to partitioning and density-based clustering algorithms, in ICEIS: Proceedings of the 19th International Conference on Enterprise Information Systems – Vol. 1, 2017. p. 89–98.
[22] P. Roy, J.K. Mandal, Performance evaluation of some clustering indices, Computational Intelligence in Data Mining - Vol. 3, Springer India, New Delhi, 2015.
[23] H. Zhang, G. Yuan, G. Ma, Basic physicochemical characteristics of fly ash from one shanghai municipal solid waste incineration (MSWI) plantin Advanced Research on Material Engineering, Chemistry, Bioinformatics Ii (2012) 362–365.
[24] Z. Yang, et al., Encapsulated behavior and extraction ability of uranium in coal ash: a quantitative investigation with SiO2-Al2O3-Fe2O3-CaO system, Fuel (2020) 259.
[25] H.J. Lee, Development of application block using geobond and ash from sewage sludge incinerator II, Journal of Korean Society of Environmental Engineers 37 (7) (2015) 412–417.
[26] E. Ghafari, D. Feys, K. Khayat, Feasibility of using natural SCMs in concrete for infrastructure applications, Construction and Building Materials 127 (2016) 724–732.
[27] H.K. Shon, et al., Preparation of titanium oxide, iron oxide, and aluminium oxide from sludge generated from Ti-salt, Fe-salt and Al-salt flocculation of wastewater, Journal of Industrial and Engineering Chemistry 15 (5) (2009) 719–723.
[28] H.Y. Hu, et al., Sintering characteristics of CaO-rich municipal solid waste incineration fly ash through the addition of Si/Al-rich ash residues, Journal of Material Cycles and Waste Management 18 (2) (2016) 340–347.
[29] M.V. Tsvetkov, et al., Influence of sodium oxide on the fusion of solid municipal waste ash, Russian Journal of Physical Chemistry B 14 (4) (2020) 647–653.
[30] A. Gupta, S.A. Begum, A comparative study on feature selection techniques for multi-cluster text data, Harmony Search and Nature Inspired Optimization Algorithms, Springer Singapore, Singapore, 2019.
[31] X. Zhu, et al., A new unsupervised feature selection algorithm using similarity-based feature clustering, Computational Intelligence 35 (1) (2019) 2–22.

The accurate production forecast of solid ashes: application and comparison of machine learning techniques

9

Abstract

Predicting industrial solid ash production in advance is helpful to develop measures for solid ash management and recycling. In this chapter, coal fly ash was used as an example to illustrate the capability of machine learning (ML) for estimating solid ash production. To do so, three ML algorithms (random forest, support vector machine, and deep neural network) were employed, and their hyperparameters were tuned to improve their modeling accuracy. In addition, the importance of the input features was analyzed using the widely-used permutation importance and the Shapley Additional Explain methods, which provides a reference for optimizing the usage of coal fly ash.

9.1 Background

Accelerating industrialization has led to the rapid development of the energy sector. As the main energy source types for coal-fired power stations, coal and lignite combustion provides large amounts of electricity, and coal consumption has increased dramatically [1]. However, the onset of the COVID-19 pandemic curbed energy demand due to behavioral changes and economic slowdown, leading to small decreases in electricity demand and coal emissions in 2020. The subsequent introduction of government economic stimulus packages and widespread vaccination caused the global economy to rebound to some extent. According to statistics, coal-fired power generation increased by 9% in 2021, reaching its highest level in history [2]. By the first half of 2021, coal market consumption had increased 11% year-on-year [3]. Coal is likely to remain a mainstay in international energy in the short term.

The huge consumption of coal has led to an increase in the production of coal fly ash (CFA), a by-product resulting from the burning of coal. In India, over 10 years (2009–2010 and 2018–2019), CFA production in the energy sector has increased by almost 76% to around 217 million metric tons [4]. The massive accumulation of CFA not only occupies land resources [5] but also poses a significant environmental threat to soil, water, and air resources. Specifically, heavy rainfall causes water to come into contact with the CFA; this process releases toxic elements that can pollute the soil or produces toxic leachate that seeps into groundwater and pollutes water resources [6]. In addition, toxic elements may be discharged into the air through smoke generation, endangering air quality and, ultimately, even

Machine Learning Applications in Industrial Solid Ash. DOI: https://doi.org/10.1016/B978-0-443-15524-6.00010-8
© 2024 Elsevier Inc. All rights reserved.

human health [7]. Given this backdrop, CFA resource utilization has received increasing attention from both academia and industry.

In recent years, CFA usage has gradually increased in various fields, notably in the construction field where CFA is the most widely used. Due to its pozzolanic activity, CFA is used as a supplementary cementitious material to partially replace cement in concrete or to prepare geopolymers [8]. CFA can also be used as a coarse or fine aggregate for asphalt pavements [9], as well as in the manufacture of ceramic glass [10], metal matrix composites, and metal coatings. In addition, given the potential uses of CFA for soil improvement and heavy metal adsorption [11], CFA also has good application prospects in agriculture.

However, CFA production varies due to a range of factors; thus, from a waste management perspective, it is beneficial to predict CFA production in advance so that appropriate treatment steps can be selected for the recycling process. In this regard, previous studies have applied a range of approaches: for example, some researchers have used neural networks in MATLAB® and linear regression statistical analysis in IBM SPSS to predict CFA generation in power plants after 5 or 10 years [12]. The average annual hazardous waste production is often predicted by multiplying the amount of industrial hazardous waste generated in the base year by the average annual growth rate index [13]. However, neither approach is ideal; the aforementioned prediction method is too simple and general, whereas the index evaluation approach is too complex and time-consuming, and the prediction accuracy cannot be guaranteed in either approach.

Given the limitations of the aforementioned methods, three different ML models are constructed in this chapter, taking installed capacity and coal consumption as features and the CFA production as a target variable. Four indicators were used to evaluate the models' feasibility. The optimal robust model, as selected by this comparison, can predict rapidly and accurately CFA production. Applying it to engineering sites could save time when planning CFA disposal and improve CFA recovery.

9.2 Dataset

9.2.1 Data collection

Data is the foundation of ML modeling and gathering a comprehensive and appropriate dataset is key to constructing a good ML model. We obtained the dataset used in this chapter by searching domestic and international databases and relevant academic websites and consulting numerous CFA-related literature and academic reports. The dataset used here was extracted from a report documenting power generation and utilization of CFA in coal-fired power plants in India in 2019–2020 [14], comprising data from 183 power plants across 17 states, as shown in Fig. 9.1.

9.2.2 Dataset analysis

Different datasets have different feature distributions. Statistical and correlation analyses are helpful to establish the optimal algorithm model and better fit the data.

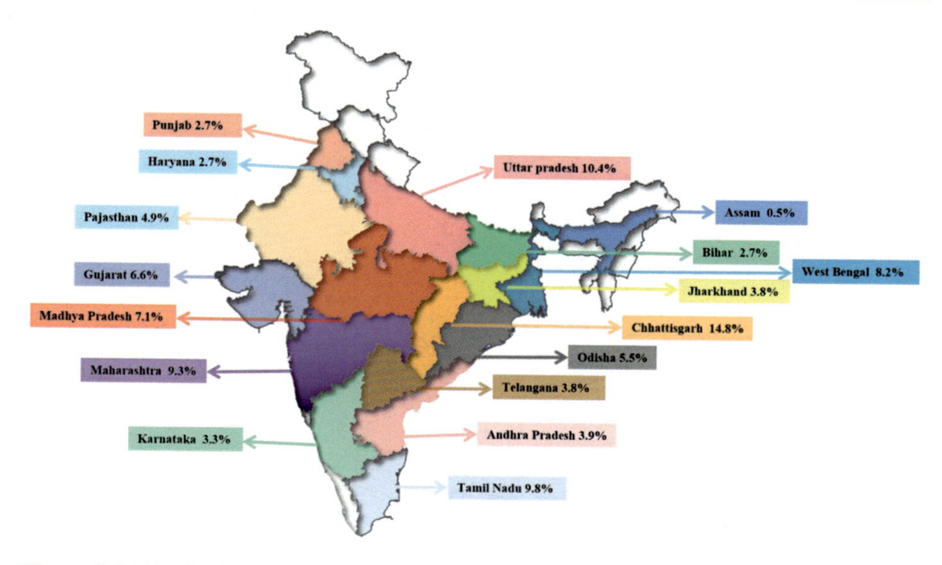

Figure 9.1 Distribution of data across India's 17 states [15].

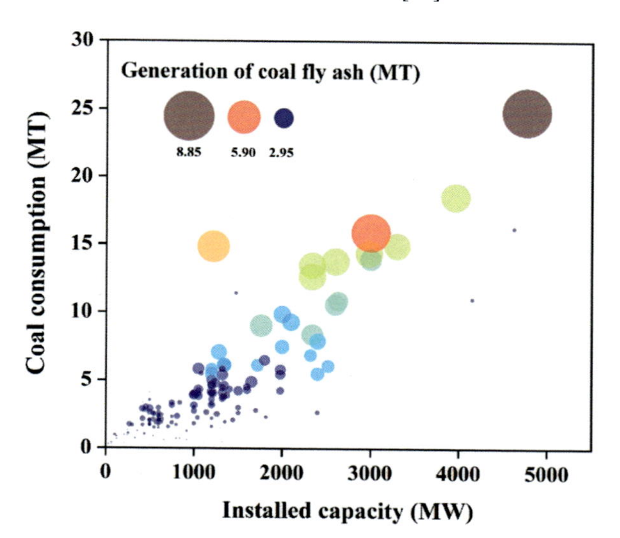

Figure 9.2 Statistical distribution of the dataset. Note that the size of the bubble represents the number of CFA produced. The brown bubble represents the maximum CFA production, and the purple bubbles represent the minimum production.

As shown in Fig. 9.2, the features and target variable of the dataset were presented and analyzed in the form of bubble graphs. The samples are mainly located in the lower-left corner of the figure. When the installed capacity varied from ~0 to 2000 MW and the coal consumption was ~0–5 MT, the CFA production was small (less than 2.95 MT). As the installed capacity and coal consumption

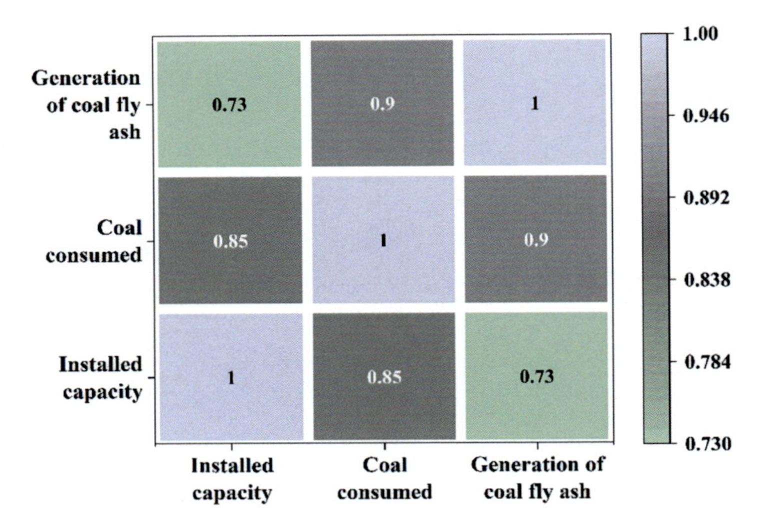

Figure 9.3 Correlation heat map. Values indicate Pearson correlation coefficients between variable pairs.

increased, so did CFA production. When the installed capacity was 4,760 MW, the CFA production reached its maximum value of 8.85 MT, and the corresponding coal consumption was around 25 MT.

The correlation between variables was further explored using Pearson correlation coefficients (R), as shown in Fig. 9.3. The R value between the coal consumption and CFA production was 0.9, higher than the R value of 0.73 identified between installed capacity and CFA production. This finding indirectly indicates that coal consumption has a greater impact on CFA production than installed capacity. In addition, the R value of 0.85 between coal consumption and installed capacity exceeds the threshold of 0.8, implying the existence of a strong correlation between these features.

9.3 Methodology

In this chapter, we used Python programming language to construct ML models with the help of the sklearn library [16]. Four evaluation indicators were used to measure the robustness of three different ML models. Finally, the optimal model was compared with the traditional production estimation method. The overall methodology workflow is illustrated in Fig. 9.4.

9.3.1 Dataset preprocessing and splitting

Feature dimension and unit transformation have a significant influence on data analysis and modeling performance. The variance ratio between the features of the

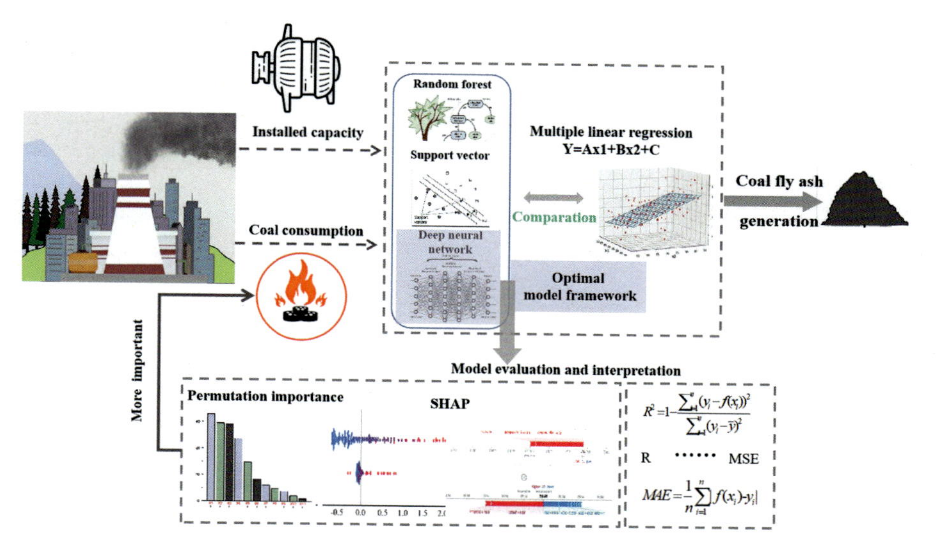

Figure 9.4 Complete diagram of methodology.

dataset in this chapter and the target variable is 200:4:2; thus, there are differences of several orders of magnitude between the variances. Feature-dominated algorithms exhibit poor modeling performance when applied to data with large variances. Accordingly, we standardized the data to eliminate dimensional effects.

After data standardization, we then divided the processed dataset into two parts: a training set and a testing set. To evaluate the impact of the dataset splitting ratio on the model's performance, we increased the size of the testing set from 10% to 45% with an interval of 5%, and R was used as the evaluation index to determine the optimal splitting ratio [17].

9.3.2 Model building and evaluation

Based on this data splitting, we then used three algorithms to build the model—random forest (RF), support vector machine (SVR), and deep neural network (DNN). Four indexes (R, R squared (R^2), mean-squared error (MSE), and mean absolute error (MAE)) were adopted to evaluate the model. A detailed explanation of ML algorithms and evaluation indexes is provided in Section 6.3.4.

To improve the model's performance, we further tuned the model hyperparameters. As DNN methods are greatly affected by their network structure and parameters [18], we used the trial-and-error method and tuned the model's hyperparameters according to recommendations in the literature [19,20]. We changed the neural network layers, learning rate, activation function, and epoch during model training and separated 10% of the data from the training set for performance validation. However, for the RF and SVR models, we used the corresponding default parameters in the sklearn integration module to ensure model performance.

During the hyperparameter tuning process, we adopted fivefold cross-validation to obtain more robust and reliable RF and SVR models. For DNN, we used the "validation_split" parameter in the Keras package to divide part of the data from the training set for model testing. Furthermore, as the result of a single random dataset split does not necessarily reflect the true performance of the model, we repeated the modeling and evaluation steps 50 times on the training and testing sets and took the average of the 50 repetitions as the final performance of the model.

9.4 Results and discussion

9.4.1 Determination of the dataset split ratio

As described earlier, to eliminate the effects of randomness, we performed 50 repeated evaluations for each dataset splitting and used the average R value to measure the model's overall performance for a specific splitting ratio. Taking RF as an example, we found that the dataset splitting ratio had little effect on performance on the training set, and the R value fluctuated slightly at around 0.98 (as shown in Fig. 9.5). In contrast, when the testing set size was 10% of the total dataset, the average R was 0.841. When the testing set size increased from 10% to 15%, the model's performance was optimal, with an average R value of 0.865. The analysis results for SVR and DNN were consistent with the above. Accordingly, a training set to testing set ratio of 0.85:0.15 was chosen as the optimal split ratio.

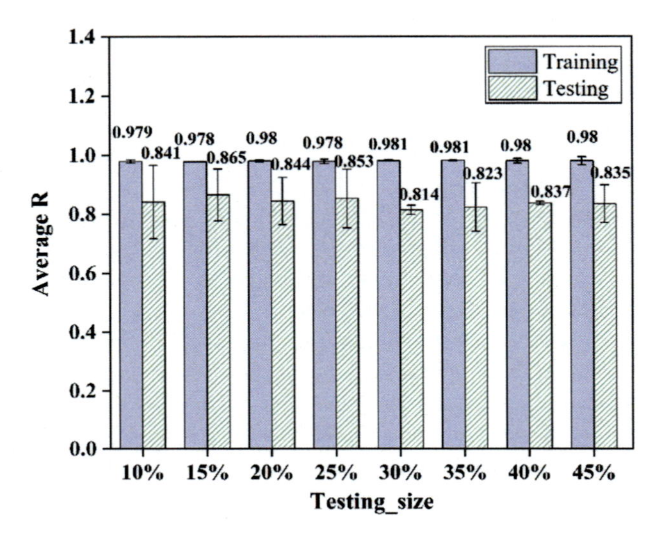

Figure 9.5 Performance of RF model under different dataset split ratios.

9.4.2 Optimal hyperparameters

In this chapter, RF and SVR, as traditional ML regression models, were trained using the default algorithm parameters in sklearn, as shown in Table 9.1. For DNN, we tuned the model's hyperparameters based on the loss values of the training set and validation set, as shown in Fig. 9.6. When epoch = 500, the loss tended to fluctuate steadily with increasing step values. The optimized DNN was then determined; as shown in Fig. 9.7, the optimal DNN model contained one input layer, five hidden layers, and one output layer, and the corresponding numbers of neurons were 2→8→32→64→16→8→1. In addition, two layers of "batch normalization" and one layer of "dropout" were included to speed up the convergence rate of the gradient-based method and prevent overfitting. Table 9.2 shows the optimal hyperparameters of the DNN model.

Table 9.1 Default hyperparameters for RF and SVR models.

RF		SVR	
Parameters	**Default value**	**Parameters**	**Default value**
n_ estimators	100	kernel	"rbf"
min_ samples_ split	2	degree	3
min_ samples_ leaf	1	gamma	scale
max_ features	"auto"	C	1
max_ depth	None	epsilon	0.1

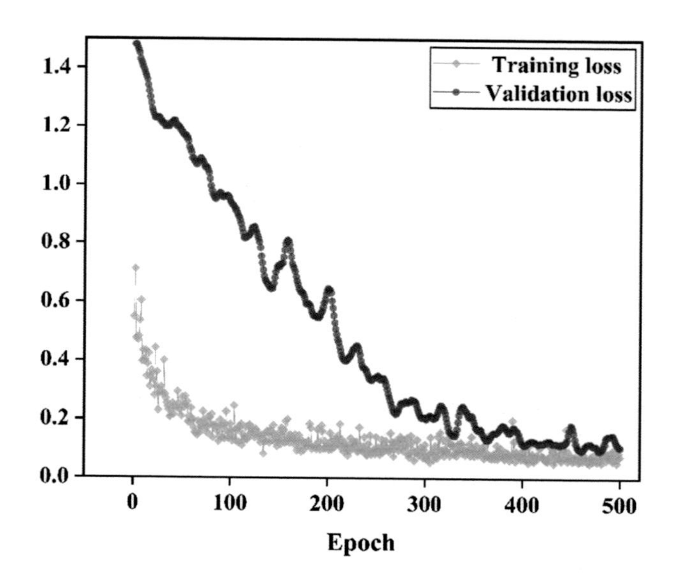

Figure 9.6 The loss during model training.

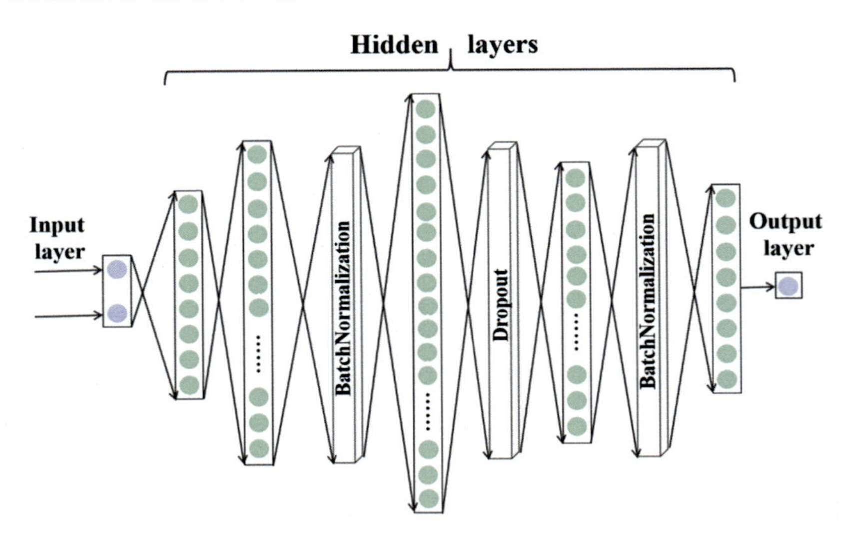

Figure 9.7 Topology structure of the optimal DNN model.

Table 9.2 Optimal hyperparameters of the DNN model.

Parameters	Option or value	Implication
Activation function	Relu	The output is no longer a linear combination of the inputs and can approximate any function.
Optimizer	Adam	A hybrid of the momentum gradient descent and RMSprop.
Learning rate	0.0005	Adjusts the weight of the neural network input.
Batch size	128	The number of samples used for training.
Epoch	500	One epoch is equal to training with all the samples in the training set.

9.4.3 Comparative analysis of model performance

The performance of RF, SVR, and DNN models was compared to select the optimal model for predicting CFA production. In Fig. 9.8, the linear fitting functions of the production as predicted by the RF model and the real production were $y = 0.878x + 0.134$ (training set) and $y = 0.861x + 0.094$ (testing set), while those of the DNN model were $y = 0.790 + 0.375$ (training set) and $y = 0.837x + 0.316$ (Testing set). All the data points were clustered around these lines, corresponding to high R and R^2 values and indicating good overall modeling performance. However, for SVR, the linear regression functions between the actual CFA generation of CFA and estimated model production were $y = 0.451x + 0.614$ and $y = 0.861x + 0.094$ on the training and testing sets, respectively. Relative to the RF and DNN models, the SVR model produced a relatively discrete data distribution for the training set and demonstrated slightly worse performance than the other two models.

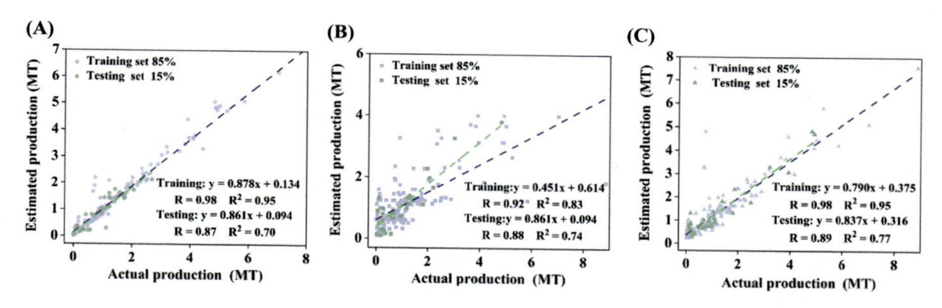

Figure 9.8 Plots illustrating the consistency of actual and predicted CFA production based on the (A) RF model, (B) SVR model, and (C) DNN model.

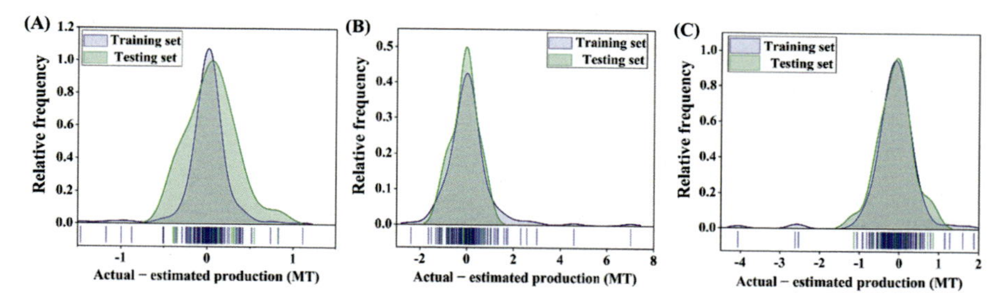

Figure 9.9 The relative frequency of difference values between the actual and estimated CFA production for the (A) RF model, (B) SVR model, and (C) DNN model.

As shown in Fig. 9.9, the differences between the actual CFA production and the estimated CFA production of the three models were small, and most of the data were concentrated around 0. However, the probability of data points appearing in the $[-0.1,0.1]$ interval in RF and DNN models was close to 0.9, whereas the equivalent value for SVR was only 0.45. In addition, for the testing set, the data points for the DNN model were more concentrated in the regions corresponding to smaller differences, thus implying that the DNN model was more robust.

Fig. 9.10 compares the performance of the three models more intuitively using four evaluation indicators. The R and R^2 values of the RF and DNN models on the training set were the same, 0.98 and 0.95, respectively, which were slightly higher than the R (0.92) and R^2 (0.83) values of the SVR model. Furthermore, the MSE and MAE values of the RF model were the smallest among the three models. In comparison, the RF model exhibited the lowest R and R^2 values on the testing set, which were 0.87 and 0.7, respectively. However, the R (0.89) and R^2 (0.77) values of the DNN model were the highest, with relatively low MSE and MAE values. Overall, the optimized DNN model was determined as the most suitable model framework for predicting CFA production in this chapter.

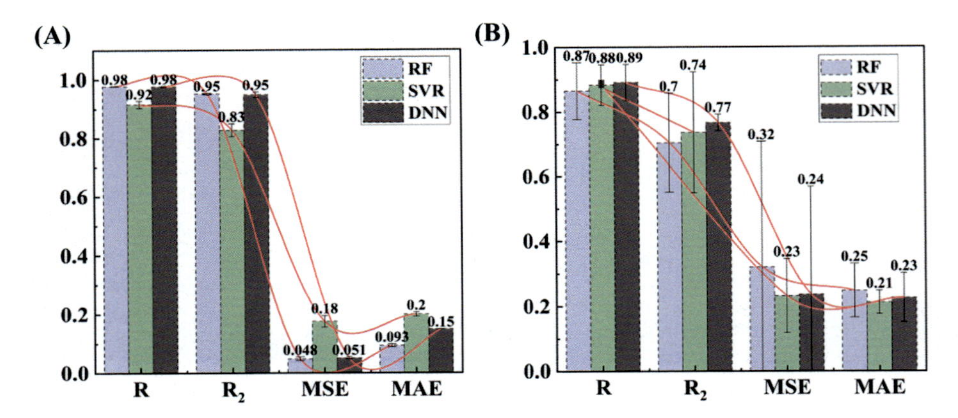

Figure 9.10 Plots showing the results of four indicators used to evaluate the performance of the three models on the (A) training and (B) testing sets.

9.5 Comparison with multiple linear regression

The multiple linear regression (MLR) approach is a traditional data analysis method that uses multiple independent variables to predict a dependent variable [21]. In this chapter, we used this method for statistical analysis; specifically, the dataset was repeatedly divided 50 times using the same ratio of 0.85:0.15, and the coefficients of the multiple linear regression equation $Y = Ax_1 + Bx_2 + C$ were established. The average results are shown in Table 9.3.

After 50 evaluations, the average R^2 and R values of the MLR model on the training set were 0.82 and 0.90, respectively, lower than the results of the three ML models tuned using fivefold cross-validation. The testing set data were then substituted into the MLR-derived equation for verification, resulting in average R and p values of 0.86 and 0.0000643805, respectively, indicating a high degree of agreement between the true values and estimated results. However, the R^2 value of 0.76 was higher than that of the RF and SVR models but lower than that of the DNN model, demonstrating that the DNN model is more suitable for CFA production prediction.

9.6 Feature importance analysis

In this section, we use the permutation importance (PI) metric provided by Python's eli5 library and "TreeExplainer" and "KernelExplainer" from the Shapley Additive Explanation (SHAP) library to analyze the importance of two features that affect CFA production. A detailed explanation of the PI and SHAP methods is provided in Section 6.4.1.

Table 9.3 The mean value of each parameter of MLR.

$Y = A \times 1 + B \times 2 + C$						
	Regression coefficient	**95% lower control limit (LCL)**	**95% upper control limit (UCL)**	**Standard error (SE)**	**t-test**	***p*-value**
Installed capacity (A)	−0.221283602	−4.6257086	−2.89796	2.5330302	−2.4167858	0.05265861
Coal consumption (B)	0.365218825	0.310336	0.404628	0.0228592	15.6097245	3.49436E-29
Constant (C)	0.173085366	0.1951878	0.318641	0.0759174	2.188839	0.041769567

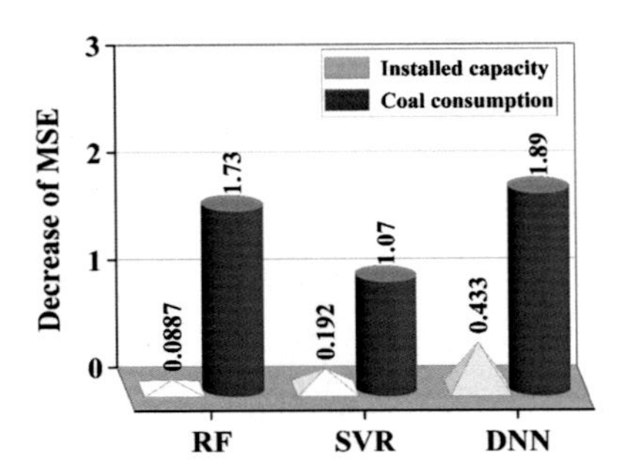

Figure 9.11 The feature importance of PI calculations.

9.6.1 Permutation importance

The model performance score decrease recorded after features are randomly rearranged determines how important the feature is [22]. Fig. 9.11 shows the PI results. After random shuffling of the coal consumption value, the MSE values of the RF, SVR, and DNN algorithms decreased by 1.73, 1.07, and 1.89, respectively. These values were generally higher than the MSE decrease values recorded when the installed capacity was randomly shuffled. All three ML models considered coal consumption as more important than installed capacity on CFA production.

9.6.2 SHAP

To further understand how features affect the model output, we used SHAP to analyze the model from global, local, and interaction perspectives. In Fig. 9.12A, the features are arranged from top to bottom on the y-axis according to their importance [23]. This indicates that coal consumption had a greater impact on the model, which was consistent with the PI results. Furthermore, for the three ML models, the higher the coal consumption, the higher the predicted CFA production. However, for installed capacity, the results were different: in the RF and SVR models, higher installed capacity values increased the predicted CFA production, but the opposite trend is recorded in the DNN model.

Fig. 9.12B used the first sample after preprocessing (feature values of 0.8059 and 1.363) as an example to illustrate the impact of features on a single prediction. The red bars represent the range in which a feature played a positive role in the model's predictions [24]. The base value was the average value of each sample's target variable, and $f(x)$ was the final predicted value of the sample, satisfying the condition $f(x) = \text{base value} + \sum\text{SHAP}$. Due to various factors, such as the modeling mechanism and data distribution, the prediction results of the three models for the same sample differ slightly. Both the coal consumption and installed capacity played a positive driving role in model prediction; however, coal consumption had a greater impact on the model prediction.

In addition to the global and local interpretation methods, Fig. 9.12 c reveals hidden relationships between features through feature interactions. When the coal consumption interacted with the installed capacity, a positive correlation with CFA

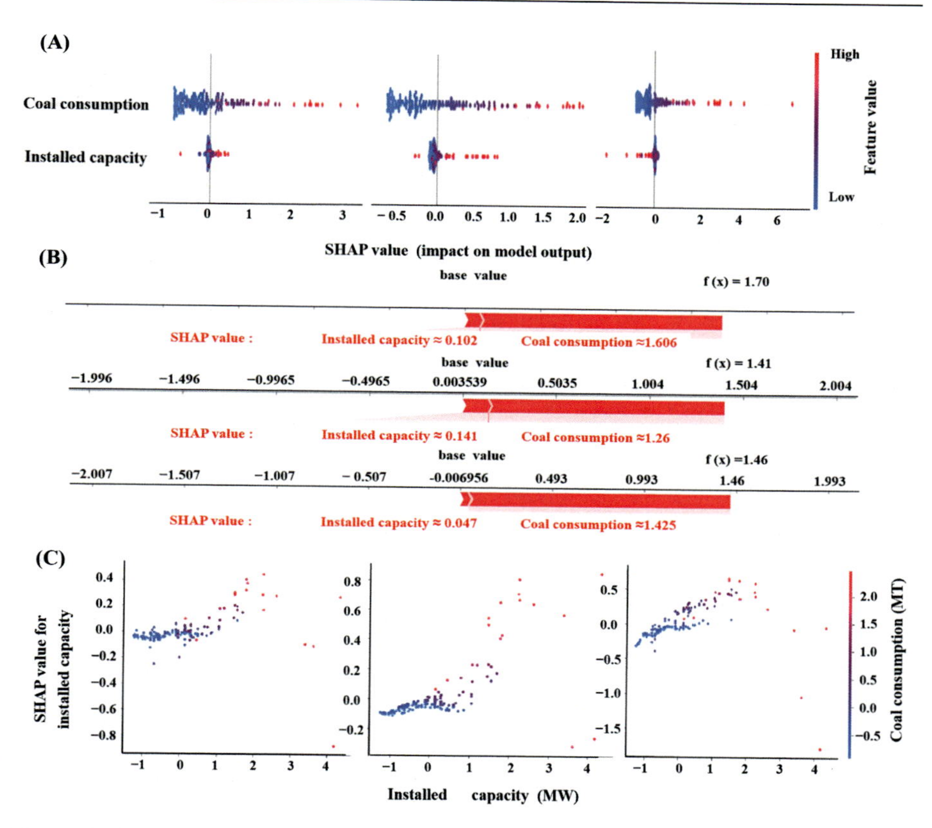

Figure 9.12 Feature importance from three perspectives using SHAP [15]: (A) global (B) local, and (C) interaction. Note that the RF, SVR, and DNN models are shown from left to right in (A) and (C) and top to bottom in (B).

production was generally recorded. When both the coal consumption and installed capacity were high, except for a few outliers, the installed capacity value had a greater impact on CFA production. In contrast, when the coal consumption and installed capacity values were relatively low, the installed capacity made little contribution to changes in the model output and even hindered prediction.

As described earlier, the impact of the installed capacity on CFA production was not always positive compared with coal consumption. In practice, the installed capacity is the designed capacity of a specific power station, and the actual capacity is affected by many external factors including coal production, the energy market, and policies. As a result, the correlation between the installed capacity and CFA production was not as strong as that between coal consumption and CFA production. In addition, due to the influence of the aforementioned factors, a power station with a high installed capacity may generate a relatively small amount of electricity, resulting in less CFA production. Models based on datasets affected by these factors may exhibit a negative impact of installed capacity on certain data samples.

9.7 Significance and outlook

High energy consumption results in increased production of solid waste such as CFA, with serious impacts on the environment and human health. The continued evolution of the concept of sustainable development has enabled more solid waste types, such as CFA, to be recycled [25]. However, the variability of CFA production can hinder the planning and design process for effective CFA disposal and utilization. Given this issue, an ML modeling framework was constructed to quickly and accurately predict CFA production based on input coal consumption and installed capacity alone. By applying this modeling framework to engineering scenarios, managers can predict CFA production in advance, which can guide the disposal and recycling of CFA, thereby maximizing its value. However, due to the small dataset size and few input features, the universality and persuasiveness of this model must be further improved. Future studies must consider other factors that influence CFA production and apply a more comprehensive dataset.

9.8 Summary

In this chapter, we constructed and evaluated three ML models, compared them with traditional methods, and determined that DNN was the optimal model for accurate CFA production prediction. Furthermore, we analyzed the feature importance from various perspectives, which provides a reference for promoting the rational use of CFA. Specific conclusions are as follows:

1. The R and R^2 values of the DNN model applied to the training set were 0.98 and 0.95, respectively, while those for the test set were 0.89 and 0.77, respectively, showing more robust performance compared with RF and SVR models.
2. The R^2 value of the traditional MLR model on the testing set was 0.76, which was higher than those of the RF and SVR models but lower than that of the DNN model.
3. Both PI and SHAP methods indicated that coal consumption had a greater positive effect than installed capacity on CFA production.

References

[1] S.Z. Naqvi, J. Ramkumar, K.K. Kar, Coal-based fly ash, in: K.K. Kar (Ed.), Handbook of Fly Ash, Butterworth-Heinemann, Oxford, UK, 2022, pp. 3−33.

[2] IEA. Electricity Market Report; IEA: Paris, France, 2022; Available online:https://www.iea.org/reports/electricity-market-report-january-2022.

[3] IEA (2021), Global Energy Review 2021, IEA, Paris https://www.iea.org/reports/global-energy-review-2021, License: CC BY 4.0.

[4] S. Arora, An Ashen Legacy: India's Thermal Power Ash Mismanagement, Centre for Science and Environment, New Delhi, India, 2020.

[5] U. Blaha, et al., Micro-scale grain-size analysis and magnetic properties of coal-fired power plant fly ash and its relevance for environmental magnetic pollution studies, Atmospheric Environment 42 (36) (2008) 8359–8370.
[6] A. Chowdhury, A. Naz, A. Chowdhury, Waste to resource: Applicability of fly ash as landfill geoliner to control ground water pollution, Materials Today: Proceedings 60 (2022) 8–13.
[7] A. Jiang, J. Zhao, Experimental study of desulfurized fly ash used for cement admixture. Proceedings of Civil Engineering in China–Current Practice and Research Report; Hindawi: Hebei, China, 2010: p. 1038–1042.
[8] R. Ragipani, et al., Selective sulfur removal from semi-dry flue gas desulfurization coal fly ash for concrete and carbon dioxide capture applications, Waste Management 121 (2021) 117–126.
[9] S. Shanmugan, et al., Enhancing the use of coal-fly ash in coarse aggregates concrete, Materials Today: Proceedings 30 (2020) 174–182.
[10] M. Zhu, et al., Preparation of glass ceramic foams for thermal insulation applications from coal fly ash and waste glass, Construction and Building Materials 112 (2016) 398–405.
[11] A. Kotelnikova, et al., Assessment of the structure, composition, and agrochemical properties of fly ash and ash-and-slug waste from coal-fired power plants for their possible use as soil ameliorants, Journal of Cleaner Production 333 (2022) 130088.
[12] N. Zahari, et al. Study on prediction fly ash generation using statistical method. in AIP Conference Proceedings, AIP Publishing LLC, 2018.
[13] I.M.W. Widyarsana, S.A. Tambunan, A.A. Mulyadi, Identification of Fly Ash and Bottom Ash (FABA) Hazardous Waste Generation From the Industrial Sector and Its Reduction Management in Indonesia, 2021.
[14] CEA, Report on Fly Ash Generation at Coal/Lignite Based Thermal Power Stations and Its Utilization in the Country for the Year 2019–2020, CEA New Delhi, India, 2020.
[15] C. Qi, et al., Comparison and determination of optimal machine learning model for predicting generation of coal fly ash, Crystals 12 (4) (2022) 556.
[16] F. Pedregosa, et al., Scikit-learn: machine learning in python, The Journal of Machine Learning Research 12 (2011) 2825–2830.
[17] C. Qi, et al., Rapid identification of reactivity for the efficient recycling of coal fly ash: hybrid machine learning modeling and interpretation, Journal of Cleaner Production 343 (2022) 130958.
[18] T. Shinozaki, S. Watanabe, Structure discovery of deep neural network based on evolutionary algorithms. in 2015 IEEE International Conference on Acoustics, Speech and Signal Processing (ICASSP), 2015.
[19] D. Beniaguev, I. Segev, M. London, Single cortical neurons as deep artificial neural networks, Neuron 109 (17) (2021) 2727–2739.e3.
[20] K.K. Panchagnula, N.V.K. Jasti, J.S. Panchagnula, Prediction of drilling induced delamination and circularity deviation in GFRP nanocomposites using deep neural network, Materials Today: Proceedings (2022).
[21] G. Wang, et al. Comparison between BP neural network and multiple linear regression method. in International Conference on Information Computing and Applications, Springer, 2010.
[22] N. Afanador, T. Tran, L. Buydens, Use of the bootstrap and permutation methods for a more robust variable importance in the projection metric for partial least squares regression, Analytica Chimica Acta 768 (2013) 49–56.

[23] J. Peng, et al., An explainable artificial intelligence framework for the deterioration risk prediction of hepatitis patients, Journal of Medical Systems 45 (5) (2021) 1–9.
[24] Y. Chen, et al., Rapid mechanical evaluation of the engine hood based on machine learning, Journal of the Brazilian Society of Mechanical Sciences and Engineering 43 (7) (2021) 1–17.
[25] C. Qi, X. Xu, Q. Chen, Hydration reactivity difference between dicalcium silicate and tricalcium silicate revealed from structural and bader charge analysis, International Journal of Minerals, Metallurgy and Materials 29 (2) (2022) 335–344.

FIELD: fast mobility evaluation and environmental index for solid ashes with machine learning

10

Abstract

Development of the combustion-based energy sector has resulted in a large amount of industrial solid ash waste. Due to the chemical composition of the raw materials, trace elements (TEs) in industrial solid ash may be released, and their potential leaching poses a severe environmental threat. How to evaluate the mobility of TEs in solid ash and assess their environmental pollution potential is a major concern in solid ash management. In this chapter, we introduced a comprehensive framework for TEs, called the Fast mobility evaluation and environmentaL inDex (FIELD), based on sequential extraction and machine learning (ML) approaches. By applying the FIELD framework to coal fly ash, we showed that the proposed framework can accurately predict TE fractions and improve environmental evaluation. Future application of the FIELD framework to real-world projects can reduce the time required for environmental assessment of TEs in industrial solid ash; thus this approach has important practical utility for solid ash management and recycling.

10.1 Background

As discussed in previous chapters, the accumulation and improper handling of industrial solid ash can cause serious air, soil, and water hazards. Therefore, solid ash is widely recycled to promote sustainable development. To date, solid ash has been extensively applied as a supplementary cementitious material because of its specific cementitious reactivity. In addition, it is also used to make ceramic glass, as a base material for roads, as coarse and fine aggregates, etc. [1].

Irrespective of whether industrial solid ash is recycled or dumped for disposal, trace elements (TEs) in the solid ash may be released into the environment when it comes into contact with water [2]. It has been reported that the concentrations of TEs in solid ash may be 4–10 times higher than their original concentrations in raw materials;[3] thus solid ash can cause widespread environmental pollution and human health problems if not properly treated. Accordingly, evaluating the mobility and environmental risk of TEs in solid ash is essential to formulate appropriate disposal strategies. Fig. 10.1 shows the potential leaching and environmental risks of industrial solid ash.

Over the past few decades, the migration capacity of TEs has been shown to depend not only on the TEs' total concentration but also on their specific chemical forms and binding states [4], which can be determined using sequential extraction methods. Of these, the Tessier [5] and Bureau Commune de Reference of the

Machine Learning Applications in Industrial Solid Ash. DOI: https://doi.org/10.1016/B978-0-443-15524-6.00004-2
© 2024 Elsevier Inc. All rights reserved.

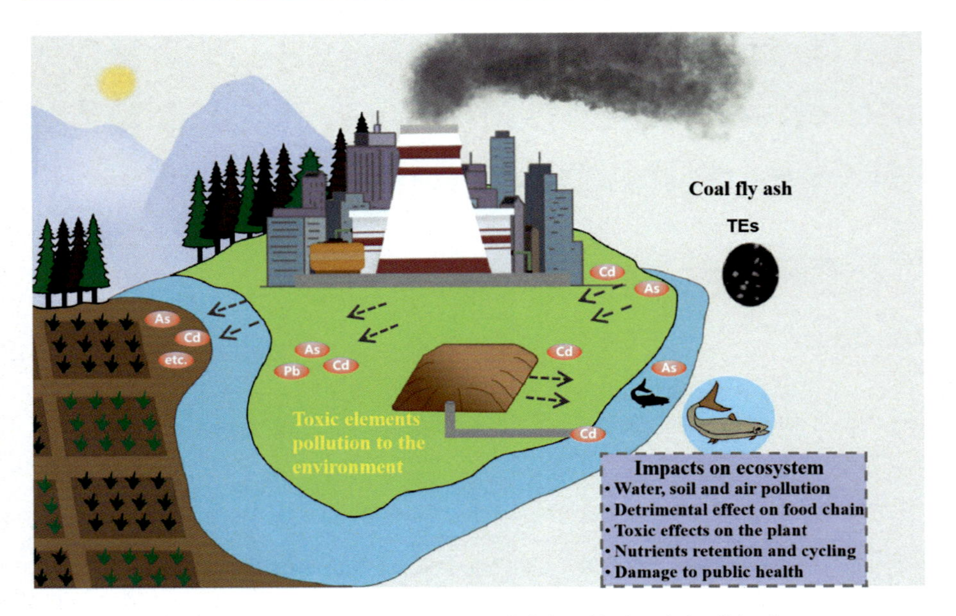

Figure 10.1 Potential leaching and environmental risks of industrial solid ash.

European Commission (BCR) [6] methods are the most widely used. Many researchers have used these methods to explore the leaching hazards of TEs in solid ash. For example, Yuan et al. [7] collected two coal fly ash (CFA) samples from a coal-fired thermal power plant located in northern China and developed an improved sequential extraction procedure to characterize the TE types in the CFA. Sočo et al. [8] used a five-step sequential extraction method to quantitatively evaluate the chemical components of copper (Cu) and zinc (Zn) in CFA in order to characterize the migration ability of metals under ambient conditions. Tian et al. [9] explored the leaching behavior of TEs such as boron (B), phosphorus (P), vanadium (V), chromium (Cr), arsenic (As), selenium (Se), molybdenum (Mo), antimony (Sb), and tungsten (W) in CFA using the sequential extraction method; the relationship between TEs in CFA was then discussed based on leaching and X-ray absorption near-edge structure results.

In addition, environmental risk assessment provides an intuitive comparison of the potential leaching risk of different industrial solid ashes (CFA, etc.). To date, the pollution load index and ecological risk index have been used to assess the environmental risks of solid ash. The pollution load index is calculated as the n-root of the product of the concentration factor for n types of TEs, where the concentration factor is the ratio of the TE's concentration in solid ash and the background value of that TE in the environment [10]. In addition to this approach, the ecological risk index also considers a trace response factor that differentiates the migration of different TEs [11].

However, the experimental sequential extraction process is time-consuming, labor-intensive, and requires highly specialized techniques and equipment. Furthermore, traditional environmental risk assessment methods only consider the total concentration of

TEs rather than TE fractions in solid ash, which may affect the effectiveness of the assessment. Given these limitations, we propose a synthesis framework called the **F**ast mob**I**lity **E**valuation and environmenta**L** in**D**ex (FIELD) in this chapter. In FIELD, ML technology was used to accurately predict TE fractions in real time based on physical and chemical properties, elemental properties, and the total TE concentrations of solid waste. In addition, considering the total concentration and TE fractions, further new environmental assessment indicators were proposed.

10.2 Methodology

The FIELD framework comprises the following four main components: (1) data collection and preprocessing; (2) deep neural network (DNN) modeling to predict TE fractions; (3) new environmental assessment indicators; and (4) black-box model interpretation. The details of these stages are described in the following sections.

10.2.1 Data collection and preprocessing

Collecting reliable data is a prerequisite for constructing ML models. Such data can be obtained from extensive laboratory experiments or literature reviews. However, for experimental data, the experimental process must be explicitly recorded and reproducible. For literature review-based data collection, careful integration of the data is required as the sequential extraction methods that were originally used may vary. The physicochemical properties of industrial solid ash, the total concentration of TEs, and the elemental properties of TEs are the main factors affecting TE fractions [12] and are taken as input features in this chapter. The percentage of each TE fraction is the target variable.

To better represent the mapping relationship between inputs and outputs, the data can be processed using one-hot encoding. In addition, dimensional processing methods such as standardization and normalization can also be used to improve the model's performance.

10.2.2 Deep neural network modeling

The DNN method is a widely used perceptron-based deep learning technique whose network layers are divided into three types: input layer, hidden layer, and output layer [13]. Unlike the perceptron approach, DNN has strong nonlinear fitting ability and can handle complex nonlinear problems efficiently [14]. Given the flexible structure and high degree of freedom of DNN, we used DNN for model construction in this chapter.

DNN modeling first requires the optimal dataset split ratio and the number of dataset split repeats required to achieve model convergence to be determined. The model is then trained and tuned for optimal performance. Finally, multiple evaluation indicators (correlation coefficient (R), coefficient of determination

(R^2), root mean squared error (RMSE), mean absolute error (MAE), mean squared error (MSE), etc.) must be used to comprehensively verify the optimal DNN model's generalization ability.

10.2.3 New environmental assessment indicators

It is widely accepted that the different TE fractions obtained from sequential extraction represent different mobility of TEs. Thus TE mobility could be entirely different in the presence of different TE fractions in solid ash, even if the total concentration is the same. Thus accounting for different TE fractions is essential to achieve an accurate mobility evaluation. In FIELD, we propose novel leaching risk assessment indexes for a single TE (EIS) and all TEs in solid ash (EIT) based on the total concentration and TE fractions, as shown in Eqs. (10.1) and (10.2).

$$\text{EIS} = \frac{\left(\sum_{i=1}^{i=n} F_i \times D_i\right) \times C_t}{C_b} \tag{10.1}$$

$$\text{EIT} = \sum_{j=1}^{j=m} EIS_j \times T_j \tag{10.2}$$

where EIS is the environmental assessment index of a single TE type, and EIT is the total environmental assessment index of all TE types. n represents the total number of TE occurrence states identified using sequential extraction; F_i represents the percentage of the *i-th* fraction; D_i represents the mobility factor for the *i-th* fraction; C_t is the total concentration of a TE in solid ash, and C_b is the background concentration value of the TE; m represents the m types of TEs observed in solid ash; and T_j is the trace response factor for the *j-th* type of TE and can be obtained from values published in previous studies. Note that the D_i term represents the migration and bioavailability of TEs of the *i-th* fraction; the value of this parameter varies from 0 to 1 and can be determined using expert knowledge and/or the analytical hierarchy process. When more representative or reasonable D_i values are obtained, the corresponding EIS of the TEs can be updated at any time.

10.2.4 Black-box model interpretation

In addition to accurate model predictions, it is essential to perform model interpretation to obtain new knowledge about potential patterns between inputs and outputs. In this chapter, we used the SHAP and permutation importance (PI) approaches to analyze the importance of the DNN model's input features from both global and local perspectives. Given that different variables have varied influences on different outputs, we also explored the feature importance under different TE fractions. For more detailed explanations of the SHAP and PI methods, see Chapter 7.

10.3 The application of FIELD to coal fly ash

10.3.1 Dataset analysis

The CFA dataset used in this chapter comes from published literature [9] [15] and contains a total of 400 samples for eight TE types (B, V, Cr, As, Se, Mo, Sb, and W). Among these samples, the physiochemical properties of CFA (16 variables), the elemental properties of TEs (32 variables), the total concentration of TEs (one variable), and the TE fraction types (five variables with one-hot coding) are feature variables. The percentages of each TE fraction (water-extractable, acid-soluble, reducible, oxidizable, and residual fractions) are the target variables, as shown in Fig. 10.2.

To improve the model's performance, we standardized the features using the sklearn package's preprocessing. scale function in Python and performed descriptive statistical analysis on the input and output variables, as shown in Figs. 10.3 and 10.4. The maximum and minimum values of the output variables were 100 and 0, respectively. The median was one-fifth of the maximum value, indicating that under different TE fractions, the percentage differences were significant but biased toward smaller values. The data points of the physiochemical properties of the CFA were evenly distributed; however, there were more outliers for the total concentrations of the eight TEs, indicating that the data were scattered.

10.3.2 Dataset splitting and repetition

Because of the significant influence of the dataset split ratio on model performance, based on suggestions in the literature, we divided the dataset into three parts in a

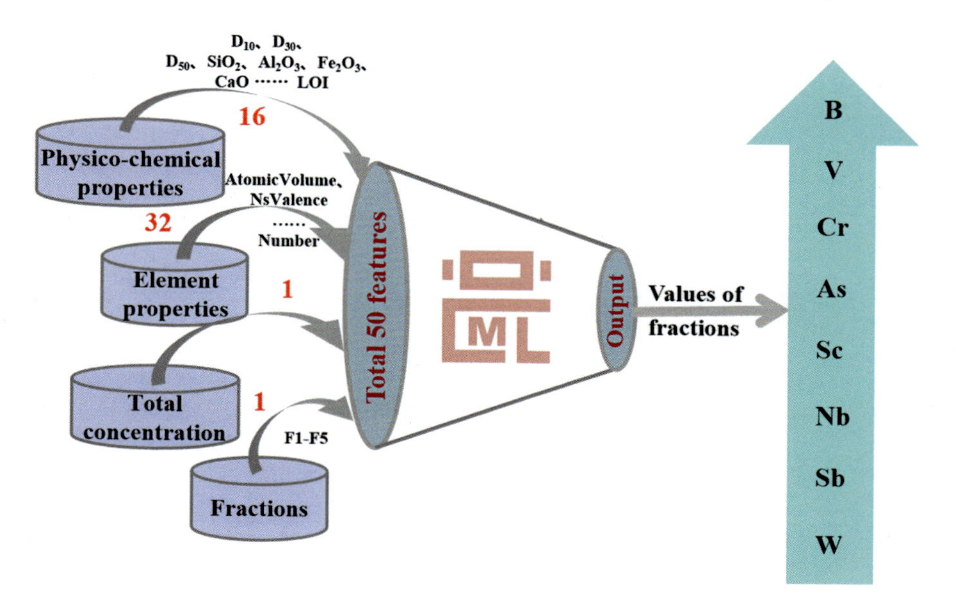

Figure 10.2 Input and output variables for the dataset.

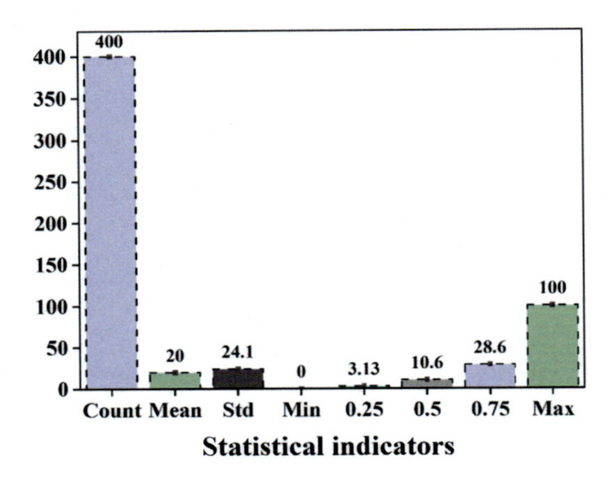

Figure 10.3 Descriptive statistical analysis of output variables.

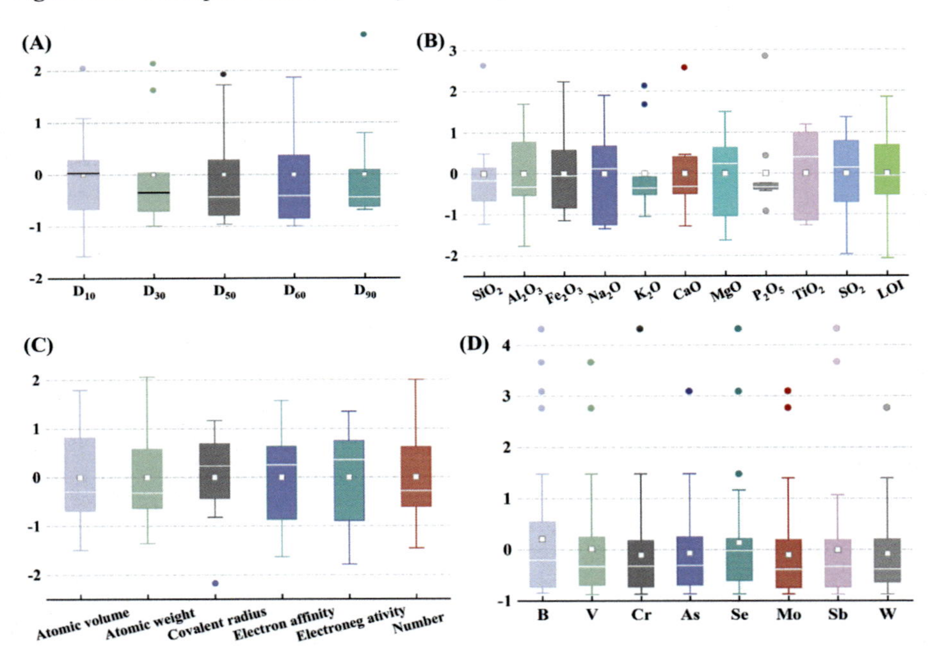

Figure 10.4 Statistical analysis of the features of the CFA dataset: (A) physical properties of CFA, (B) chemical properties of CFA, (C) element properties of TEs, and (D) total TE concentrations.

ratio of 6:2:2: [16] 60% of the data was used for DNN model training, 20% for model validation, and the remaining 20% for generalization testing.

To reduce the impact of randomness, we repeated the dataset splitting process multiple times. As shown in Fig. 10.5, to achieve a balance between model convergence and computing resource usage, the model was trained and evaluated 20 times, and the mean value of the evaluation index was taken as the final performance of the model. Note that this process was also repeated 20 times for each subsequent problem involving dataset splitting.

10.3.3 Hyperparameter tuning

The architecture and hyperparameters of a DNN model have a huge impact on its performance. Accordingly, we used trial and error and tuning approaches based on the recommendations of published literature. After determining a set of hyperparameters, we adjusted the model by observing the loss values of the training and validation sets in the hyperparameters tuning process (Fig. 10.6). When the optimal parameters had been determined for the DNN model (Table 10.1), the problem of overfitting was effectively controlled (Fig. 10.6).

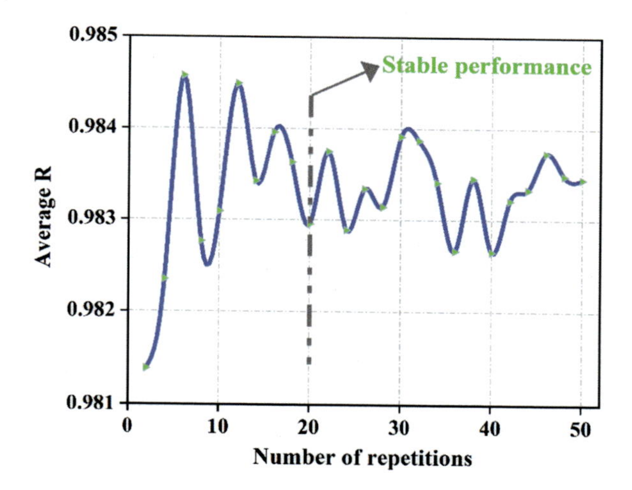

Figure 10.5 Number of repetitions of model convergence.

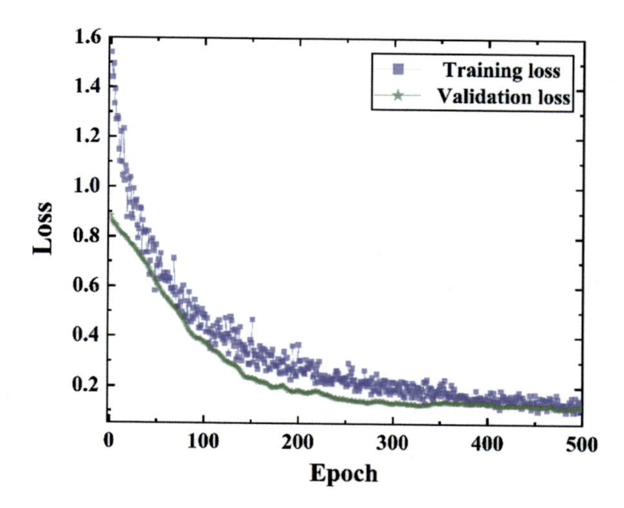

Figure 10.6 The loss values of the training and validation sets during model training.

Table 10.1 The optimal DNN architecture and hyperparameter values.

Parameter	Value	Implication	Reference
Optimizer	Adam	The first-order optimization algorithm replaces the traditional stochastic gradient descent method	Wang and Yao [17]
Activation	ReLu	Realizes the nonlinearity between neuron inputs and outputs	
Hidden layer neurons	150–200–100-50	Basic computing unit	Liu, et al. [18]
Dropout	0.25–0.35–0.25	Neurons are discarded according to a certain probability to prevent overfitting	Phornchaicharoen and Padungweang [19]
Batch_size	128	The number of data samples captured in training	Pyrkov, et al. [20]
Epochs	500	Number of input repetitions of all the data into the network for complete forward calculation and back propagation	
Learning rate	0.0005	Adjusts the input weight	Shokr, et al. [21]

10.3.4 Performance evaluation

Fig. 10.7 shows the average performance of the model on three subsets. The mean R values of the training, validation, and testing sets were 0.98,0.91, and 0.92, respectively. The corresponding mean R^2 values were 0.97, 0.82, and 0.83, and the mean MSE and MAE values were all below 0.3. These values collectively indicate that the optimal DNN model had good robustness and generalization ability.

Fig. 10.8 further compares the consistency between the estimated TE fractions and true TE fractions. In Fig. 10.8A, the linear fitting equations of the data in the training, validation, and testing sets were $y = 0.92776x + 2.16915$, $y = 0.60725 + 6.54343$, and $y = 0.73018x + 1.6926$, respectively, with all corresponding R^2 values above 0.8. In

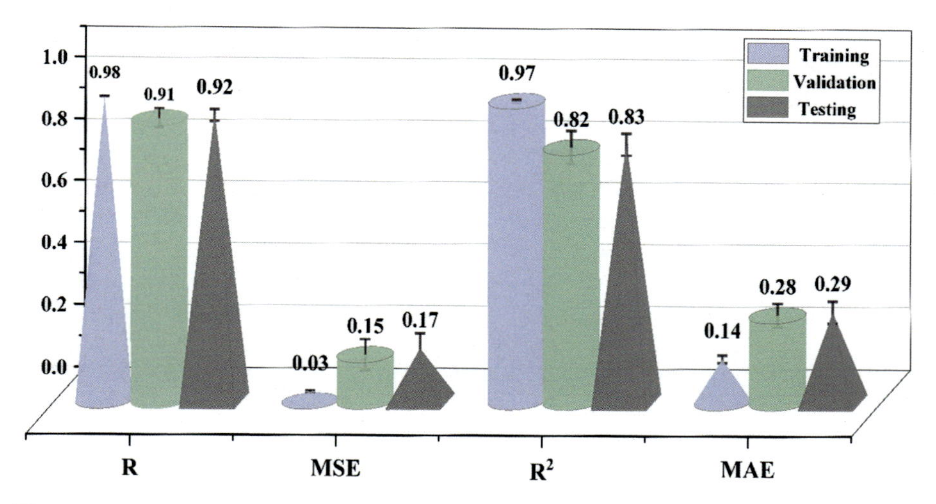

Figure 10.7 Model performance evaluation using four indicators.

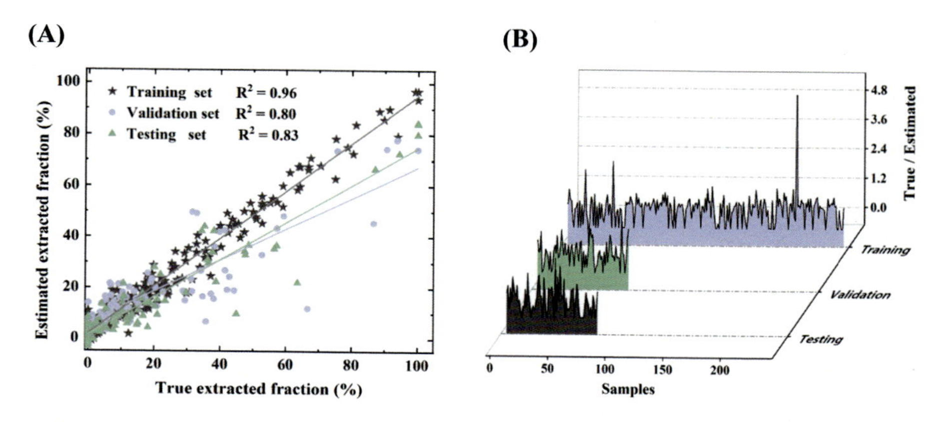

Figure 10.8 Comparison of the true and estimated TEs fractions: (A) linear fitting and (B) true/estimated fractions by set.

Fig. 10.8B, except for a few outliers, the ratios of true TE fractions to estimated TE fractions were generally close to 1. The earlier analyses all indicate that the true TE fraction values were highly consistent with the estimated values from the model.

10.3.5 Environmental index of trace elements

Based on the total concentration of TEs and the fraction mobility, we proposed two new environmental assessment indexes: EIS and EIT. The mobility and bioavailability of TEs in the five fractions in sequential extraction order show a gradually decreasing trend; [22] therefore, the mobility factor D_i values of the water-extractable, acid-soluble, reducible, oxidizable, and residual fractions were set as 1, 0.8, 0.6, 0.4, and 0.2. The background concentration C_b values of B, V, Cr, As, Se, Mo, Sb, and W were chosen as 46.9, 66, 40, 9, 0.51, 1.11, 0.68, and 3.34 mg kg^{-1}, based on suggestions from the literature [23].

As shown in Fig. 10.9A, the EIS values were generally smaller than the values of the traditional environmental assessment method C_t/C_b because $\sum_{i=1}^{i=n} F_i \times D_i \leq 1$. Furthermore, the most hazardous elements were considered differently by two assessment methods. Based on the traditional C_t/C_b method, Sb was considered the most hazardous, because higher Sb concentrations promote leaching into the environment. In contrast, the EIS method considers Se to be the most hazardous element.

Further analysis of the first YAN CFA sample showed that the total concentrations of Sb and Se were 66.67 mg kg^{-1} and 22.57 mg kg^{-1} compared with background concentration values of 0.68 mg kg^{-1} and 0.51 mg kg^{-1}, respectively. According to traditional evaluation methods, the relative accumulation of Sb was greater than that of Se. However, the water-extractable, acid-soluble, reducible, oxidizable, and residual fractions of Sb in the YAN sample represented 3.03%, 5.63%, 0%, 0%, and 91.34%, respectively. Thus more than 90% of Sb is in the most stable residual state. In contrast, in this sample, most Se was contained in the more mobile fractions. Specifically, the water-extractable fraction contained 24.24%, while the corresponding values for the acid-soluble, reducible, and oxidizable fractions were 23.36%, 12.45%, and 39.95%. Xiong et al. [24]. found that 70% of Se in CFA would be removed under strongly acidic conditions ($pH < 1$). With increasing pH, the low proton level under alkaline conditions promotes the oxidation of Se, thereby promoting the dissolution of Se zero-valent oxyanions into the leachate. Thus the high mobility of Se makes it a more serious environmental threat than Sb, as evidenced by the EIS value for this element. The same phenomenon was also observed when comparing B and As in the YAN samples.

Based on our literature review, we found that T_j values for Hg, Cd, As, Cu, Pb, Ni, Cr, V, Zn, and Mn were available. Using the TEs in this chapter, EIT calculations were performed for As, Cr, and V by taking their T_j values 10, 2, and 2, respectively. As shown in Fig. 10.9B, according to the traditional $\sum(\frac{C_{ti}}{C_{di}}) * T_{ji}$ measure, the CFA samples with the top five environmental risk rankings are YEM, PAN, ZHA, SHU, and YAN. Based on the EIT approach, the PAN sample had the highest environmental risk, followed by IND, FAR, ZHA, and JAP. The environmental risk levels of the CFA samples obtained by the two methods differed significantly as the traditional evaluation methods consider only the total TE concentrations and ignore the chemical fractions or mobility.

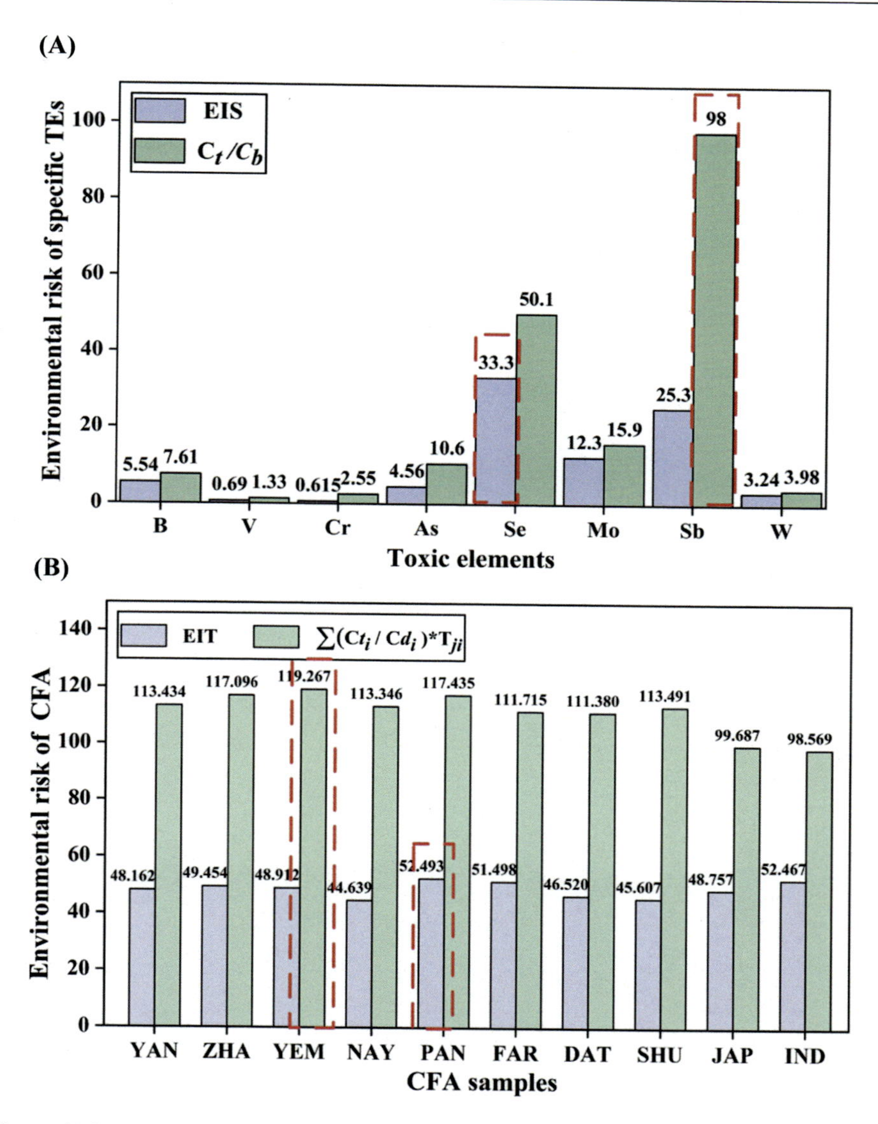

Figure 10.9 Plots comparing the novel environmental indicators and conventional evaluation method (A) on the first CFA sample and (B) on all CFA samples. In Fig. 10.9B, the abbreviations on the *x*-axis represent different CFA samples. Note that the red dotted boxes highlighted the most environmentally risky TEs and CFA samples calculated by EIS, EIT, C_t/C_b, and $\sum(\frac{C_{ti}}{C_{di}}) * T_{ji}$ respectively. Only As, Cr, and V were used during the calculation of EIT considering the availability of T_j. The environmental risk was mainly contributed (≥90% in all cases) by As due to its relatively high accumulation and toxicity.

Furthermore, $\sum(\frac{C_{ti}}{C_{di}}) * T_{ji}$ identified that the YEM sample presented the highest environmental risk, mainly due to the contribution of As, which represented 108.274 of the total risk value of 119.267. However, As accounted for only a small proportion of the high-mobility fractions, with 2.553% in the water-extractable fraction, 0.213% in the acid-soluble fraction, and 32.766% in the reducible fraction. This may be due to the high alumina and iron oxide contents in the YEM samples, which cause adsorption and precipitation to become the main mechanisms that control the leaching of As [25]. In comparison, the As content in PAN was more mobile (2.340% in the water-extractable fraction, 5.106% in the acid-soluble fraction, and 47.234% in the reducible fraction). Therefore, the environmental risk of the PAN sample should be greater than that of the YEM sample when the total concentrations in the samples are similar. Overall, the earlier analysis confirms that the new environmental assessment indexes proposed in the FIELD framework are superior to conventional assessment methods.

10.3.6 Model interpretation and knowledge discovery

To further evaluate the model, we used PI and SHAP to assess the importance of features. Fig. 10.10A and B illustrate the normalized feature importance of the CFA

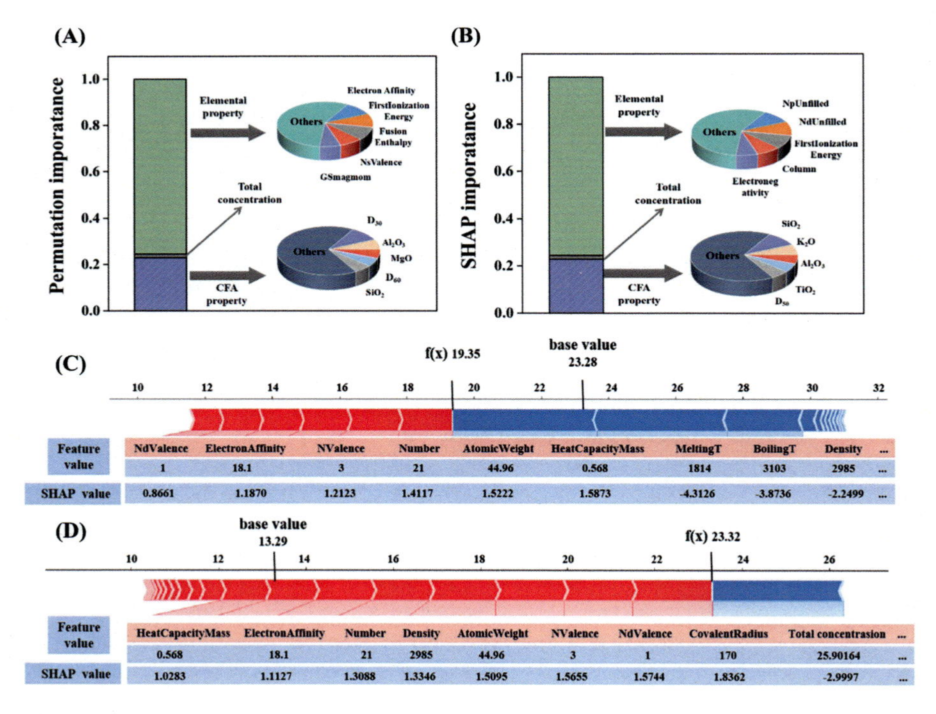

Figure 10.10 Feature plots based on the optimal DNN model: (A) permutation importance for all samples, (B) SHAP importance for all samples, (C) feature importance for the DAT water-extractable fraction, and (D) feature importance for the DAT acid-soluble fraction.

physicochemical properties and elemental properties. Note that the TE fractions after one-hot coding are not included in this figure, but the TE fraction is still the most important factor directly affecting the fraction percentage, even if the other input features are the same.

As shown in Fig. 10.10A and B, in comparison to the physicochemical properties and total concentration of CFA, the TEs' elemental properties have a more significant impact on the fraction percentage with PI and SHAP scores of 0.744 and 0.755, respectively. A comprehensive analysis showed that electron affinity, first ionization energy, and the number of unfilled p- and d-orbitals were the most important elemental properties affecting the TE fractions, while SiO_2 content, Al_2O_3 content, and particle size D_{30} were the most important physicochemical properties. Notably, there were some differences in the importance rankings of the elemental properties and physicochemical properties as determined by the two methods; this difference may be due to the different internal mechanisms in the calculation process [26].

In Fig. 10.10C and D, the feature importance values of the water-extractable and acid-soluble fractions in separate DAT samples were taken as examples for local analysis. This analysis further confirmed that the influence of each feature on the output variable was different. For example, the contribution of the atomic weight to the water-extractable fraction was negative, but this feature had a positive impact on the prediction of the acid-soluble fraction.

Fig. 10.11 further illustrates the feature importance ranking for five different TE fractions. The boiling T, melting T, density, total concentration, and MgO content were the five most important variables for the water-extractable fraction. For the residual fraction, the boiling T, covalent radius, atomic volume, total concentration, and SiO_2 content had the strongest effect on the TE fraction percentages. Each feature variable

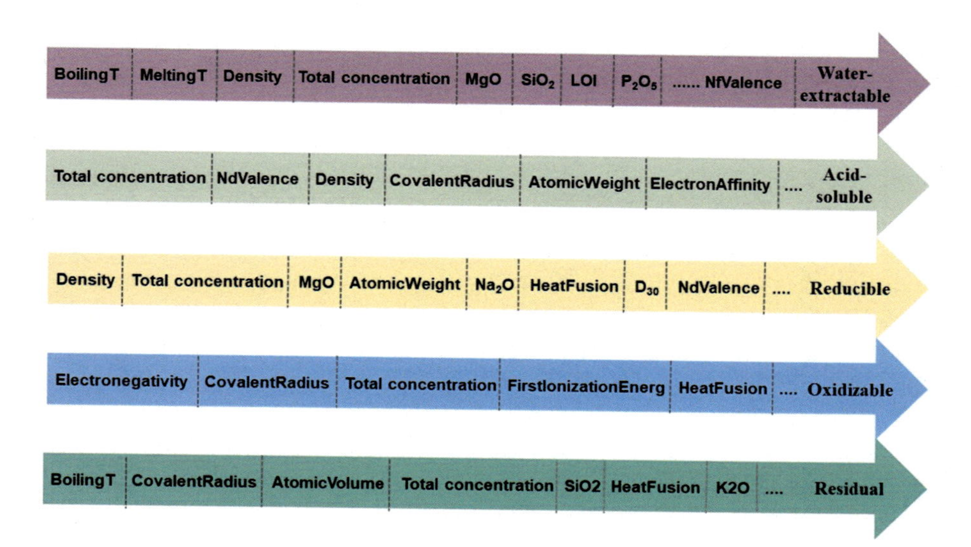

Figure 10.11 Variable ranking for specific TEs fraction.

had different effects on different TE fractions, but the total concentration (which consistently ranked in the top five) was a common important input feature to all the fractions percentage prediction. According to Jin et al. [27], the total concentration directly affects the leaching characteristics and bioavailability of As in CFA.

10.4 Summary

In this chapter, we proposed a comprehensive framework for environmental assessment. Specifically, a DNN model was constructed to predict TE fractions in solid ash. Based on the total concentration and fraction, two novel environmental evaluation indicators, EIS and EIT, were proposed. The case study of applying the FIELD framework to CFA analysis shows that:

1. The model achieved average R^2 values of 0.97, 0.82, and 0.83 on the training, validation, and testing sets, respectively, indicating that the optimal DNN model had high robustness and generalization ability.
2. The EIS calculation results showed that Se in the YAN samples had higher mobility and posed a greater environmental threat compared to other TEs; the EIT results indicated that the PAN samples posed a higher environmental risk due to their high concentrations and mobility of As.
3. The model interpretation shows that the elemental properties had the greatest impact on TE fractions. In addition, the feature importance varied for different TE fractions.

References

[1] C. Qi, et al., Rapid identification of reactivity for the efficient recycling of coal fly ash: hybrid machine learning modeling and interpretation, Journal of Cleaner Production 343 (2022) 130958.

[2] M. Ahmaruzzaman, A review on the utilization of fly ash, Progress in Energy and Combustion Science 36 (3) (2010) 327–363.

[3] G. Akar, et al., Leaching behavior of selected trace elements in coal fly ash samples from yenikoy coal-fired power plants, Fuel Processing Technology 104 (2012) 50–56.

[4] H. Li, et al., Comprehensive assessment of environmental and health risks of metal (loid) s pollution from non-ferrous metal mining and smelting activities, Journal of Cleaner Production 375 (2022) 134049.

[5] A. Tessier, P.G. Campbell, M. Bisson, Sequential extraction procedure for the speciation of particulate trace metals, Analytical Chemistry 51 (7) (1979) 844–851.

[6] A. Ure, et al., Speciation of heavy metals in solids and harmonization of extraction techniques undertaken under the auspices of the BCR of the Commission of the European Communities, International Journal of Environmental Analytical Chemistry 51 (1–4) (1993) 135–151.

[7] C.-G. Yuan, Leaching characteristics of metals in fly ash from coal-fired power plant by sequential extraction procedure, Microchimica Acta 165 (1) (2009) 91–96.

[8] E. Sočo, J. Kalembkiewicz, Investigations of sequential leaching behaviour of Cu and Zn from coal fly ash and their mobility in environmental conditions, Journal of Hazardous Materials 145 (3) (2007) 482–487.

[9] Q. Tian, et al., Distributions and leaching behaviors of trace elements in fly ash, Acs Omega 3 (10) (2018) 13055–13064.

[10] S. Caeiro, et al., Assessing heavy metal contamination in Sado Estuary sediment: an index analysis approach, Ecological Indicators 5 (2) (2005) 151–169.

[11] L. Hakanson, An ecological risk index for aquatic pollution control. a sedimentological approach, Water Research 14 (8) (1980) 975–1001.

[12] M. Anju, D.K. Banerjee, Comparison of two sequential extraction procedures for heavy metal partitioning in mine tailings, Chemosphere 78 (11) (2010) 1393–1402.

[13] V. Gholami, M.J. Booij, Use of machine learning and geographical information system to predict nitrate concentration in an unconfined aquifer in Iran, Journal of Cleaner Production 360 (2022) 131847.

[14] P. Shah, et al., Deep neural network-based hybrid modeling and experimental validation for an industry-scale fermentation process: identification of time-varying dependencies among parameters, Chemical Engineering Journal 441 (2022) 135643.

[15] Y. Jiang, et al., Topological representations of crystalline compounds for the machine-learning prediction of materials properties, Npj Computational Materials 7 (1) (2021) 28.

[16] M. Sarmadi, et al., Modeling, design, and machine learning-based framework for optimal injectability of microparticle-based drug formulations, Science Advances 6 (28) (2020) eabb6594.

[17] Y. Wang, Y. Yao, Breast lesion detection using an anchor-free network from ultrasound images with segmentation-based enhancement, Scientific Reports 12 (1) (2022) 14720.

[18] P. Liu, S. Joty, H. Meng, Fine-grained opinion mining with recurrent neural networks and word embeddings, in: Proceedings of the 2015 Conference on Empirical Methods in Natural Language Processing, 2015.

[19] A. Phornchaicharoen, P. Padungweang, Face recognition using transferred deep learning for feature extraction, in: 2019 Joint International Conference on Digital Arts, Media and Technology with ECTI Northern Section Conference on Electrical, Electronics, Computer and Telecommunications Engineering (ECTI DAMT-NCON), 2019.

[20] T.V. Pyrkov, et al., Extracting biological age from biomedical data via deep learning: too much of a good thing? Scientific Reports 8 (1) (2018) 5210.

[21] E. Shokr, A. De Roeck, M.A. Mahmoud, et al., Modeling of charged-particle multiplicity and transverse-momentum distributions in pp collisions using a DNN, Scientific Reports 12 (1) (2022) 8449.

[22] L. Chen, et al., Heavy metals chemical speciation and environmental risk of bottom slag during co-combustion of municipal solid waste and sewage sludge, Journal of Cleaner Production 262 (2020) 121318.

[23] L. Dou, T. Li, Regional geochemical characteristics and influence factors of soil elements in the pearl river delta economic zone, China, International Journal of Geosciences (2015) 12. Vol. 06 No. 06.

[24] X. Xiong, et al., Potentially trace elements in solid waste streams: fate and management approaches, Environmental Pollution 253 (2019) 680–707.

[25] T. Wang, et al., Leaching characteristics of arsenic and selenium from coal fly ash: role of calcium, Energy & Fuels 23 (6) (2009) 2959–2966.

[26] X. Mi, et al., Permutation-based identification of important biomarkers for complex diseases via machine learning models, Nature Communications 12 (1) (2021) 3008.

[27] Y. Jin, et al., Evaluation of bioaccessible arsenic in fly ash by an in vitro method and influence of particle-size fraction on arsenic distribution, Journal of Material Cycles and Waste Management 15 (4) (2013) 516–521.

Identifying the amorphous content in solid ashes: a machine learning approach using an international dataset

Abstract

This chapter presents a case study of applying machine learning (ML) modeling to the identification of amorphous phases in solid ash. Taking coal fly ash (CFA) as an example, we show that ML algorithms are robust in amorphous phase prediction based on the chemical composition of CFA. Random forest regression models were built with the optimization of artificial bee colony. This chapter evaluated the model using correlation coefficient, r-square, root mean square error, and mean absolute error, giving results of testing set of 0.773, 0.477, 6.542, and 5.279. Feature importance and permutation importance were used to investigate feature contribution. It is proved that the established model had good robustness and generalization capability, which can effectively determine the potential of CFA as supplementary cementitious materials to promote the cleaner production of cement or geopolymer resources.

11.1 Background

Rapid economic development and industrial production have driven energy resource demands [1]. Given its characteristics of high yield and high consumption, coal is widely used in power production and industrial production. About 40% of the world's electricity comes from coal [2]. Coal fly ash (CFA) is generated as a by-product in the power generation process in thermal power plants [3], with a mass of around 800 million tons per year produced [4]. Extensive fly ash emissions not only occupy land resources and seriously damage the environment but also endanger human health [5].

The resource utilization of CFA has attracted increasing attention from industry and academia. CFA is used to prepare ceramic glass [2], backfill mining areas, pave roads, and as an adsorbent for heavy metal ions in polluted water [6]. Researchers also have attempted to use carbohydrate-based composite slurry to remediate soil [7] or recover superplastic materials and rare earth elements from CFA leachate [3]. CFA is widely recycled as an auxiliary cementitious material (SCM) for the clean production of concrete or geopolymer resources [8] due to the widespread adoption of geopolymers as a replacement for Portland cement [9].

CFA can be used as an SCM because of its reactivity [10]. Mineral composition analysis of CFA indicates that it contains three states: crystalline phases, amorphous phases, and a small amount of incompletely burned carbon [11]. The amorphous phases are widely believed to be the main cause of CFA reactivity [12]. Different

Machine Learning Applications in Industrial Solid Ash. DOI: https://doi.org/10.1016/B978-0-443-15524-6.00016-9
© 2024 Elsevier Inc. All rights reserved.

combustion conditions, cooling methods, and coal sources can affect the amorphous phase content and, thus, the reactivity of CFA, which has led to widespread discussion both within China and globally. Some researchers have combined particle size analysis and Fourier transform infrared spectroscopy to propose a method for rapidly predicting CFA reactivity from the phase chemistry of aluminosilicate glasses [13]. Based on the physical and chemical properties of CFA, researchers have constructed a comprehensive evaluation index and equation for the suitability of CFA in the manufacture of high-strength geopolymers [14]. These widely used methods are mostly based on empirical knowledge and instrumental measurements, which are time-consuming, complex, and costly. Thus, developing rapid and convenient CFA reactivity detection methods remains an area of significant research interest.

ML methods have been widely applied to many fields of recycling and sustainability, [15]. However, there is little research to date on the use of ML for the rapid determination of CFA suitability as an SCM. In this chapter, a random forest (RF) regression model optimized using the artificial bee colony (ABC) algorithm was established to explore the intrinsic correlation between CFA's reactivity and its chemical composition [16]. This model enables rapid reactivity identification by predicting CFA amorphous content based on chemical composition without requiring the use of X-ray diffraction (XRD) and Rietveld methods. The model established in this chapter helps to facilitate efficient CFA recycling.

11.2 Modeling methodology for the amorphous phase

The modeling methodology applied in this chapter can be divided into four stages: (1) preprocess the dataset and split it into a training set and testing set; (2) use a combination

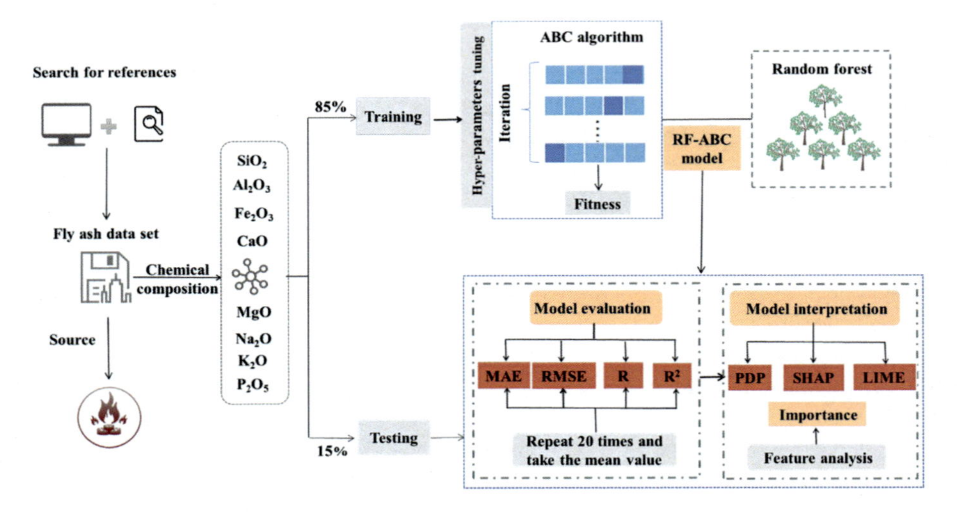

Figure 11.1 The ML modeling workflow used in this chapter.

of the RF and ABC algorithms to train the model and optimize the hyperparameters; (3) use the testing dataset to evaluate the model; and (4) feature analysis and model interpretation based on the complete dataset. The workflow is shown in Fig. 11.1.

11.2.1 Dataset

Before modeling, it is essential to clean and preprocess the dataset, which has a significant effect on the model's performance. The dataset used in this chapter contained 123 samples that were obtained through an extensive literature search and analysis using Scopus and Elsevier databases. Due to the incompleteness of the raw data, we used mean

Table 11.1 Part of the processed dataset.

Chemical composition (features)								Reactivity of CFA (target)
SiO_2	Al_2O_3	Fe_2O_3	CaO	MgO	Na_2O	K_2O	P_2O_5	Amorphous (%)
50.58	34.94	3.93	1.65	1.26	0.91	3.02	0.7	61
50.7	36.4	3.9	1.4	1.2	1	2	0.7	65.2
59	19.7	6.8	3.6	2.7	0.7	1.8	0.7	48
47.5	27.3	14.3	4.25	1.48	0.74	0.54	0.91	74.2
53.3	32.5	3.1	6.9	0.9	0.27	0.59	0.1	81.7
54.4	32.1	7.49	1.06	0.75	0.14	0.22	0.09	75.8
67.3	22.5	3.74	1	0.53	0.5	2.11	0.09	78.4
71.2	24.7	1.16	0.08	0.12	0.01	0.53	0.04	62.8
51.43	23.59	15.33	1.75	1.15	0.36	0.84	1.32	59.16
57.3	29.6	3.89	4.79	0.89	0.2	0.86	0.531	73.1

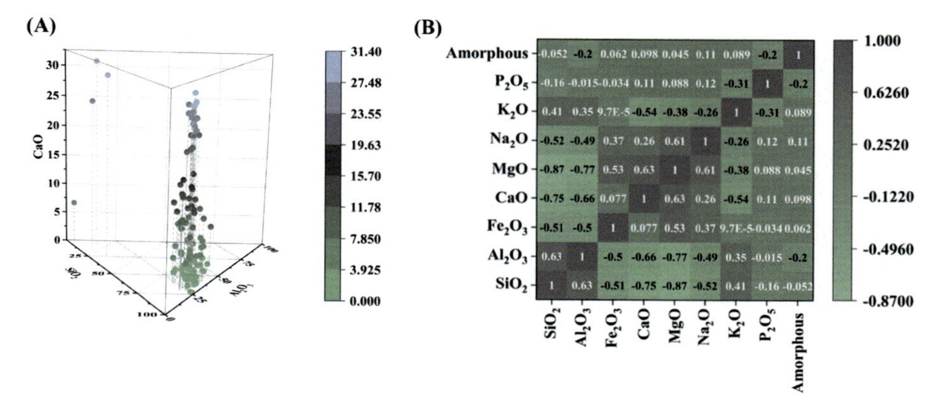

Figure 11.2 Statistical distribution of the variables. (A) Three-dimensional phase diagram, with the data distribution of the $SiO_2 - Al_2O_3 - Fe_2O_3 - CaO$ system displayed in scatter form. (B) Heat map showing the correlation coefficients between the eight chemical compositions and the amorphous phase.

values to fill the missing values to improve the data completeness. Part of the processed dataset is shown in Table 11.1. The dataset contained eight input features (SiO_2, Al_2O_3, Fe_2O_3, CaO, MgO, Na_2O, K_2O and P_2O_5, and SiO_2, Al_2O_3, Fe_2O_3, and CaO). The distribution of all chemical components and their correlations are shown in Fig. 11.2.

11.2.2 Random forest-artificial bee colony model

A hybrid modeling methodology was proposed to predict the mass fraction of the amorphous phase according to the chemical composition of CFA. The hybrid method used the RF technique to model the nonlinear relationship between the input features and target variables. The ABC algorithm was then used to adjust the RF hyperparameters and improve the model's generalization ability [17].

The first step in the modeling process involved splitting the preprocessed dataset into training and testing sets in a certain proportion. We varied the training set proportion from 55% to 90% in 5% increments and evaluated the modeling performance using the R coefficient. The formula of R is as follows:

$$R(X, Y) = \frac{Cov(X, Y)}{\sqrt{Var[X]Var[Y]}}$$

where $R \in [-1, 1]$, $Cov(X, Y)$ is the covariance of X and Y, $Var[X]$ is the variance of X, and $Var[Y]$ is the variance of Y.

As shown in Fig. 11.3, the training set displayed strong stability as the proportion of the training set increased with values mainly concentrated around 0.97; however, the change in the training set proportion had a greater impact on the testing performance. As the training set proportion increased from 55% to 65%, the R of the testing set

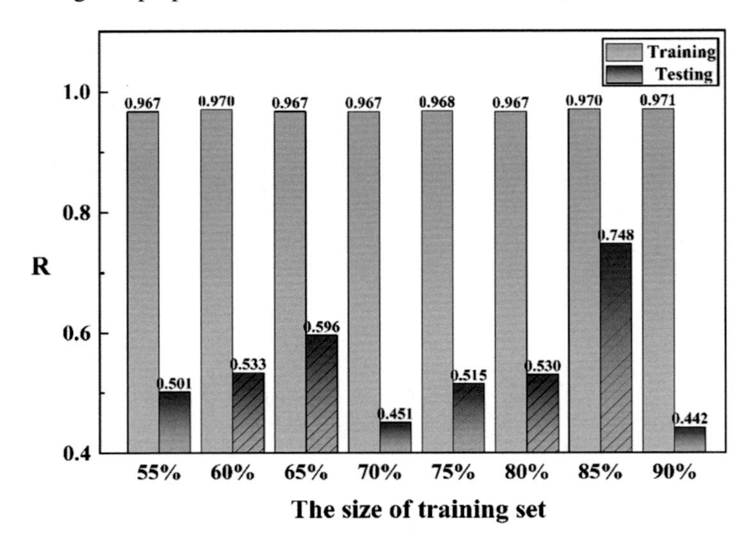

Figure 11.3 Sensitivity study of the size of the training set.

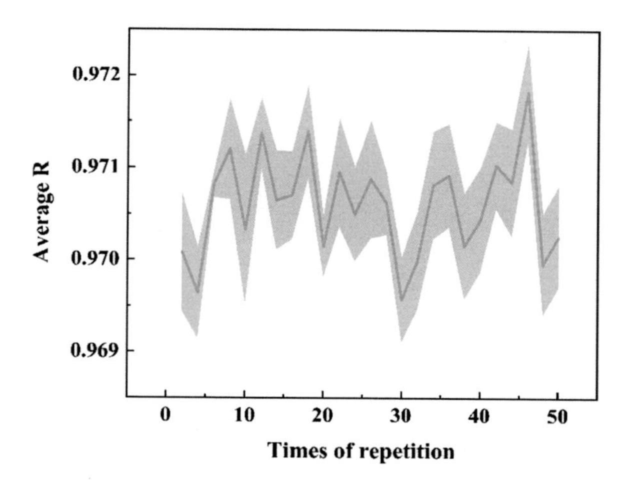

Figure 11.4 Plot illustrating the number of repetitions required to obtain stable fitness for one set of hyperparameters.

increased from 0.501 to 0.596. When the proportion of the training set was 70%, the R of the test set decreased to 0.451. Subsequently, there was an upward trend: when the training set proportion was 85%, the R value of the testing set reached its maximum value of 0.748, indicating relatively optimal model performance. When the size of the training set was further increased from 85% to 90%, the R value for the training set reached its maximum value; however, the R value of the testing set decreased from 0.748 to 0.442. The low model performance can be potentially attributed to the use of a training set that was too small, but the use of a training set that was too large resulted in overfitting and reduced model generalization. Therefore, a compromise training set proportion of 85% was selected based on the analysis earlier.

The ABC algorithm was used to optimize the hyperparameters due to its significant advantages in solving multiobjective problems. A population size of 200 was chosen, and the maximum number of iterations was set as 50 [18]. To avoid randomness in the results, fivefold cross-validation was used. The mean value of R was used as the fitness value for each set in Fig. 11.4; as shown, when the random dataset was repeatedly split 20 times, the performance of the model tended to be stable, and the average R value fluctuated slightly around 0.970. Therefore, the number of repetitions was determined to be 20 times.

11.2.3 Performance evaluation

After hyperparameter optimization, the trained model was then evaluated on the testing set. In this chapter, in addition to the aforementioned correlation coefficient R, three additional evaluation indicators (mean absolute error (MAE), root mean squared error (RMSE), and R squared (R^2)) were used to evaluate the RF model's generalization ability and ensure that the evaluation results were reasonable [19]. The MAE calculates the absolute mean of the differences between the predicted and true values for all samples

[20], while the RMSE squares the residuals between the predicted and true values for each sample, averages them, and then takes the square root of the result [21]. The smaller the value of these two indicators, the better the fitting result. The R^2 parameter represents the coefficient of determination, and its value reflects the proportion of the total variance explained by the variance in the predicted data [22]. The value range of R^2 is [0,1]: the closer its value is to 1, the better the prediction. The MAE, RMSE, and R^2 indexes can be calculated as follows:

$$MAE = \frac{1}{n}\sum_{i}^{n} |y_i - \hat{y}_i|$$

$$RMSE = \sqrt{\frac{1}{n}\sum_{i}^{n} (y_i - \hat{y}_i)^2}$$

$$R^2 = \frac{\sum_i^n (\hat{y}_i - \overline{y}_i)^2}{\sum_i^n (y_i - \overline{y}_i)^2}$$

where n is the number of samples, yi and $\hat{y}_i$ are the true and predicted values of a sample, respectively, and $\overline{y}_i$ is the mean of the samples.

11.3 Results and discussion

In this section, we defined the hyperparameters of the RF-ABC model and compared the model performance before and after tuning. The interpretation of the model was also investigated in detail.

11.3.1 Hyperparameter tuning and model performance

The hyperparameter tuning process of the ABC algorithm was repeated 20 times with 50 iterations in each ABC optimization step. The identified optimal hyperparameters are shown in Table 11.2.

The optimal RF model was then constructed using the optimized hyperparameters, and the performance of the optimal model was compared with the model before tuning, as shown in Figs. 11.5 and 11.6. The performance of the optimal RF model decreased slightly on the training set but increased on the testing set. After

Table 11.2 Optimized hyperparameters and their explanation.

Hyperparameters	Optimized	Explanation
n_estimators	50	The number of weak learners in the integration algorithm
max_depth	11	The maximum depth of the tree, beyond which branches are cut
min_samples_split	4	The minimum number of samples that a node can split
min_samples_leaf	1	The minimum number of samples that a leaf node contains
max_features	3.58421507	Limit of the number of features considered when branching

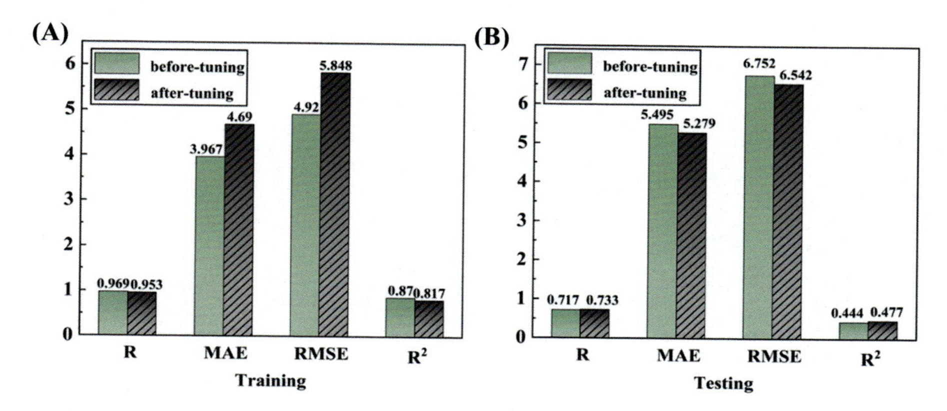

Figure 11.5 Comparison of model performance before and after hyperparameter tuning on (A) the training set and (B) the testing set.

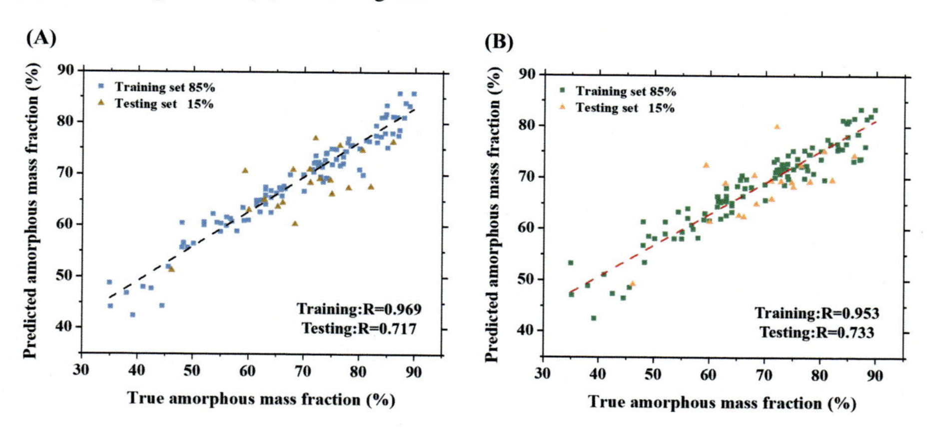

Figure 11.6 Plot illustrating the fitting between the predicted and real values of the amorphous phase (A) before tuning and (B) after tuning.

the hyperparameter tuning process, the correlation coefficient R values for the training set and testing set were 0.953 and 0.733, respectively, with corresponding MAE values of 4.690 and 5.279. Although the R^2 value on the testing set was relatively low (0.477), it was still determined to fall within an acceptable range for the interpretation of the model. Overall, the model's performance was relatively good.

11.3.2 Feature analysis and model interpretation

11.3.2.1 Feature contribution

The RF model provides variable importance measures to determine which features are more closely related to the target variable [19]. The feature contribution can be also calculated using the permutation importance, which iteratively removes each variable in turn and compares the model predictions [23]. The feature importance scores were

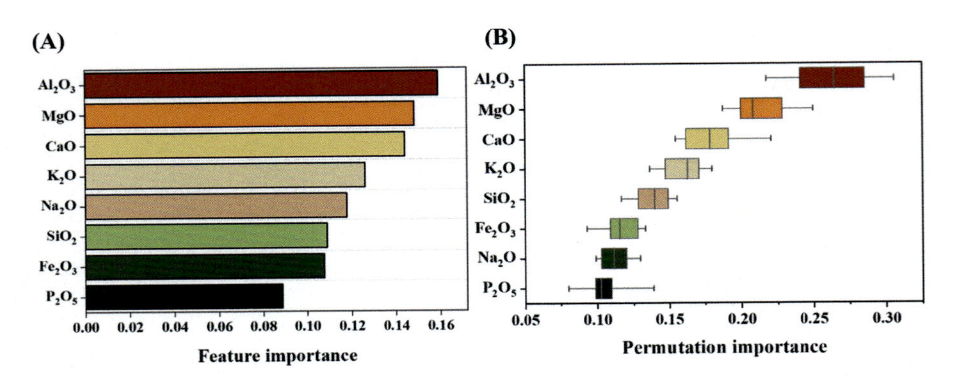

Figure 11.7 Plots illustrating the importance of variables: (A) feature importance and (B) permutation importance.

calculated using the aforementioned methods and sorted as shown in Fig. 11.7. The feature importance values of Al_2O_3, MgO, and CaO were 0.159, 0.148, and 0.144, respectively, and their corresponding permutation importance values were 0.2642, 0.2139, and 0.1790. These three chemical compositions ranked in the top three in both ranking diagrams, indicating that these compounds contributed the most to the content of the amorphous phase. P_2O_5 had the lowest scores, indicating it had the least influence on the amorphous phase. These findings are consistent with Jin et al. (2020), who demonstrated that the reactivity of CFA arises primarily from the hydration of SiO_2 and Al_2O_3 in the vitreous phase under certain alkaline conditions [11]. Using XRD analysis, Tennakoon et al. (2015) also found that the $Na_2O-Al_2O_3-SiO_2-H_2O$ gel combination was controlled by SiO_2 and Al_2O_3 availability in the reaction environment [24].

A small amount of MgO will form more of the vitreous phase to promote the alkali–silica hydration reaction [25]. Li and Xu [26] highlighted that higher Ca content can also increase the network improver content, which is likely to enhance the structure of CFA hydrate and improve its reactivity. The aforementioned experimental results prove that Al_2O_3, MgO, and CaO were important variables affecting the reactivity of CFA, consistent with our findings from the feature importance analysis.

11.3.2.2 Partial dependence plot

Feature importance and permutation importance have certain limitations: these metrics can reflect the overall importance of features but not the patterns of how the input features affect the output variable [27]. To address this limitation, we use the partial dependence plots (PDP) method. Fig. 11.8 shows the PDP between the target response and eight groups of single features. The response fluctuations clearly reflect the amorphous content change with increasing mass fractions of the various chemical components. For example, as the mass fractions of Al_2O_3, CaO, and MgO increased, the amorphous phase content, which is used here to represent the reactivity of CFA, initially increased and then decreased before finally converging at a relatively stable value. When the mass fractions of these three chemical components were around 20%, 15%, and 2%, respectively,

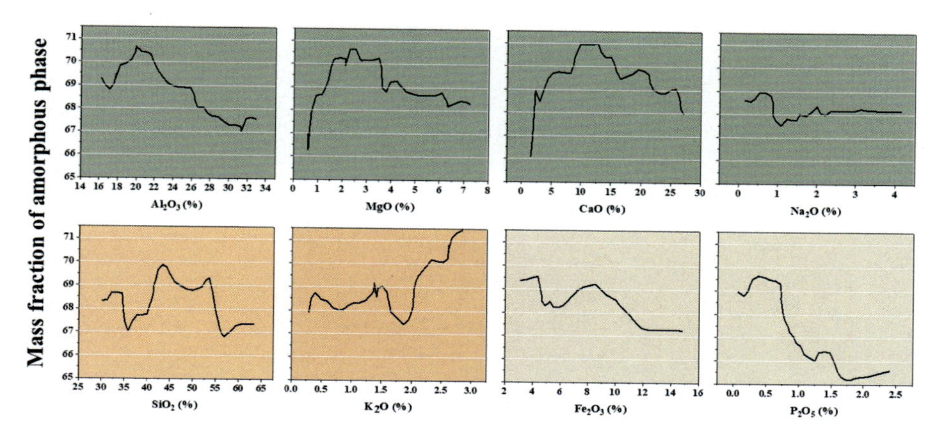

Figure 11.8 Partial dependence plots of single features.

the amorphous content reached its maximum value. In addition, the impact on the amorphous phase remained stable once the mass fraction of Na_2O exceeded 2%.

The interaction of chemical components within CFA ultimately affects its reactivity. Therefore, Al_2O_3, MgO, and CaO were selected to explore binary interactions, and complex interactions between Al_2O_3, MgO, and CaO were observed. High-amorphous content values occur under the following three conditions:

1. When the mass fraction of Al_2O_3 was ~20% and the MgO fraction was in the 2%–4% range.
2. When CaO was in the 10%–15% range and the mass fraction of Al_2O_3 was ~20%.
3. When the mass fraction of MgO was 2%–4% and CaO was 10%–20%.

The earlier analysis illustrated that complex interactions occurred among these chemical components, including coupled interactions in the amorphous phase. Only when the mass fraction of each chemical component was in its optimal range (e.g., $10\% < CaO < 15\%$) were high amorphous phase contents observed. The results were consistent with the conclusions in the published literature. For example, Cho [28] found that the proportion of components forming the glass network frame (SiO_2, Al_2O_3, and Fe_2O_3) and network modifier components (CaO, MgO, Na_2O, and K_2O) significantly affected CFA reactivity. Fernandez-Jimenez [29] observed that in the case of different activated alumina content values, the reaction degree of FA was different even though the active silica content in CFA was very similar.

11.3.2.3 SHapley Additive explanation and Local Interpretable Model-Agnostic Explanations

In addition to the PDP approach described earlier, the SHapley Additive explanation (SHAP) and Local Interpretable Model-Agnostic Explanations (LIME) methods were used to further explain the model in this section. In Fig. 11.9A, the features were sorted in descending order of importance [12]. When the SHAP value is >0, the feature positively affects the predicted value; conversely, values <0 have a negative impact on the model prediction [30]. In general, Al_2O_3, MgO, and CaO had the greatest effect on the

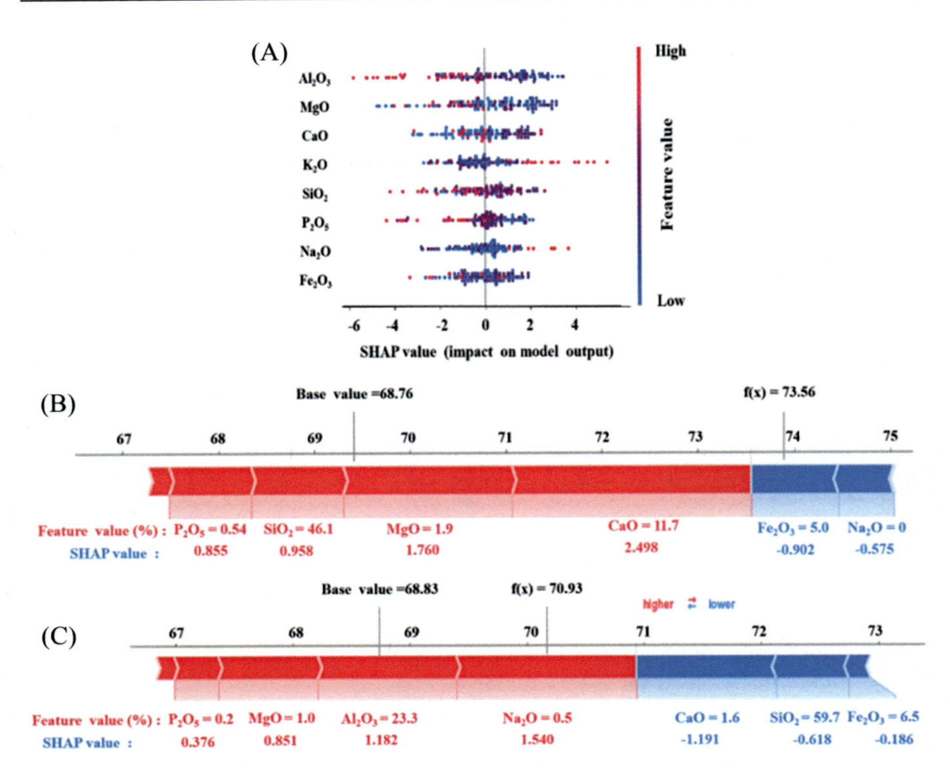

Figure 11.9 Model interpretation using SHAP: (A) overall interpretation. Each point represents a sample, and the color represents the largeness of a feature value. (B) Local interpretation of the 5th sample, and (C) Local interpretation of the 95th sample. Red indicates feature contributions that increase the score, whereas blue indicates features that decrease the score.
Source: Reproduced with permission from Elsevier.

amorphous content, and Fe_2O_3 had the least effect. This is broadly consistent with the theory that CFA reactivity is controlled by the alkalinity of the glass phase (($CaO + MgO + Al_2O_3$)/SiO_2) [31]. High K_2O content values promoted amorphous content formation within the CFA; however, a negative effect on the amorphous content was observed in the presence of higher Al_2O_3 mass fractions. This is because high Al_2O_3 content makes the hydration process too fast and affects the long-term hydration of CFA [32].

Some studies suggest that Class F fly ash containing more than 70% $SiO_2-Al_2O_3-Fe_2O_3$ has stronger volcanic ash reactivity [33]. Additionally, other studies have suggested that SiO_2 and Al_2O_3 significantly affect the reactivity of CFA because silica reacts with the calcium hydroxide released from the hydration of calcium oxide to produce calcium hydrate [14], which is similar to the cement hydration process. In contrast to the conventional view that SiO_2 and Fe_2O_3 have significant effects on reactivity [11], the findings in Figs. 11.7 and 11.9A imply that Fe_2O_3 content has a relatively small effect on CFA reactivity.

In addition to the general analysis of the model, the local interpretation of the different samples is shown in Fig. 11.9B and C. The base value is the mean value of all the target variables, and *f*(*x*) is the predicted value of the mass fraction of the amorphous phase in the sample. The difference between the base value and *f*(*x*) represents the marginal contribution, with each characteristic's marginal contribution indicated by an arrow. For the 5th CFA sample, the SHAP values for CaO, MgO, SiO_2, and P_2O_5 were 2.498, 1.760, 0.958, and 0.855, respectively; thus, these four chemical components contributed positively to the 5th CFA sample's amorphous content. In contrast, Fe_2O_3 and Na_2O had SHAP values of -0.902 and -0.575, which correlated negatively with the CFA's amorphous content. For the 95th CFA sample, low CaO levels and high SiO_2 levels had negative effects on the model's results, whereas high Al_2O_3 levels and low Na_2O levels had positive effects on the model's results.

Fig. 11.10 illustrates the decision-making process of the black box model for the selected sample points using the LIME method. LIME not only provides prediction intervals for the target variables [34] but also graphically visualizes the mass fraction variation ranges for all characteristics and the corresponding positive and negative effects on the model's output [35]. The LIME analysis was performed for samples 5 and 95 in the CFA dataset, corresponding to predicted amorphous content values of 73.56 and 70.93, respectively. For the fifth CFA sample, there was a significant negative effect on the amorphous content when the mass fraction of Al_2O_3 exceeded 28.37, while the amorphous content was higher when the mass fractions of CaO were ~5.6%–14.78%. In contrast, the amorphous content decrease recorded for the 95th CFA sample mainly occurred because the mass fraction of CaO was below 2.22%. The amorphous content increased when the mass fraction of Al_2O_3 was between ~19.35% and 23.5%.

The influence of chemical composition on the amorphous content of different CFA samples varied, as shown in Fig. 11.10. To achieve efficient CFA recovery, chemical compositional differences between CFA samples must be considered. The hydration reactivity can be improved in the presence of an appropriate combination of chemical compositions; this finding is similar to the conclusion that increased Ca^{2+}, Na^{+}, and K^{+} content values can increase CFA reactivity.

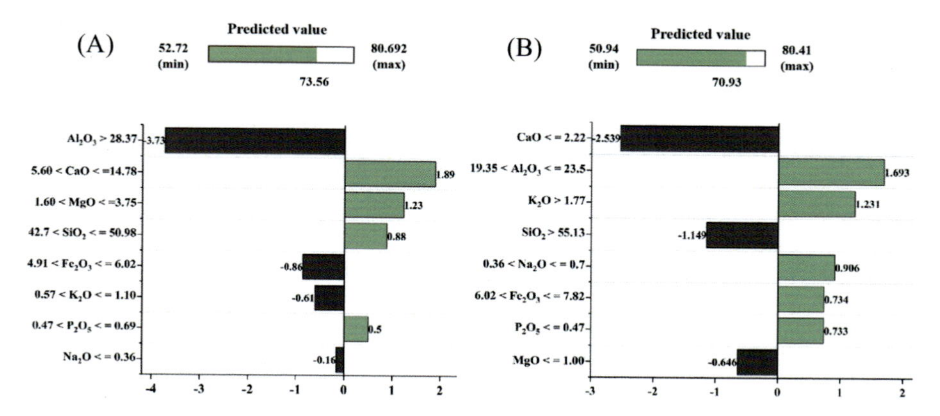

Figure 11.10 Local interpretation based on LIME for (A) the 5th sample in the dataset and (B) the 95th sample in the dataset.

11.3.3 Study significance and outlook

Compared with the traditional classification of fly ash of Classes C and F based on CaO content, the RF-ABC regression model developed in this chapter is based only on the CFA chemical composition and does not require complex XRD-based analysis of the amorphous phase, thus allowing fast and accurate CFA screening. By applying this model to engineering sites, it can effectively assess the potential of CFA as SCMs, promote the recycling of CFA resources, and prevent the environmental pollution and human health impacts caused by mass CFA accumulation. This model can also offer secondary sorting opportunities for unqualified CFAs determined by traditional methods, promote reuse, and create added value for power plants. Further analysis of the model shows that Al_2O_3, MgO, and CaO contribute significantly to the amorphous content of CFA and can be considered key factors in model construction. When the mass fractions of Al_2O_3, MgO, and CaO were within their ideal ranges, higher-amorphous content values were observed with high prediction accuracy. The combination of various CFA samples with different chemical compositions can be potentially used to improve overall CFA reactivity and contribute to the recycle of CFA resources.

References

[1] T. Ju, et al., On the state of the art of crystalline structure reconstruction of coal fly ash: a focus on zeolites, Chemosphere 283 (2021) 131010.

[2] Z. Yang, et al., Utilization of waste cooking oil for highly efficient recovery of unburned carbon from coal fly ash, Journal of Cleaner Production 282 (2021) 124547.

[3] M. Mathapati, et al., A review on fly ash utilization, Materials Today: Proceedings 50 (2022) 1535–1540.

[4] M. Ganesapillai, et al., Sustainable recovery of plant essential Nitrogen and phosphorus from human urine using industrial coal fly ash, Environmental Technology & Innovation 24 (2021) 101985.

[5] C. Yang, et al., Study on the correlation between Fe/Ti forms and reaction activity in high-alumina coal fly ash, Science of the Total Environment 792 (2021) 148419.

[6] H. Zhou, et al., Utilization of coal fly and bottom ash pellet for phosphorus adsorption: sustainable management and evaluation, Resources, Conservation and Recycling 149 (2019) 372–380.

[7] R. Terzano, et al., Copper stabilization by zeolite synthesis in polluted soils treated with coal fly ash, Environmental Science & Technology 39 (16) (2005) 6280–6287.

[8] J.-H. Park, et al., Exploration of the potential capacity of fly ash and bottom ash derived from wood pellet-based thermal power plant for heavy metal removal, Science of the Total Environment 740 (2020) 140205.

[9] A.L. Almutairi, et al., Potential applications of geopolymer concrete in construction: a review, Case Studies in Construction Materials 15 (2021) e00733.

[10] V.J. Azad, O.B. Isgor, Modeling chloride ingress in concrete with thermodynamically calculated chemical binding, International Journal of Advances in Engineering Sciences and Applied Mathematics 9 (2) (2017) 97–108.

[11] Y. Jin, et al., Structure refinement of fly ash in connection with its reactivity in geopolymerization, Waste Management 118 (2020) 350–359.

[12] X. Wen, et al., Quantifying and comparing the effects of key risk factors on various types of roadway segment crashes with LightGBM and SHAP, Accident Analysis & Prevention 159 (2021) 106261.
[13] Z. Zhang, H. Wang, J.L. Provis, Quantitative study of the reactivity of fly ash in geopolymerization by FTIR, Journal of Sustainable Cement-Based Materials 1 (4) (2012) 154–166.
[14] Z.T. Yao, et al., A comprehensive review on the applications of coal fly ash, Earth-Science Reviews 141 (2015) 105–121.
[15] H.Y. Ismail, et al., Modelling of yields in torrefaction of olive stones using artificial intelligence coupled with kriging interpolation, Journal of Cleaner Production 326 (2021) 129020.
[16] Y. Song, et al., Machine learning enables rapid screening of reactive fly ashes based on their network topology, ACS Sustainable Chemistry & Engineering 9 (7) (2021) 2639–2650.
[17] C. Qi, Q. Chen, S. Sonny, Kim, Integrated and intelligent design framework for cemented paste backfill: a combination of robust machine learning modelling and multi-objective optimization, Minerals Engineering 155 (2020) 106422.
[18] A. Afzal, et al., Battery thermal management: an optimization study of parallelized conjugate numerical analysis using cuckoo search and artificial bee colony algorithm, International Journal of Heat and Mass Transfer 166 (2021) 120798.
[19] W. Huo, et al., Performance prediction of proton-exchange membrane fuel cell based on convolutional neural network and random forest feature selection, Energy Conversion and Management 243 (2021) 114367.
[20] F.H.F. Leite, et al., Surgical planning of horizontal strabismus using multiple output regression tree, Computers in Biology and Medicine 134 (2021) 104493.
[21] T. Chai, R.R. Draxler, Root mean square error (RMSE) or mean absolute error (MAE)? – arguments against avoiding RMSE in the literature, Geoscientific Model Development 7 (3) (2014) 1247–1250.
[22] E. Kasuya, On the use of R and R squared in correlation and regression, Ecological Research 34 (1) (2019) 235–236.
[23] N.L. Afanador, T.N. Tran, L.M.C. Buydens, Use of the bootstrap and permutation methods for a more robust variable importance in the projection metric for partial least squares regression, Analytica Chimica Acta 768 (2013) 49–56.
[24] C. Tennakoon, et al., Influence and role of feedstock Si and Al content in geopolymer synthesis, Journal of Sustainable Cement-Based Materials 4 (2) (2015) 129–139.
[25] Z. Wang, et al., Hydration properties of alkali-activated fly ash/slag binders modified by MgO with different reactivity, Journal of Building Engineering 44 (2021) 103252.
[26] Z. Li, G. Xu, X. Shi, Reactivity of coal fly ash used in cementitious binder systems: a state-of-the-art overview, Fuel 301 (2021) 121031.
[27] D. Wang, et al., A machine learning framework to improve effluent quality control in wastewater treatment plants, Science of the Total Environment 784 (2021) 147138.
[28] Y.K. Cho, S.H. Jung, Y.C. Choi, Effects of chemical composition of fly ash on compressive strength of fly ash cement mortar, Construction and Building Materials 204 (2019) 255–264.
[29] A. Fernández-Jimenez, et al., Quantitative determination of phases in the alkali activation of fly ash. Part I. Potential ash reactivity, Fuel 85 (5) (2006) 625–634.
[30] S. Zheng, et al., Corporate environmental performance prediction in China: an empirical study of energy service companies, Journal of Cleaner Production 266 (2020) 121395.

[31] E. Sakai, et al., Hydration of fly ash cement, Cement and Concrete Research 35 (6) (2005) 1135–1140.
[32] G. Li, Y. Sun, C. Qi, Machine learning-based constitutive models for cement-grouted coal specimens under shearing, International Journal of Mining Science and Technology 31 (5) (2021) 813–823.
[33] İ. Acar, Characterization and Utilization Potential of Class F Fly Ashes (2013).
[34] J. Peng, et al., An explainable artificial intelligence framework for the deterioration risk prediction of hepatitis patients, Journal of Medical Systems 45 (5) (2021) 61.
[35] P. Pandey, A. Rai, M. Mitra, Explainable 1-D convolutional neural network for damage detection using lamb wave, Mechanical Systems and Signal Processing 164 (2022) 108220.

The reactivity classification of coal fly ash based on the random forest method

12

Abstract

A rapid reactivity classification method is crucial to promote the recycling of solid ash as a supplementary cementitious material. On this basis, we establish a reactivity classification model in this chapter, taking coal fly ash (CFA) as an example. The CFA reactivity classification was achieved using the random forest algorithm with finely tuned hyperparameters. The accuracy, recall rate, precision, and area under curve (AUC) metrics were used to evaluate model performance. The accuracy of the optimized model was 85.45%, the recall rate was 97.56%, the precision was 84.29%, and the AUC value was 0.924. Furthermore, feature importance, partial dependence plot (PDP), and SHapley Additive exPlanation (SHAP) methods were used to measure the contribution of features. The feature importance and SHAP results indicate that K_2O, Fe_2O_3, Na_2O, and Al_2O_3 had a strong influence on the model. The PDP analysis further demonstrated that the probability of high- reactivity CFA increased as the percentages of K_2O and Na_2O increased and decreased as the Al_2O_3 percentage increased. The earlier results prove that the established model has good robustness and generalization ability and can be used to rapidly classify the reactivity of CFA and determine whether it has the potential to be used as a supplementary cementitious material.

12.1 Background

Due to urbanization, China's electricity demand is gradually increasing; however, the country's current power supply is still largely dependent on coal-fired power generation. By the end of 2021, the installed capacity of power generation in China had reached 2.38 billion kilowatts; of this total, thermal power generation accounted for 1.30 billion kilowatts, of which 1.11 billion kilowatts was provided by coal power 0. Coal-fired power generation produces large amounts of solid waste such as fly ash, which accounts for more than 50% of the total solid waste generated by coal-fired power plants [1]. The consumption of 1 ton of raw coal can generate ~250–300 kg of fly ash. Therefore, China's annual fly ash generated by coal-fired power generation is around 600 million tons, accounting for approximately 50% of the world's total fly ash production, and fly ash has become the largest single pollution source in China's industrial solid waste production 0.

Machine Learning Applications in Industrial Solid Ash. DOI: https://doi.org/10.1016/B978-0-443-15524-6.00009-1
© 2024 Elsevier Inc. All rights reserved.

If a large amount of fly ash is stacked in the open, it will be entrained by air movement, causing air pollution and health risks if inhaled by humans. If fly ash is treated using traditional landfill approaches, its accumulation over time will take up large areas of land. In addition, it may also cause river blockages, and toxic substances in fly ash can contaminate soil and groundwater in the form of leachate [1]. Thus how fly ash can be safely and efficiently utilized has become an important concern for the power generation industry.

Fly ash has been extensively used as an auxiliary cementitious material due to its hydration reactivity [2]. The use of high-quality fly ash and high-efficiency water-reducing agent composite to produce high-grade concrete is a relatively mature technical process for efficient fly ash utilization, and fly ash concrete made has the advantages of good workability, strong pumpability, improved impact resistance, and enhanced frost resistance. The reactivity of fly ash derives from the presence of an amorphous phase in amorphous minerals [3] and determines whether it can be used as a cementitious material. Different coal sources, combustion conditions, and cooling time will affect the fly ash's amorphous phase and, thus, the reactivity of the fly ash [4]. At present, the fly ash amorphous phase is mainly measured using X-ray diffraction (XRD) [5] or the crystal structure analysis method for amorphous phase analysis [6]. For example, Tang used Rietveld full-spectrum fitting to quantitatively study fly ash phases [7]. In addition, Liu et al. used XRD and Fourier transform infrared spectroscopy (FTIR) approaches to analyze the mineral phase composition of fly ash [8]. However, the aforementioned methods have the disadvantages of complex operation and high costs. Therefore, the industry urgently requires an efficient reactivity classification method to guide the usage of fly ash.

Given the industry difficulties described earlier, based on the intrinsic correlation between fly ash's reactivity and its chemical composition, this chapter establishes a random forest (RF) classification model and improves the model's accuracy through hyperparameter tuning. This model eliminates the need for complex amorphous phase analysis using XRD and crystal structure analysis and can quickly classify the reactivity of fly ash. In addition, the model comprehensively analyzes the influence of each fly ash sample's chemical composition on its reactivity, which provides an important reference for efficient fly ash recycling.

12.2 Basics of machine learning modeling

12.2.1 Random forest algorithm

The RF approach is a supervised ensemble learning model for classification and regression [9]. It is essentially an ensemble learning algorithm [10] that can process the prediction results of multiple independent weak learners, that is, single tree models. The predictions of a single tree model may be relatively good but will likely overfit some of the data. The RF method overcomes this limitation by combining numerous tree models, each of which individually might suffer variably from overfitting. Therefore, RF algorithms have been widely used in research on fly ash and related fields. For example,

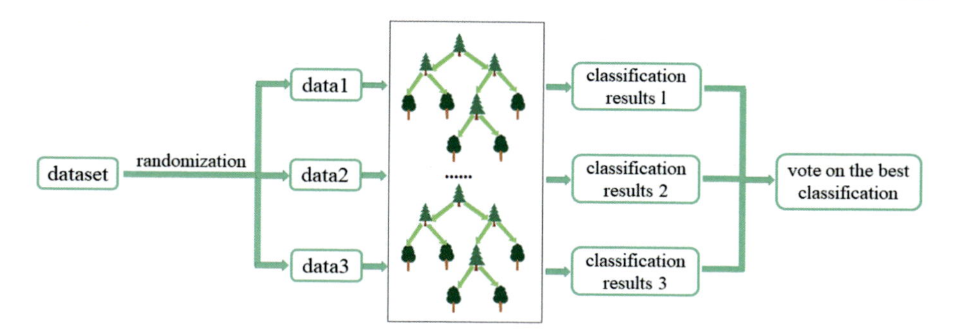

Figure 12.1 Schematic diagram of RF classification.

Sihag et al. used the RF method to evaluate the effect of fly ash on soil infiltration properties [11], Khan et al. used the RF method to simulate the wear depth of environmentally friendly fly ash concrete [12], and Jiang et al. used the RF algorithm to predict the compressive strength of fly ash concrete [13].

The basic concept of RF classification is as follows. First, k subsets are selected from the original training set using bootstrap sampling, each with the same subset size; then, k decision tree models are built for each of the k subsets to obtain k classification results. Finally, each record is voted on to decide its final classification based on the k classification results (Fig. 12.1) [14].

12.2.2 Feature importance

Understanding how the model works and identifying the features that have a greater impact are important aspects of how machine learning (ML) can help with knowledge discovery [15]. As an example, if a single decision tree is used, the default approach is to separately calculate how much purity is added to each variable splitting node. In contrast, the RF algorithm calculates feature importance based on the depth of the split point where the variable is located [16]; from there, it determines which features are closely related to the target variable [17].

Suppose there are J features $X_1, X_2, \ldots, X_j$, I decision trees, and c categories; the Gini index score $\mathrm{VIMj}^{(\mathrm{Gini})}$ is then calculated for each feature X_j, that is, the average change in node split impurity of the j-th feature in all decision trees of the RF. The formula for calculating the Gini index of the i-th tree node q is shown in Eq. (12.1).

$$\mathrm{GI}_q^{(i)} = \sum_{c=1}^{|c|} \sum_{c' \neq c} p_{qc}^{(i)} p_{qc'}^{(i)} = 1 - \sum_{c=1}^{|c|} \left(p_{qc}^{(i)}\right)^2 \tag{12.1}$$

where p_{qc} denotes the proportion of category c in node q. The importance of feature X_j at node q of the i-th tree, that is, the change in the Gini index before and after branching of node q, is calculated as shown in Eq. (12.2).

$$VIM_{jq}^{(Gini)(i)} = GI_q^{(i)} - GI_l^{(i)} - GI_r^{(i)} \tag{12.2}$$

where $GI_l^{(i)}$ and $GI_r^{(i)}$ denote the Gini indices of the two new nodes after branching. Assuming that there are i trees in the RF, the Gini index score $\mathrm{VIMj}^{(\mathrm{Gini})}$ is calculated as shown in Eq. (12.3).

$$VIM_j^{(Gini)} = \sum_{i=1}^{I} VIM_j^{(Gini)(i)} \tag{12.3}$$

12.2.2.1 Partial dependence plot

Partial dependence plots (PDPs) show the marginal effect of one or two features on the prediction results of an ML model and demonstrate whether the relationship between the feature and the target variable is linear, monotonic, or more complex [18]. If the feature analyzed is not correlated with other features and is independent, the partial dependency graph perfectly illustrates how the feature affects the prediction on average. In contrast, when the features are correlated with each other, the partial dependency graph requires additional validation of the relationship between features and predictions. PDPs show how features affect the model's prediction compared to feature importance. Broadly, the greater the observed variation of the PDP curve, the more important a feature is, and the flatter the curve, the less important the feature is.

If the relationship between y and feature X_I is studied, the partial dependence graph is a function of X_I and the model's predicted value, which fits a model *mymodel*, and X_i^k represents the i-th feature of the k-th sample in the training set, as shown in Eq. (12.4).

$$\mathrm{f}(\mathrm{X}_1) = \frac{1}{n}\sum_{k=1}^{n} mymodel\left(X_1, X_2^k, X_3^k, \cdots X_n^k\right) \tag{12.4}$$

12.2.3 SHapley Additive exPlanation

In addition to the PDP approach described earlier, SHAP is also a popular model interpretation tool [19]. Compared with other model interpretation tools, SHAP is more versatile and can be used for both global and local interpretation.

The basic concept of SHAP involves calculating the marginal contribution of a feature when it is added to the model. The mean value is then calculated considering the different marginal contributions of the feature over all feature sequences. SHAP differs from model interpretation tools that use a parameter's magnitude or a positive or negative measure of an indicator's contribution to a model. In this approach, the combined contribution of each indicator is calculated in each sample from the SHAP value. If an indicator shows a consistent trend over most of the samples, this indicates that the model identifies this indicator as contributing significantly to its predictions [20].

12.2.4 Model evaluation

A model built based on a training set must be evaluated using data from the test set. In this chapter, we use the accuracy rate, recall rate, receiver operating characteristic curve (ROC), and AUC to measure the model performance. The aforementioned four evaluation metrics are used to evaluate the model's generalization ability and confirm the reliability of the evaluation results.

Among these metrics, the accuracy rate refers to the proportion of the samples that are correctly predicted among all the samples. The samples can be divided into four cases based on their real and predicted values: true positive (TP), false positive (FP), true negative (TN), and false negative (FN). Accuracy values closer to 1 indicate better prediction effects, as shown in Eq. (12.5).

$$Accuracy = \frac{TP + FN}{TP + FN + FP + TN} \tag{12.5}$$

The precision, which indicates the proportion of samples correctly classified as positive cases, is calculated as shown in Eq. (12.6). The recall, also called the sensitivity or check-all rate, measures the correctly classified proportion of all the samples that should be classified as true, as shown in Eq. (12.7).

$$Precision = \frac{TP}{TP + FP} \tag{12.6}$$

$$Recall = \frac{TP}{TP + FN} \tag{12.7}$$

The AUC is the area under the ROC curve. This metric represents the working characteristics of a subject and is used to measure the effectiveness of ML for "binary classification." The ROC curves are plotted with the false positive rate (FPR) on the horizontal axis and the true positive rate (TPR) on the vertical axis, with different cutoff point values selected. The higher the TPR value, the better the prediction for positive values; analogously, the higher the FPR value, the worse the model's performance in predicting negative values. Therefore, the closer the curve is to the (0, 1) point, the higher its corresponding AUC value.

$$TPR = \frac{TP}{TP + FN} \tag{12.8}$$

$$FPR = \frac{FP}{TN + FP} \tag{12.9}$$

12.3 Fly ash reactivity classification modeling

12.3.1 Dataset collection

Data collection and preprocessing are necessary steps before modeling, and the data quality has a substantial impact on model performance [21]. In this chapter, data from 123 CFA samples were obtained through an extensive search and analysis of literature conducted using the Scopus and Elsevier databases. Due to the incompleteness of the original data, we used the average values of the each column to fill in the missing values and improve the usability of the data [22]. The aforementioned studies mainly used X-ray fluorescence spectroscopy (XRF) for oxide content analysis of the CFA samples and XRD and Rietveld refinement to characterize the amorphous phase in CFA. Considering the percentage content of each chemical component of the coal ash samples and the data availability, a total of eight major chemical components were selected as input features.

The eight main chemical components of CFA in the dataset were SiO_2, Al_2O_3, Fe_2O_3, CaO, MgO, Na_2O, K_2O, and P_2O_5. In this chapter, these components were used as features to characterize the CFA reactivity. Considering the EN 450−1 and IS 3812−1 standards in fly ash reactivity classification, an amorphous content value of 82% was used as the threshold value to classify the CFA's reactivity into high-reactivity (22 samples) and low-reactivity CFA (101 samples). The high-reactivity CFA was represented by a value of "1," whereas the low-reactivity CFA was represented by a value of "0" in the dataset. A sample of the dataset is shown in Table 12.1.

12.3.2 Dataset analysis

Fig. 12.2 shows the distribution of the input features in the dataset. As shown, the distributions of the chemical components in the dataset significantly differ. For example, Al_2O_3 conforms to a relatively clear Gaussian distribution, whereas the values for CaO are strongly skewed toward the left side of the plot.

In addition, Pearson correlation coefficients were calculated for the whole dataset, as shown in Fig. 12.3A. The different circle sizes and colors used in the figure correspond to the magnitude of the correlation coefficients, where larger correlation coefficient values indicate a stronger linear correlation between the variables [23]. As shown, SiO_2 and Al_2O_3 have the strongest positive correlation, whereas SiO_2 and MgO have the strongest negative correlation. Note that variable pairs with small correlation coefficients (around zero) only indicate a weak linear correlation between variables and do not necessarily imply there is nonlinear correlation between the variables.

In the existing literature, $SiO_2-Al_2O_3-Fe_2O_3$ three-phase diagrams are often used to analyze CFA samples [24]. Fig. 12.3B shows the $SiO_2-Al_2O_3-Fe_2O_3$ three-phase diagram for the CFA samples. As shown, the classification of high-reactivity and low-reactivity CFA is extremely nonlinear and cannot be reasonably classified using this diagram type. Therefore, using ML classification algorithms is a potentially viable alternative option for CFA reactivity classification.

12.3.3 Machine learning classification

The quality of the data affects the performance of the ML model and ultimately determines the limits of the model's precision and reliability. In this chapter, the data were transformed using normalization processing [25], and features with zero variance were removed. An optimal division ratio of 3:7 was determined by

Table 12.1 Parts of the dataset.

No	SiO_2	Al_2O_3	Fe_2O_3	CaO	MgO	Na_2O	K_2O	P_2O_5	Reactivity
0	50.58	34.94	3.93	1.65	1.26	0.91	3.02	0.69	0
1	50.70	36.40	3.90	1.40	1.20	1.00	2.00	0.69	0
2	59.00	19.70	6.80	3.60	2.70	0.70	1.80	0.69	0
3	47.50	27.30	14.30	4.25	1.48	0.74	0.54	0.91	0
4	53.30	32.50	3.10	6.90	0.90	0.27	0.59	0.10	0
……	……	……	……	……	……	……	……	……	……
118	53.20	26.00	8.60	2.40	1.60	0.50	2.70	0.30	1
119	28.50	17.90	8.40	27.30	3.80	0.20	1.00	0.30	0
120	48.20	25.90	8.80	2.30	1.50	0.50	2.60	0.30	0
121	50.80	33.40	6.40	2.40	0.80	0.40	0.70	0.30	0
122	41.70	29.00	3.80	10.00	2.40	0.50	0.80	1.50	0

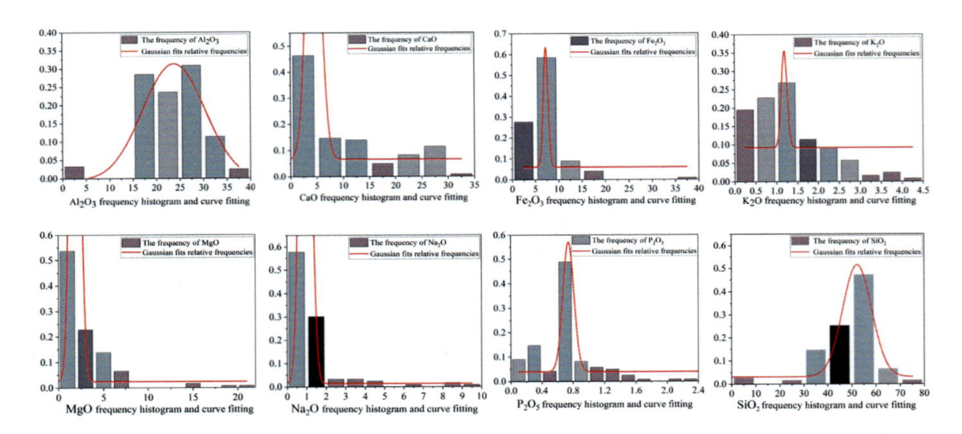

Figure 12.2 Data distribution of input features.

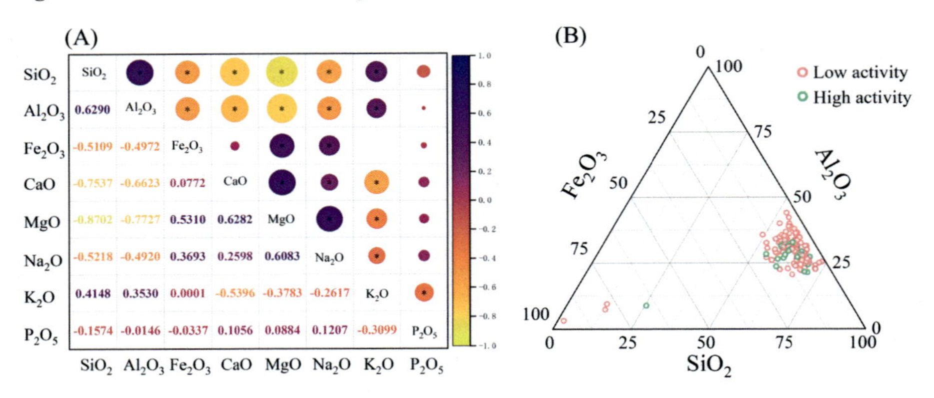

Figure 12.3 Dataset preanalysis: (A) correlation coefficient graph and (B) $SiO_2-Al_2O_3-Fe_2O_3$ three-phase diagram.

convergence analysis [26]; thus, the 123 samples in the dataset were randomly divided in a 3:7 ratio to generate the testing set and the training set, respectively. The eight chemical oxides described earlier were used as input indicators, and the RF method was used to build the CFA reactivity classification model. In addition, 10-fold cross-validation was used as the validation method for the two classifiers mentioned earlier.

12.4 Tuning of model parameters and analysis of evaluation results

In this chapter, six hyperparameters of the RF algorithm were tuned for model optimization: the number of base evaluators (n_estimators), the maximum depth of the tree (max_depth), the minimum number of leaf node samples (min_samples_leaf), the minimum number of samples required to split the internal nodes (min_samples_split), the maximum number of features considered in constructing the optimal model of decision tree (max_features), and the division criterion (criterion). Based on the influence of these parameters on the model's evaluation performance on unknown data, the parameters with greater influence were first tuned, and the results of the previous tuning step were then used in the next tuning step, and finally, the optimal value of each parameter was determined.

In the following text, the fine-tuning of the number of base evaluators (i.e., n_estimators) is explained as an example. In this chapter, we evaluated the influence of n_estimators based on the learning curve and determined whether the parameter could always increase the model's prediction accuracy. The default value of n_estimators is 100. To optimize model fitting, the initial parameter range was set as 0–200 with a step of 10, as shown in Fig. 12.4A. The classification model obtained the highest accuracy of 85.38% at the 71st tree. Based on the initial learning curve, the range was then further refined to a smaller interval (60–80) with a reduced step to optimize this parameter. After optimization, the accuracy of the model was still found to be 85.38% when the other parameters were set to default values, as shown in Fig. 12.4B.

The aforementioned tuning steps were then repeated to optimize the other five hyperparameters using the learning curve, with the resulting optimized hyperparameter values shown in Table 12.2. The corresponding accuracy of the model was 85.45%.

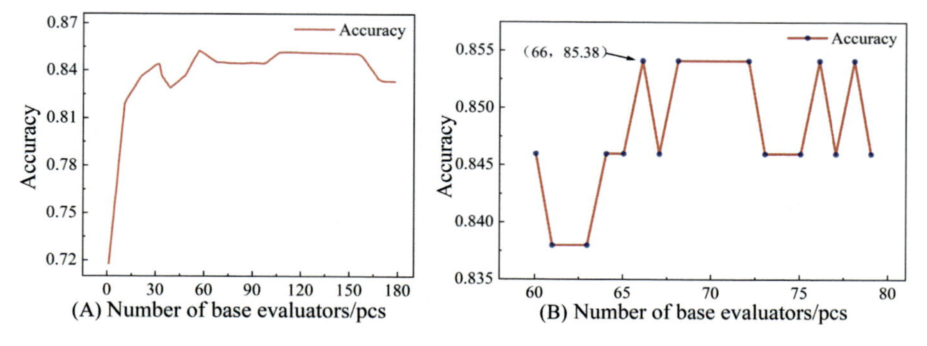

Figure 12.4 Learning curves of the n_estimators parameter: (A) preliminary division plot and (B) refined division plot.

Table 12.2 Hyperparameter tuning for the RF models.

Parameters	Default value	Range	Optimum value
Number of base evaluators	100	0–200	66
Maximum depth of decision tree	none	1–20	6
Minimum number of samples of leaf nodes	1	1–10	1
Minimum number of samples for splitting internal nodes	2	2–20	2
Maximum number of features	auto	1–30	3
Division criteria	'gini'	'gini', 'entropy'	'gini'

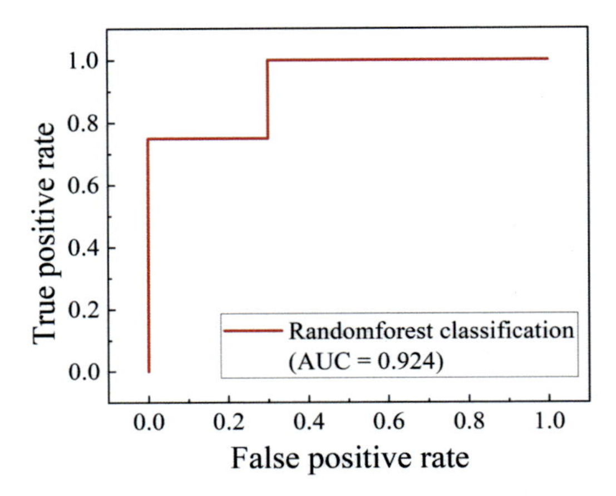

Figure 12.5 ROC curve of the optimal RF model.

In addition, this chapter also evaluated the model using two further evaluation indicators, namely recall and precision, and obtained a recall rate of 97.56% and a precision rate of 84.29%. The ROC graph and the AUC value were also evaluated, as shown in Fig. 12.5. The obtained AUC value of 0.924 indicates that the classification model is highly robust. Overall, the aforementioned results imply that the constructed RF model could achieve good predictive performance on the reactivity classification of the CFA samples.

12.5 Postanalysis of the model

In this chapter, the importance of eight input features was investigated. As the effect of randomness induced by dataset splitting can lead to differences in the importance scores calculated each time, the importance calculation was repeated 10 times to minimize the effect of randomness. The importance scores of the eight features are shown in Fig. 12.6, and their values are as follows: K_2O—0.1779, Fe_2O_3—0.1490, Na_2O—0.1417, Al_2O_3—0.1228, CaO—0.1164, MgO—0.1116, SiO_2—0.1011, and P_2O_5−0.0795. Overall, K_2O, Fe_2O_3, Al_2O_3, and Na_2O had the highest importance scores, indicating that they contribute the most to amorphous phase content, whereas P_2O_5 contributed the least. This conclusion is consistent with the findings of Quan et al. [27].

Fig. 12.7 shows the PDPs of the eight input features, where the fluctuations of the curves in the graph clearly reflect the trend of fly ash reactivity as the content of a single feature changes. For example, the probability of high-reactivity CFA decreased as the percentages of MgO and Al_2O_3 increased, whereas its probability initially increased and then decreased as the percentages of Fe_2O_3 and CaO

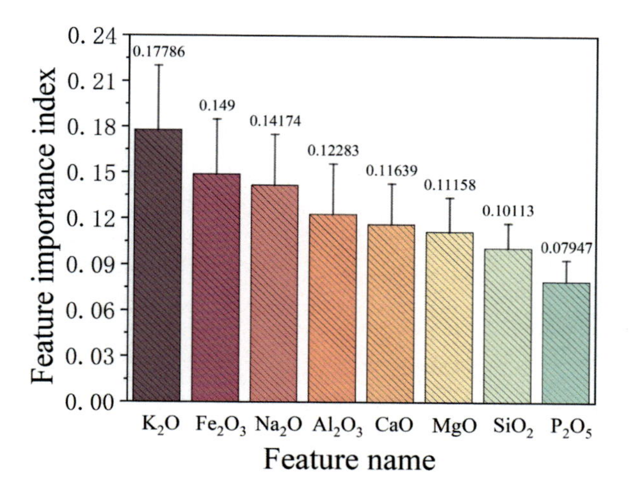

Figure 12.6 Feature importance scores.

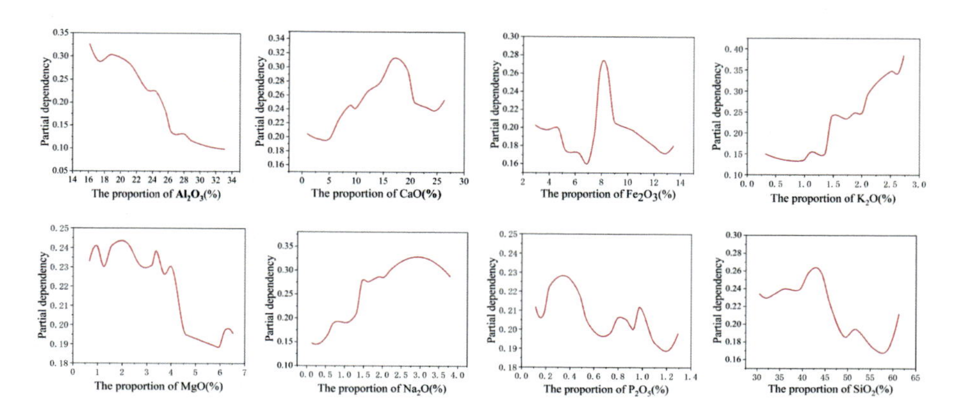

Figure 12.7 Partial dependency diagram for eight elements.

increased. The PDP results are also in good agreement with the conclusions published in existing literature [28].

The global interpretation results for the entire dataset achieved using SHAP are shown in Fig. 12.8. As shown, K_2O, Al_2O_3, and Na_2O had a greater influence on the classification model for low-reactivity CFA (class 0), whereas SiO_2, Fe_2O_3, and CaO had a moderate influence. In contrast, P_2O_5 and MgO had the smallest influence. Consistent results were observed for high-reactivity CFA (class 1), indicating

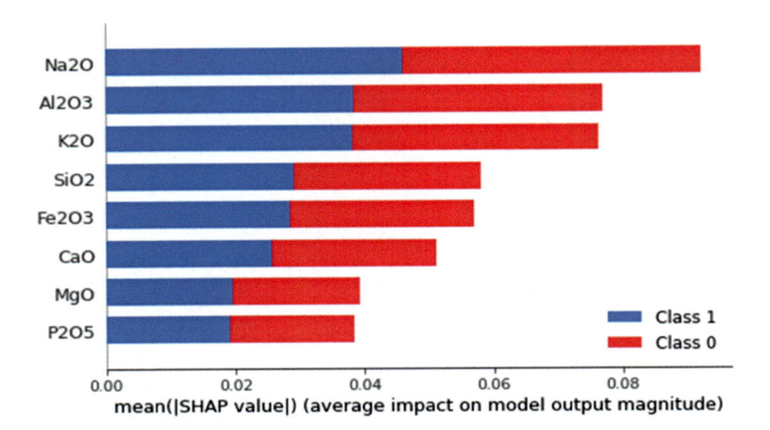

Figure 12.8 SHAP values for global interpretation.

that these trends are globally applicable for the interpretation of the whole model. Overall, these results are in agreement with the results from the feature importance analysis and PDPs, further indicating the consistency and continuity of the different feature analysis methods used in this chapter.

12.6 Summary

1. In this chapter, the cross-validation method was selected to validate the model accuracy, and the optimal hyperparameter combination for the RF classification algorithm was determined based on learning curves. Based on four evaluation metrics, the accuracy of the model was 85.45%, its recall was 97.56%, its precision was 84.29%, and the corresponding AUC value was 0.924, indicating robust performance of the predictive model.
2. The feature importance, PDP, and SHAP methods were used for model interpretation. The feature importance yielded the importance scores of eight features as K_2O—0.1779, Fe_2O_3—0.1490, Na_2O—0.1417, Al_2O_3—0.1228, CaO—0.1164, MgO—0.1116, SiO_2—0.1011, and P_2O_5—0.0795. The PDP analysis revealed that the probability of high-reactive fly ash decreased with the increasing MgO and Al_2O_3, whereas it initially increased and then decreased with increasing Fe_2O3 and CaO content. The global SHAP interpretation revealed that K_2O, Al_2O_3, and Na_2O influenced the model to a greater extent, followed by SiO_2, Fe_2O_3, and CaO.
3. Overall, in this chapter, an RF classification model was successfully established. Comprehensive analysis indicated that this RF model has good generalization ability and robustness for CFA reactivity classification. This classification model can be applied to real-world applications, which will greatly decrease the need for complex amorphous phase analysis by XRD and crystal structure analysis. Using this approach, CFA reactivity can be rapidly classified, which can greatly improve production efficiency and save material and labor costs. In addition, the model comprehensively analyzes the influence of each chemical component of fly ash on its reactivity, thus providing useful constraints for the efficient recycling of fly ash.

References

[1] C. Zhicheng, W. Zhefeng, F. Xiong, Coal fly ash utilization of thinking, Journal of Green Environmental Protection Building Materials 08 (2021) 60–61.

[2] H. Yafei, L. Keqing, H. Bin, et al., Effect of composite cementitious system on strength of wet shotcrete with tailings and its ratio optimization, Journal of Central South University (Science and Technology) 52 (11) (2021) 3999–4009.

[3] S. Alterary Seham, H. Marei Narguess, Fly ash properties, characterization, and applications: a review, Journal of King Saud University - Science (2021) 101536.

[4] F. Tournier Robert, I. Ojovan Michael, Undercooled phase behind the glass phase with superheated medium-range order above glass transition temperature, Physica B: Condensed Matter (2020) 412542.

[5] G.V.P.B. Singh, K.V.L. Subramaniam. Quantitative XRD study of amorphous phase in alkali activated low calcium siliceous fly ash. Construction and Building Materials 124 (2016) 139–147.

[6] M. Sivasubramanian, R.R. Saravanan, L. Animenan, et al., Crystal structure analysis, hirshfeld surface analysis, spectral investigations (FT-IR, FT-R), DFT calculations, ADMET studies and molecular docking of 3H-Methyl-1H-pyrazole-1-carboxamide (3MPC), Journal of the Indian Chemical Society 99 (4) (2022) 100402.

[7] X. Tang, Quantitative study of fly ash physical phase based on Rietveld full spectrum fitting, Non-Ferrous Metallurgy Energy Conservation 31 (01) (2015) 48–51.

[8] L. Xing, T. Yixuan, H. Jinfeng, et al., Study on the evaluation of fly ash activity for geopolymers, Materials Guide 36 (02) (2022) 102–108.

[9] Q. Chongchong, Y. Xingyu, L. Guichen, et al., A new generation of artificial intelligence in mine filling review and prospect the application, Journal of Coal 46 (02) (2021) 688–700.

[10] B. Kang, C. Seok, J. Lee, A benchmark study of machine learning methods for molecular electronic transition: tree-based ensemble learning versus graph neural network, Bulletin of the Korean Chemical Society 43 (3) (2022) 328–335.

[11] P. Sihag, B. Singh, S. Gautam, et al., Evaluation of the impact of fly ash on infiltration characteristics using different soft computing techniques, Applied Water Science 8 (6) (2018) 187.

[12] M.A. Khan, F. Farooq, M.F. Javed, et al., Simulation of depth of wear of eco-friendly concrete using machine learning based computational approaches, Materials 15 (1) (2022) 58. Available from: http://doi.org/10.3390/ma15010058.

[13] Y.M. Jiang, Y. Li H, Y.S. Zhou, Compressive strength prediction of fly ash concrete using machine learning techniques, Buildings-Basel 12 (5) (2022) 16.

[14] F. Quannan, W. Jianbin, Z. Jianping, et al., A review of random forest methods, Statistics and Information Forum 26 (03) (2011) 32–38.

[15] Q. Yang, L. Sheng, J. Liang, et al., Bridge anomaly monitoring data recognition method based on statistical feature mixture and random forest importance ranking, Chinese Journal of Sensor Technology 35 (06) (2022) 756–762.

[16] D. Rengasamy, M. Jimiama, A. Kumar, et al., Feature importance in machine learning models: a fuzzy information fusion approach, Neurocomputing 511 (2022) 163–174.

[17] T.Z. Yinguo, L. Xiaobo, Yuxiang, et al., Hyperspectral image classification based on feature importance, Infrared Technology 42 (12) (2020) 1185–1191.

[18] T. Parr, J.D. Wilson, Partial dependence through stratification, Machine Learning with Applications 6 (2021) 100146.

[19] A.B. Parsa, A. Movahedi, H. Taghipour, et al. Toward safer highways, application of XGBoost and SHAP for real-time accident detection and feature analysis. Accident Analysis & Prevention 136 (2020) 105405.
[20] S.C. Matias, J.M. Scavuzzo, M.N. Campero, et al., Feature importance: opening a soil-transmitted helminth machine learning model via SHAP, Infectious Disease Modelling 7 (1) (2022) 262–276.
[21] W. Mengting, C. Qiusong, Q. Chongchong, Slope safety and stability evaluation and protective measures based on machine learning, Chinese Journal of Engineering Science 44 (02) (2022) 180–188.
[22] K.S. Ashwin, Statistics of spatial averages and optimal averaging in the presence of missing data, Spatial Statistics 25 (2018) 1–21.
[23] Y. Xiao, L. Ya, W. Hairui, et al., Based on Pearson correlation coefficient of the hybrid domain feature selection method of the rolling bearing, Journal of Chemical Industry Automation and Instrumentation 49 (03) (2022) 308–315.
[24] R.X. He, Mine power plant fly ash physical and chemical properties analysis, Value Engineering 35 (31) (2016) 96–97.
[25] Y. Chunlin, H. Mingqing, Q. Haoyu, Quality classification of fractured rock mass based on multivariate data normalization treatment, China Mining Industry 31 (08) (2022) 158–164.
[26] S. Zhu, X. Gao, Z. Zhang, et al., Infrared spectrum dataset partition ratio and pretreatment method research, Journal of Analytical Chemistry 50 (09) (2022) 1415–1429.
[27] Q. Liu, Z. Bai, D. Wang, et al., Chemical composition, physicochemical properties and Application analysis of fly ash in China, China Non-Metallic Minerals Industry Guide 01 (2021) 1–9.
[28] C. Qi, W. Mengting, J. Zheng, et al., Rapid identification of reactivity for the efficient recycling of coal fly ash: hybrid machine learning modeling and interpretation, Journal of Cleaner Production 343 (2022) 130958.

Forecasting the uniaxial compressive strength of solid ash-based concrete

13

Abstract

This chapter presents a method for estimating the compressive strength of concrete made using slag–fly ash–superplasticizer. Random forest and particle swarm optimization algorithms were combined to construct a prediction model and perform hyperparameter tuning. The correlation coefficient (R), the explanatory variance score (EVS), the mean absolute error (MAE) and the mean square error (MSE) were used to evaluate the performance of the model. R = 0.954, EVS = 0.901, MAE = 3.746, and MSE = 27.535 of the optimal RF-PSO model on the testing set indicated the high generalization ability. In addition, principal component analysis was also attempted to reduce the dimensionality of the input features. Finally, feature analysis was performed to promote knowledge discovery in a complex slag–CFA–superplasticizer system.

13.1 Background

Concrete is a widely used building material composed of cement, coarse and fine aggregates, admixtures, and water. The strength of concrete is mainly determined by cement; however, cement is costly and large amounts of CO_2 are produced during its usage, thus posing a serious threat to the environment given its extremely widespread usage in construction [1]. Cement manufacturing techniques are constantly being improved to reduce their CO_2 emissions [2], but a growing body of research suggests that using alternative materials may be a more appropriate and sustainable approach. As a result, green building materials are being used more widely; for example, it is becoming increasingly popular to partially replace Portland cement with solid waste materials such as blast furnace slag (BFS) and fly ash (FA) [3].

BFS is a by-product of smelting pig iron in blast furnaces at around 1,500°C. According to recent statistics, China's BFS production of more than 139 million tons accounted for 32% of the world's total in 2017 [4]. The massive accumulation of BFS poses a potential threat to the environment. The main components of BFS contain SiO_2, CaO, and Al_2O_3 [5], which have pozzolanic activity and are comparable to Portland cement. Razak and Babu [6] demonstrated that BFS is a good substitute for Portland cement as it can reduce the heat of hydration and improve durability while maintaining the concrete's strength. High levels of BFS are also often used in

Machine Learning Applications in Industrial Solid Ash. DOI: https://doi.org/10.1016/B978-0-443-15524-6.00008-X

© 2024 Elsevier Inc. All rights reserved.

concrete formulations for offshore applications as they reduce chlorine penetration, thus ensuring better protection of their internal reinforcements in a marine setting [7].

The main sources of FA are coal-fired power plants and urban central heating boilers, which account for 60%–88% of the by-products of pulverized coal combustion [8]. With the increasing development of the electric power industry, FA emissions have increased year on year and have now become one of the main industrial waste types produced in China. The accumulation of FA not only wastes land resources but also causes harm to vegetation and human beings. Based on the pozzolanic effect, FA can be used as a binder or raw material for clinker production and can partially replace cement in concrete production [8]. Compared with traditional Portland cement, FA-based concrete has low water requirements, low heat of hydration, low risk of early cracking, and high strength gains in the late period [9].

Concrete strength is key to effectively using BFS and FA as supplementary cementitious materials (SCMs). To date, many researchers have investigated the mechanical properties (e.g., elastic modulus, compressive strength, and tensile strength) of BFS-based and FA-based concrete. For example, Oner et al. [10] prepared 28 different formulations of concrete mixtures in the laboratory and found that up to 40% of the cement could be replaced by FA without reducing the mechanical properties of the resulting concrete. Chidiac et al. [11] studied the mechanical properties of concrete mixed with abrasive BFS (GGBFS) and found that, using the same composition of cementitious materials, when the GGBFS content reached 50%, the 28-day compressive strength of the concrete was equivalent to that of ordinary Portland cement concrete [12]. When the GGBFS content exceeds 50% [7], its 28-day strength may be less than that of ordinary Portland cement concrete. Prang et al. [13] demonstrated that increasing the Blaine value rate of BFS can increase the compressive strength of the resulting concrete. However, the aforementioned studies required extensive experiments, which are time-consuming, and costly, thus hindering the efficient design of concrete preparation using SCMs. Accordingly, a more efficient strength prediction method is urgently required.

In recent years, the rapid development of artificial intelligence has made it feasible to accurately predict concrete strength [14]. For example, Nazanin et al. [15] used radial basis function, multilayer perceptron, support vector regression, adaptive network-based fuzzy inference system, and deep neural network methods to accurately predict the elastic modulus and compressive and tensile strength of FA-based concrete. In addition, Choudhary et al. [16] combined artificial neural networks with sequential feature selection to accurately predict the compressive strength of FA-based concrete. They also found that cement, silica fume, FA, and water had the greatest impact on compressive strength. Although the studies cited earlier applied these methods successfully, few other studies to date have applied machine learning (ML) to investigate the strength of concrete made with BFS, FA, and superplasticizers. Furthermore, the impact of dimensionality reduction on ML performance in this context has not been fully explored.

In this chapter, we combined the random forest (RF) and particle swarm optimization (PSO) algorithms to predict the compressive strength of concrete containing BFS, FA, and superplasticizers. Specifically, we used RF for nonlinear relational mapping from input to output and PSO for hyperparameter tuning. Additionally, principal component analysis (PCA) was also tested for dimensionality reduction to improve the model's performance. Finally, we conducted a sensitivity analysis on the input features to provide a reference for future studies investigating how to improve the strength of concrete.

13.2 Materials and method

The method adopted in this chapter was divided into the following main components (Fig. 13.1). First, the data were collected from previous studies through a literature search, and the PCA algorithm was attempted for dimensionality reduction. The initial and dimensionally reduced datasets were then randomly divided, and the optimal splitting ratio was determined according to the RF model performance. Based on this, PSO was used to tune the model's hyperparameters with different indicators used for evaluation. Finally, feature sensitivity analysis was performed using the optimized RF model.

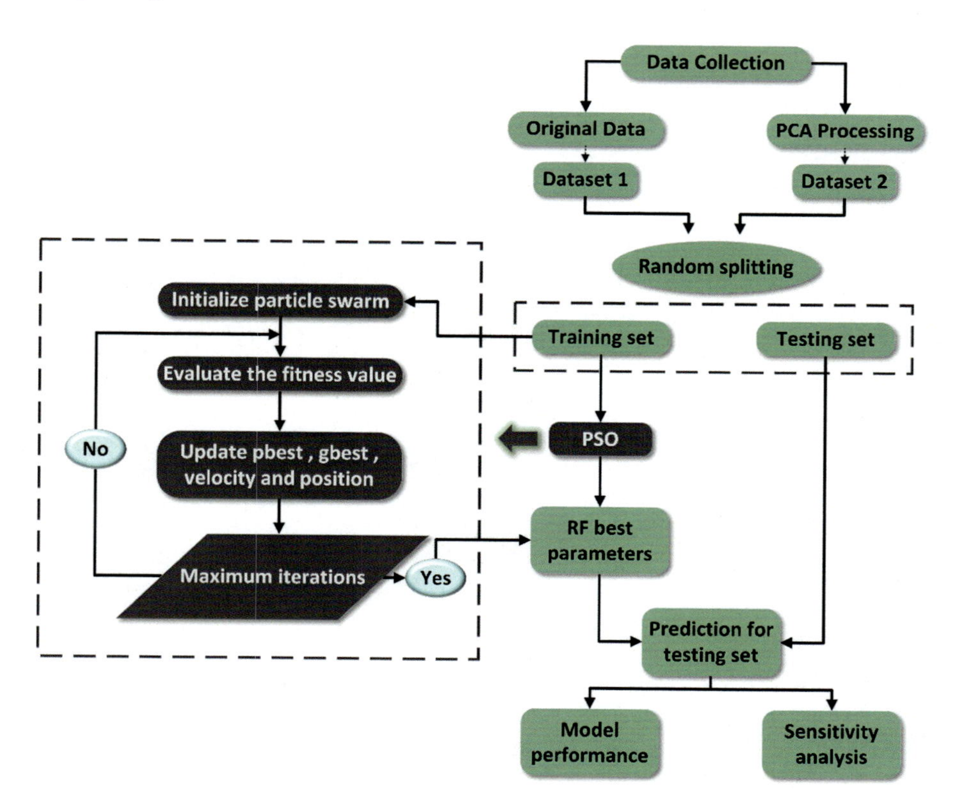

Figure 13.1 Methodology for strength prediction.

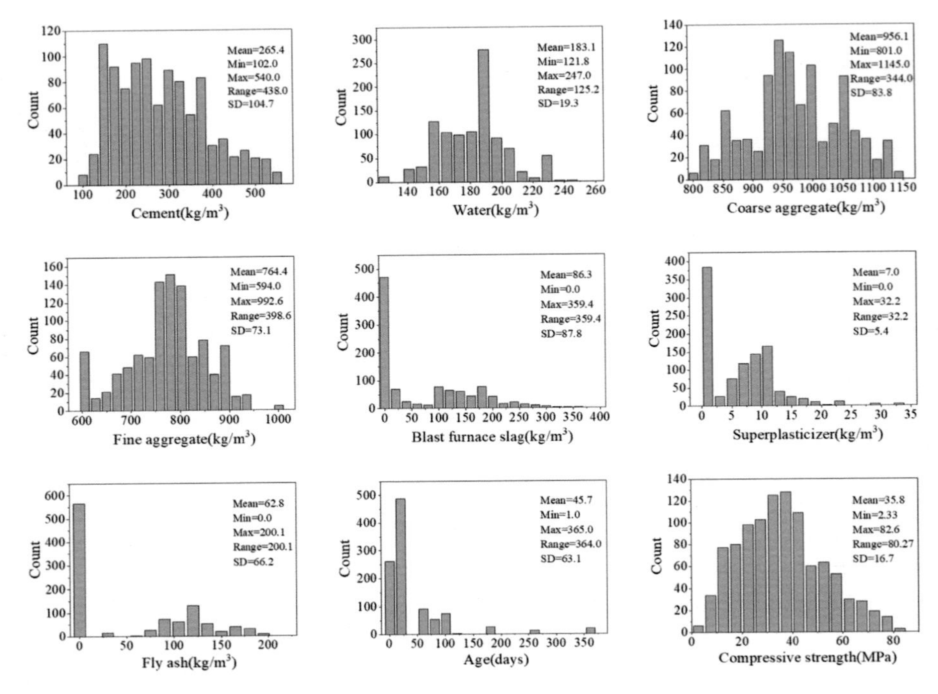

Figure 13.2 Statistical analysis of features and target variables in the dataset.

13.2.1 Dataset preparation

The data were collected from widely used and representative studies [17], comprising a total of 1,030 samples. The input variables consisted of cement (kg/m^3), blast furnace slag (kg/m^3), fly ash (kg/m^3), water (kg/m^3), superplasticizer (kg/m^3), coarse aggregate (kg/m^3), fine aggregate (kg/m^3), and age (days). The compressive strength of the concrete at a given age was the target variable, which was obtained by performing typical laboratory compressive test procedures on bulk samples. Statistical analyses for each parameter in the dataset are shown in Fig. 13.2.

The PCA method was used to reduce the dimension of eight input features to evaluate the impact of dimensionality reduction on model performance. To project the original feature to the most informative dimension to the maximum possible extent, 95% of the information was retained, resulting in five input variables being obtained. We then used initial dataset 1, with eight input variables, and dataset 2, whose dimension was reduced, to construct the RF models separately and compared their performance.

13.2.2 Dataset splitting

As described in Section 7.3.2, raw datasets must be split into training and testing sets. The training set trains the model by determining the mapping from input to output values, while the testing set is used to assess the model's generalization ability on unknown data. Given that the split ratio of the dataset will affect the model's

Table 13.1 Value range of the hyperparameters of the RF models based on two datasets.

Hyperparameters	Explanation	Type	Tuning range
Max_depth	The maximum depth of each decision tree	Integer	1–15
Number_DT	The number of decision trees in the forest	Integer	50–2,000
Min_samples_split	The minimum number of samples required to split an internal node	Integer	2–15
Min_samples_leaf	The minimum number of samples at the leaf node	Integer	1–15
Max_features	The number of features to be used when looking for the optimal split	Float	0.4–1

performance [18], the size of the testing set was gradually increased from 10% to 65% of the total data in 5% increments, and the correlation coefficient (R) was used for evaluation. In addition, to avoid randomness due to the splitting step, each split ratio was repeated 50 times, and the average R was then taken as the model's performance under this split ratio.

13.2.3 Hyperparameter tuning

In contrast to model parameters, hyperparameters are established before the model's learning process begins and cannot be modified during training [19]. The hyperparameter selection affects the model's performance [20]. In this chapter, PSO was used to tune the RF hyperparameters for the models corresponding to the two datasets. Table 13.1 describes the meanings and value ranges of the various hyperparameters used. The correlation coefficient R was used for the fitness function, and the maximum number of iterations was set to 50. A detailed description of the principles of PSO is provided in Section 7.3.4.4.

13.2.4 Performance evaluation

In ML modeling, evaluation indicators are required to measure the model's performance. This chapter evaluated the model performance using R, the explained variance score (EVS), mean absolute error (MAE), and mean squared error (MSE). For the specific formula and explanation of the indicators, see Section 7.4.2.

13.3 Results and discussion

13.3.1 Best split ratio of datasets

To improve the model's performance, we adjusted the split ratio of the two datasets. As shown in Fig. 13.3, when the size of the testing set increased from 10% to 15%,

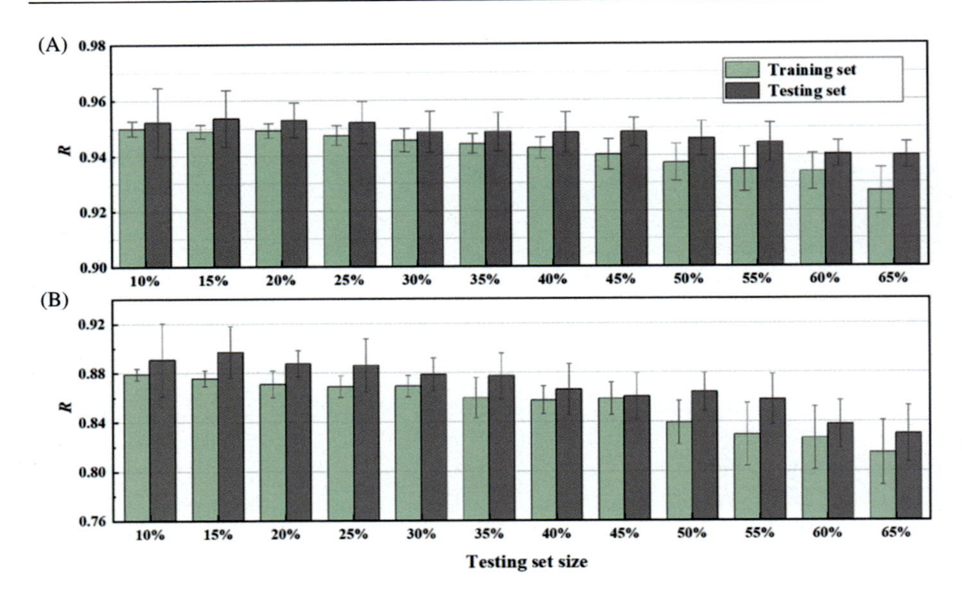

Figure 13.3 Selection of the optimal split ratios of (A) dataset 1 and (B) dataset 2.

the performance of the models based on the two datasets was greatly improved: the mean values of *R* on the testing set reached their maximum values of 0.958 and 0.895 for datasets 1 and 2, respectively. However, with further increases in the size of the testing set, the average *R* values gradually decreased. Although there were some small fluctuations, the *R* values did not exceed 0.958 and 0.895 for the two models. As a result, the size of the testing set was set to 15%.

13.3.2 Determination of the optimal hyperparameters

In this chapter, we used the PSO algorithm to optimize the hyperparameters of the models for the two datasets. As shown in Fig. 13.4, as the number of iterations increased for dataset 1, the average *R* value of the model increased from 0.9515 in the first generation to 0.9534 in the 15th generation and then tended to be stable. For dataset 2, the *R* value of the model tended to be stable after five iterations ($R = 0.862$); however, the *R* value of the model increased once more when the number of iterations reached 25, and the global optimal position was updated. With increasing iterations, the average *R* value then remained unchanged at 0.863 until 32 iterations. In summary, optimal and stable model performance can be obtained when 50 iterations are selected. The corresponding hyperparameters obtained from this process are shown in Table 13.2.

13.3.3 Selection of the optimal RF-PSO model

To select the best concrete strength prediction method, two RF-PSO models were constructed based on dataset 1 (model 1) and dataset 2 (model 2) and then

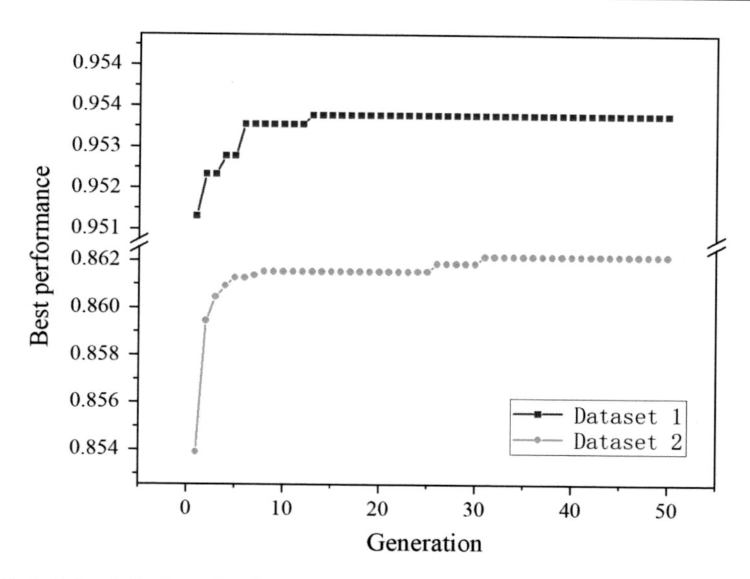

Figure 13.4 Using PSO iterative for hyperparameter optimization. Note that the y-axis uses the R value as the metric.

Table 13.2 Optimal RF hyperparameters for datasets 1 and 2.

Hyperparameters	Dataset 1	Dataset 2
Max_depth	15	15
Number_DT	1457	356
Min_samples_split	2	2
Min_samples_leaf	1	1
Max_features	0.466	0.978

compared. As shown in Fig. 13.5, four evaluation indicators were used for comparative analysis. For model 1, the R, EVS, MAE, and MSE values on the training set were 0.954, 0.901, 3.746, and 27.535, respectively. In contrast, the predictive performance of model 2 ($R = 0.864$, EVS = 0.740, MAE = 6.130, and MSE = 72.351) after PCA treatment was worse than that of model 1. The aforementioned results showed that PCA did not improve model performance; thus, the application of PCA was not appropriate for the specific dataset used in this chapter.

We further evaluated the performance of the models in terms of the observed and predicted compressive strength ratios ($CS_{\_obe}/CS_{\_pre}$). As shown in Fig. 13.6, for the two models, over 80% of $CS_{\_obe}/CS_{\_pre}$ observations were in the range of $\sim 0.8-1.2$, indicating good RF-PSO model performance. In addition, the observed $CS_{\_obe}/CS_{\_pre}$ values plot slightly to the right of 1, indicating that the compressive strength values predicted by the RF_PSO model were slightly lower than the actual values recorded in the original dataset.

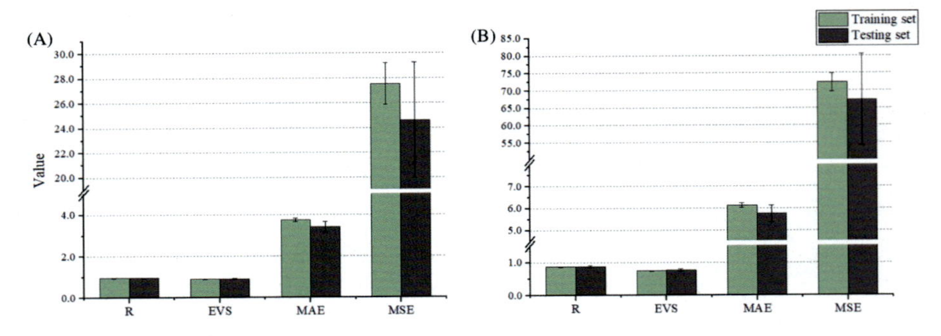

Figure 13.5 Performance evaluation of (A) model 1 and (B) model 2 using four indicators.

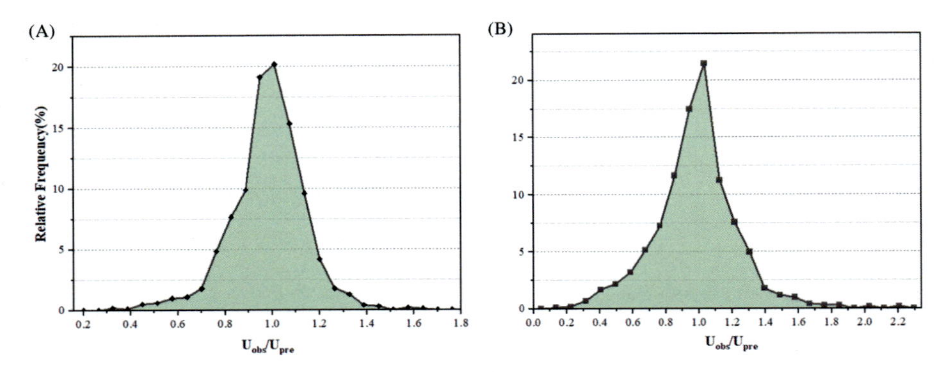

Figure 13.6 Relative frequencies of observed and predicted compressive strength ratios $CS_{\_obe}/CS_{\_pre}$ of (A) model 1, (B) model 2.

In Fig. 13.7, the data points are evenly distributed around the $y = x$ line, i.e., the perfect regression line. For Model 1, the difference between the predicted and measured compressive strength of concrete was 20.68% below 1 MPa and 23.98% above 5 MPa. For model 2, the error was 12.91% below 1 MPa and 45.54% above 5 MPa. The R^2 value of model 1 was 0.92, higher than the equivalent 0.77 value for model 2. These indicators demonstrate that model 1 outperformed model 2, with high consistency between the predicted and actual strength; thus, model 1 was selected as the final model for our chapter.

13.3.4 Feature sensitivity analysis

In addition to constructing a prediction model with good performance, it is also important to determine which variables have an impact on the compressive strength of concrete. In this section, feature sensitivity analysis was performed on the input variables of model 1 using the built-in RF feature importance function in Python's sklearn package. The calculated scores were normalized, as shown in Fig. 13.8.

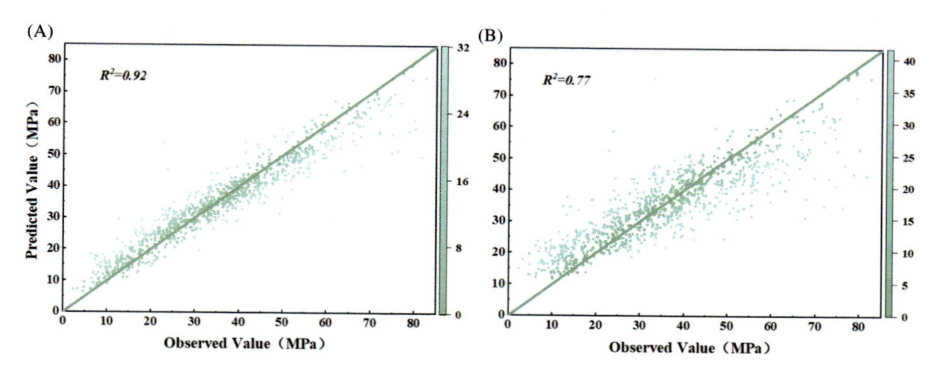

Figure 13.7 Comparison between the predicted and actual strength values for (A) model 1 and (B) model 2.

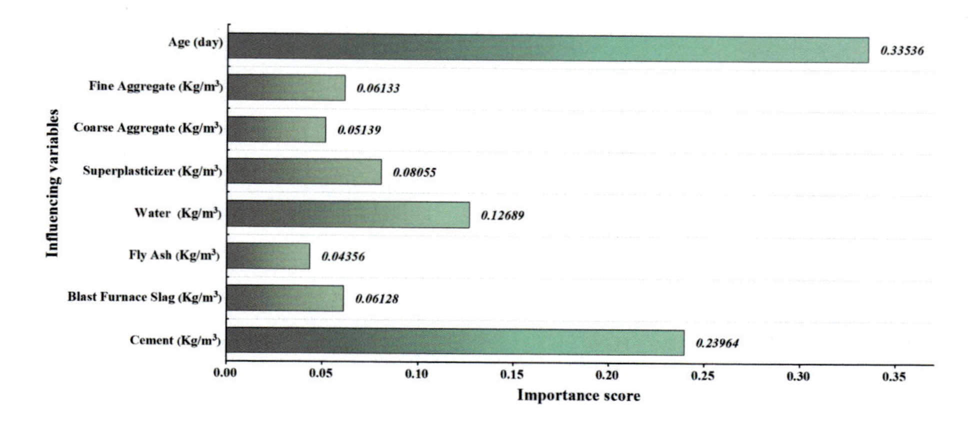

Figure 13.8 Sensitivity analysis of input features.

The top three most important features were curing age, cement content, and water content, which contributed 33.536%, 23.964%, and 12.689%, respectively. In comparison, FA, coarse aggregate, BFS, and fine aggregate contents all have relatively smaller influences on compressive strength with importance scores of 4.356%, 5.139%, 6.128%, and 6.133%, respectively.

Despite the significantly lower contribution of components such as FA in this chapter, Cho et al. [21] found that after 91 days of curing, the strength of FA-based concrete will increase due to the pozzolanic effect. Moreover, the combination of FA and BFS has been found to offset the short-term strength loss of FA-based concrete while maintaining long-term performance [22]. BFS also can improve the mechanical properties of concrete by reducing its porosity and increasing its elasticity to weak acids and salts. Jain et al. found that the compressive strength of concrete was highest when BFS replaced 10% of cement [23]. In addition, many studies have shown that adding superplasticizers to concrete can reduce the amount of water required for mixing, thereby reducing the water–cement ratio and porosity and improving the performance of the concrete [24,25]. To improve the compressive strength of concrete efficiently, the amount of superplasticizer is generally limited to within 3% [26].

13.4 Summary

In this chapter, the RF and PSO algorithms were combined to construct a rapid prediction model for concrete strength. PCA dimension reduction was also attempted to improve the performance of the model. The final model suitable for our method was selected using four evaluation indexes and a variety of comparison methods, and a feature sensitivity analysis of this model was performed. This chapter's main conclusions are as follows:

1. The predictive performance of model 2 ($R = 0.864$, EVS = 0.740, MAE = 6.130, and MSE = 72.351) after PCA treatment was worse than that of model 1. As a result, PCA was not appropriate for the current BFS–FA–superplasticizer concrete dataset.
2. After PSO tuning, the R, EVS, MAE, and MSE values of the optimal model (based on dataset 1) were 0.954, 0.901, 3.746, and 27.535, respectively, indicating that the model could accurately predict concrete strength.
3. Feature sensitivity analysis shows that the curing time had the most significant impact on concrete compressive strength, followed by cement, water, superplasticizers, fine aggregate, BFS, coarse aggregate, and FA content, which also have an important influence in determining the strength of concrete.

References

[1] I. Vázquez-Rowe, et al., Production of cement in Peru: understanding carbon-related environmental impacts and their policy implications, Resources, Conservation and Recycling 142 (2019) 283–292.
[2] S.T. Chopperla, et al., Development of an efficient procedure for sustainable low carbon cement manufacturing process, Applied Mechanics and Materials (2015). Trans Tech Publ.
[3] P. Lehner, P. Konečný, P. Ghosh, Variation of durability and strength parameters of pumice based mixtures, Materials 14 (13) (2021) 3674.
[4] S.K. Tripathy, et al., Utilisation perspective on water quenched and air-cooled blast furnace slags, Journal of Cleaner Production 262 (2020) 121354.
[5] V. Václavík, et al., The use of blast furnace slag, Metalurgija 51 (4) (2012) 461–464.
[6] H. Abdul Razak, D. Venkatesh Babu, Fresh, strength and durability characteristics of binary and ternary blended self compacting concrete, International Journal of Engineering and Advanced Technology 9 (2) (2019).
[7] I. Yuksel, Blast-furnace slag, Waste and Supplementary Cementitious Materials in Concrete (2018) 361–415. Elsevier.
[8] A. González, R. Navia, N. Moreno, Fly ashes from coal and petroleum coke combustion: current and innovative potential applications, Waste Management & Research 27 (10) (2009) 976–987.
[9] R. Elmrabet, A. El Harfi, M. El Youbi, Study of properties of fly ash cements, Materials Today: Proceedings 13 (2019) 850–856.

[10] A. Oner, S. Akyuz, R. Yildiz, An experimental study on strength development of concrete containing fly ash and optimum usage of fly ash in concrete, Cement and Concrete Research 35 (6) (2005) 1165–1171.

[11] S. Chidiac, D. Panesar, Evolution of mechanical properties of concrete containing ground granulated blast furnace slag and effects on the scaling resistance test at 28 days, Cement and Concrete Composites 30 (2) (2008) 63–71.

[12] R. Majhi, A. Nayak, B. Mukharjee, Development of sustainable concrete using recycled coarse aggregate and ground granulated blast furnace slag, Construction and Building Materials 159 (2018) 417–430.

[13] P. Subpa-asa, et al., Evaluation of the prediction and durability on the chloride penetration in cementitious materials with blast furnace slag as cement addition, Construction Materials 2 (1) (2022) 53–69.

[14] I. Nunez, et al., Estimating compressive strength of modern concrete mixtures using computational intelligence: a systematic review, Construction and Building Materials 310 (2021) 125279.

[15] N. Fasihihour, et al., Experimental and numerical model for mechanical properties of concrete containing fly ash: systematic review, Measurement 188 (2022) 110547.

[16] D. Choudhary, Learning algorithms using BPNN & SFS for prediction of compressive strength of ultra-high performance concrete, Journal of Machine Learning Research: JMLR 4 (2) (2019) 27.

[17] D.-C. Feng, et al., Machine learning-based compressive strength prediction for concrete: an adaptive boosting approach, Construction and Building Materials 230 (2020) 117000.

[18] Q.-F. Li, Z.-M. Song, High-performance concrete strength prediction based on ensemble learning, Construction and Building Materials 324 (2022) 126694.

[19] T. Yu, H. Zhu, Hyper-parameter optimization: a review of algorithms and applications (2020). arXiv Preprint arXiv:2003.05689.

[20] M. Claesen, B. De Moor, Hyperparameter search in machine learning (2015). arXiv Preprint arXiv:1502.02127.

[21] J. Li, Regression and classification in supervised learning, Proceedings of the 2nd International Conference on Computing and Big Data (2019) 99–104. Association for Computing Machinery: Taichung, Taiwan.

[22] G. Li, X. Zhao, Properties of concrete incorporating fly ash and ground granulated blast-furnace slag, Cement and Concrete Composites 25 (3) (2003) 293–299.

[23] K.L. Jain, L.S. Rajawat, G. Sancheti, Mechanical properties of ground granulated blast furnace slag made concrete, IOP Conference Series: Earth and Environmental Science (2021). IOP Publishing.

[24] S.Y. Zhang, Y.F. Fan, N.N. Li, The effect of superplasticizer on strength and chloride permeability of concrete containing ggbfs, Advanced Materials Research (2013). Trans Tech Publ.

[25] J. Gołaszewski, J. Szwabowski, Influence of superplasticizers on rheological behaviour of fresh cement mortars, Cement and Concrete Research 34 (2) (2004) 235–248.

[26] A.M. Zeyad, A. Almalki, Influence of mixing time and superplasticizer dosage on self-consolidating concrete properties, Journal of Materials Research and Technology 9 (3) (2020) 6101–6115.

Challenges and future perspectives of machine learning in industrial solid ashes management

14

Abstract

Although we have presented several case studies related to machine learning (ML)- based industrial solid ash management, there remain many challenges and new future application prospects for ML in industrial solid ash management. Some of these applications may rely on new monitoring technologies throughout the solid ash process; however, other key aspects will likely include the creation of new ML algorithms, data integration, and the design of new application methodologies. In this chapter, the current challenges facing ML in industrial solid ash management will be discussed and possible development trends will be described. Herein, we aimed to shed light on some key areas within this topic rather than speculating on all the possibilities.

14.1 Major challenges of current studies

14.1.1 Multidisciplinary collaboration

One of the key challenges in the application of ML to solid ash management is the inherently multidisciplinary nature of the problem. Engineers, practitioners, mathematicians, data scientists, and ML researchers must all find a common language to clarify their goals, conduct rigorous analysis and training, and avoid common pitfalls, incorrect usage of methods, and misinterpretations. The No Free Lunch theorem confirms that no single ML method is superior to others a priori [1]; therefore, a thorough understanding of the data types to be used and problems to be solved—an area where practitioners and data scientists should collaborate closely—is essential to develop efficient ML models.

Further evidence for the current lack of multidisciplinary collaboration is that data scientists in solid ash management are placing increasing emphasis on new and more complex ML algorithms, for example, ensemble learning based on hybrid metapredictors, hyperparameter optimization with many training subcycles, and sophisticated feature engineering. This overcomplexity challenges the real-world application of ML models. Cross talk is required to understand industry needs and what can be achieved with current methods; this discussion is essential for the development of efficient and robust ML models for solid ash management.

Machine Learning Applications in Industrial Solid Ash. DOI: https://doi.org/10.1016/B978-0-443-15524-6.00007-8
© 2024 Elsevier Inc. All rights reserved.

14.1.2 The "independent" testing set

When we apply certain ML methods, we normally estimate the method's generalization capability using an independent testing set as in the case studies presented in this book. In a normal ML application procedure, we split the entire dataset randomly into two or three subsets (training/testing or training/validation/testing). By doing so, we sample data points from each scenario or category of the whole dataset into the training and testing sets. Assuming we have a solid ash dataset containing data from 10 different origins, in most cases, the training and testing sets would contain some data points from each origin, assuming that the number of data points from each source is not very low. Therefore, ML models trained with a training set prepared from such a dataset might have good generalization capability on the testing set given that the ML models have already learned the underlying knowledge from samples from all data origins, as described earlier. However, the generalization capability that matters is the trained model's predictive performance on unknown data points that it has not previously encountered. In other words, trained ML models should be tested on completely new scenarios, which is a significant aspect of real-world applications of ML modeling.

Thus the preparation of a complete "independent" testing set is of vital significance for evaluating generalization capability. There are two approaches suggested for the "independent" test set preparation, and we will use the previous examples as an illustration. The first approach is to collect independent data points from new scenarios. For example, the generalization of the ML model discussed earlier could be effectively verified if the predictive performance was tested on data points from an 11th origin. If the collection of new data points is not possible, an alternative approach can be used based on dataset splitting; specifically, we could use data points from nine of the aforementioned origins for model training and then use all the data points from the remaining origin for independent testing. In this scenario, the ML models would not observe any of the data points from the 10th origin during the training process, and, thus, the predictive performance of the 10th origin would be a strong indicator of the model's generalization capability. A detailed description of the implementation of the second approach can be found in the literature, such as in [2].

14.1.3 Findable, accessible, interoperable, and reusable data

The present data in solid ash management rarely meet all the guiding principles for data management, that is, the data are findable, accessible, interoperable, and reusable (FAIR) [3]. In essence, FAIR requires that the (metal) data be accessible in a global, persistent, and transparent way and should be provided with a license for use and a detailed source. To fully meet the FAIR principles, we must not only create and use standard ways to report data, but we must also provide access to tools, protocols, codes, and parameters for data reproducibility. Therefore, developing a user-friendly ecosystem for the sharing and access of FAIR datasets is fundamental and the key challenge for unlocking the true potential of data-driven approaches in solid ash management. Keeping reproducibility and data sharing in mind at the outset of research, rather than in retrospect, is essential to meeting the FAIR principles of a dataset.

14.1.4 The reliability of machine learning models on single prediction

The reliability of a trained ML model in solving a particular problem is an important concern in ML modeling. Instead of estimating the general reliability of ML models using approaches such as the correlation coefficient between the predicted and true values or the classification accuracy, the reliability of ML models on a given case can instead be verified by evaluating whether the prediction is sensitive to the addition of this case to the training set. Taking classification problems as an example, the classifier's decision is reliable on a given case if the decision does not depend on the addition of this case to the training set, irrespective of whether this new case is labeled correctly or not. This leads to a widely accepted approach to assess the reliability of an ML method for a given case. We can verify the reliability by labeling the case with all the possible decisions and then adding the case to the training set and rerunning the model training process. If the new decision made by the updated ML model is not very different from its original decision, we can assume that the classification of that case is relatively trustworthy. However, a decision that is sensitive to the addition of new cases to the training set is not reliable. The aforementioned concept has been described in detail in [4].

14.1.5 Gray-box and white-box machine learning

The ultimate goal of science, which remains especially true for ML applications in solid ash management, is to not only achieve better predictive performance but also to be able to explain results. At present, one of the main drawbacks of ML modeling, and probably the aspect that causes the most doubt, is the "black box" nature of most advanced ML algorithms. For example, why does a particular ML algorithm achieve better performance? What are the critical input variables for a typical problem on a global scale? How will the input influence the target? Why do various inputs affect the output in different ways? What input value ranges and what parts of the input space are most crucial for predictive performance? Why do ML models have good predictive performance in some instances and large discrepancies in others? Thus interpretable and explainable ML models, constructed using gray-box or even white-box ML algorithms, are the key to further promoting our understanding of mechanisms underlying data science.

14.1.6 Implementing verified machine learning models in existing solid ash management systems

The ultimate goal of the current investigation into the application of ML modeling in industrial solid ash management is to promote improved efficiency. The predictive performance of ML models has been well tested and verified in certain scenarios, as shown in the case studies in this book, with considerable effort devoted to these topics. However, it remains highly challenging to embed ML models into existing industrial solid ash management systems: how to do so and to what extent ML models can support practitioners is currently unknown.

14.2 Future perspectives

In the following section, we present some possible future opportunities and research directions associated with ML-based solid ash management. To organize this text concisely, we group the future perspectives into three categories: data preparation, ML algorithms and modeling methodology, and application scenarios. These aspects are detailed as follows.

14.2.1 Data preparation

High-quality datasets form the basis of ML applications in industrial solid ash management. It can be said with confidence that more emphasis should always be placed on improving data rather than developing algorithms. The following future research directions are suggested:

- For better development of ML models in this field, high-quality testing data, especially independent testing data as described in Section 7.1.1, are essential, especially given the need for broadly standardized criteria for algorithm evaluation.
- The high-throughput data generation approaches that have been widely employed in various fields—for example, biological and medical sciences—may have the potential to reshape the data generation landscape in solid ash management. Compared with conventional one-by-one approaches, high-throughput technology can rapidly generate large, high-quality databases at a low cost. For solid ash management, high-throughput data generation comprises at least three possible aspects: high-throughput computations/simulations, high-throughput experiments, and high-throughput monitoring.
- The integration of smart sensors within current industrial solid ash management systems provides a significant opportunity to generate extra information throughout the solid ash management workflow, that is, from solid ash generation to solid ash recycling. Based on sensor-provided information, new application scenarios could be designed.
- At present, the metadata associated with industrial solid ash management can be collected in multiple ways. For example, there are different methods to determine the chemical composition of industrial solid ashes. However, there is a current lack of clear guidance on comparing and transforming metadata from various methods. Integrating metadata from various sources represents a good opportunity for future research.
- Currently, there are no automated and streamlined data-processing tools. This type of tool could further promote data preparation. For example, with this tool, data preparation could be performed not only by well-trained researchers and engineers but also by practitioners with little or even no training.
- As ML is heavily dependent on data, especially high-quality data, there is a great need to develop new methods to effectively collect unbiased solid ash data. Furthermore, techniques are also required that can improve data quality by generating more, higher-quality, representative data, in which active learning might play an important role.
- International database collaborations involving a sufficiently large community of researchers, a sufficient variety of data instances, and a highly regarded collaboration leader should be promoted.
- Stringent requirements for data collection and transparent use of statistical techniques are expected to be imposed. This can be further supported by increasing pressure from publishers and grant agencies to ensure the public release of any scripts or datasets relating to a study.

14.2.2 Machine learning algorithms and modeling methodology

The rapid development of algorithms in the fields of ML and AI will continue to provide new algorithms for application to industrial solid ash management. However, some research directions can be actively pursued:

- With the development of increasingly advanced algorithms, model performance will suffer when the algorithms are more complex than the data or the algorithms are excessively complicated for predictions on existing datasets. In this scenario, algorithms that can automatically generate new data and add the new instances to the model training process would be desirable.
- The balance between data availability and the predictive performance of ML models must be maintained, especially during the early stages of ML application in solid ash management when datasets are limited. To address this, metalearning, which learns knowledge both within and across problems, represents a potentially promising solution. Another possible solution to limited data availability is the use of the Bayesian framework. A customized ML application strategy should be designed for small datasets, such as the one proposed in [5].
- Easy-to-use platforms that can support ML implementation, including ML algorithm selection, automatic hyperparameter tuning, etc., are essential to the widespread industrial application of ML in solid ash management. Existing tools, such as Google AutoML, can be considered good starting points; however, the development of more tailored automatic ML implementation tools for solid ash management represents an important future research direction.
- Integrating ML and other methods into the modeling of dynamic systems, for example, identifying system errors in solid ash management, should also be examined.
- The successes of ML in solid ash management rely on sufficient labeled data for the model training step. However, it is much more expensive to collect and label sufficient data, meaning this approach is often not practical in engineering situations. Transfer learning may represent a potential solution to this issue; this approach can take full advantage of the domain knowledge in one engineering scenario and reuse it in other related but different scenarios [6]. With the help of transfer learning theories, ML applications in solid ash management are expected to expand from academic research to engineering scenarios.
- With increasing data availability, the use of more advanced ML algorithms should be attempted in solid ash management, such as unsupervised and semisupervised learning, recurrent or graph-based neural networks, simultaneous training of ML model types (multitasking), combining input data with different structures (multimodal design), ML-assisted data generation (active learning), and retraining ML models used in one area with new data from another related area (transfer learning).

14.2.3 The application of machine learning algorithms in solid ash management

The rapid increase of industrial solid ash production calls for more efficient recycling and management strategies, in which ML modeling will play a critical role. However, efforts must also be devoted to designing ML application scenarios in solid ash management.

- Following years of experimental investigation, we have abundant knowledge of solid ash recycling and management, such as how chemical oxides in solid ash influence its reactivity. Thus innovative methods that can integrate expert knowledge into the ML modeling process are attracting increasing attention. Algorithm design is not the sole focus of ML; the design of modeling methodologies for different application scenarios is also an important aspect.

Coupling data science with theoretical and experimental approaches could lead to innovative solid ash management methods.

- ML has significant potential to identify system errors, equipment faults, and control errors, which represent the current bottlenecks for solid ash management systems.
- The current applications of ML in solid ash management are fragmented: for example, while one study may focus on reactivity, another may focus on heavy metal leaching, ash generation rate, or origin identification. While each resulting ML model can tackle one specific problem well, this approach neglects other related problems. Thus a shift in emphasis is required from ML modeling for single problems to integrated modeling encompassing multiple problems.
- The costs and potential profits of different solid ash management strategies should be assessed with the help of ML modeling. By doing so, solid ash management strategies can be optimized considering dynamic changes in ash generation, environmental regulations, and market value.

References

[1] D.H. Wolpert, W.G. Macready, No free lunch theorems for optimization, IEEE Transactions on Evolutionary Computation 1 (1997) 67–82.

[2] C. Qi, et al., Towards Intelligent Mining for Backfill: a genetic programming-based method for strength forecasting of cemented paste backfill, Minerals Engineering 133 (2019) 69–79.

[3] L. Reiser, et al., FAIR: a call to make published data more findable, accessible, interoperable, and reusable, Molecular Plant 11 (9) (2018) 1105–1108.

[4] M. Kukar, Estimating the Reliability of Classifications and Cost-Sensitive Combining of Different Machine Learning Methods (Ph.D. thesis) {in Slovene}, Faculty of Computer and Information Science, 2001, University of Ljubljana, Ljubljana, Slovenia.

[5] P. Bednyakov, et al., Physics and applications of charged domain walls, NPJ Computational Materials 4 (1) (2018) 65.

[6] S.J. Pan, Q. Yang, A survey on transfer learning, IEEE Transactions on Knowledge and Data Engineering 22 (10) (2010) 1345–1359.

Appendix

```
# Clustering analysis
```

```
1. # Original data distribution before clustering
2. import numpy as np
3. import matplotlib.pyplot as plt
4. from sklearn import metrics
5. from sklearn.cluster import KMeans
6. import pandas as pd
7. from sklearn.decomposition import PCA
8. data = pd.read_csv('C:/Users/lenovo/Desktop/cluster.csv')
9. X = data.iloc[:,:-1].values
10. y = data.iloc[:,-1].values
11. # Plot the data distribution figures of the original data
12. plt.figure()
13. plt.scatter(X[:,1], X[:,5],marker='o', facecolors='none',
    edgecolors='k', s=30)
14. x_min, x_max = min(X[:,1]) - 1, max(X[:,1]) + 1
15. y_min, y_max = min(X[:,5]) - 1, max(X[:,5]) + 1
16. plt.title('Input data')
17. plt.xlim(x_min, x_max)
18. plt.ylim(y_min, y_max)
19. plt.xticks(())
20. plt.yticks(())
21. plt.show()
22. plt.scatter(X[y == 0, 0], X[y == 0, 2], s = 30, c = 'yellow',
    label = 'SSA') # Label 0
23. plt.scatter(X[y == 1, 0], X[y == 1, 2], s = 30, c = 'cyan',
    label = 'CFA') # Label 1
24. plt.scatter(X[y == 2, 0], X[y == 2, 2], s = 30, c = 'magenta',
    label = 'MSWBA') # Label 2
25. plt.scatter(X[y == 3, 0], X[y == 3, 2], s = 30, c = 'orange',
    label = 'MSWFA') # Label 3
```

```
1. # Data distribution after clustering
2. import numpy as np
3. import matplotlib.pyplot as plt
4. from sklearn import metrics
5. from sklearn.cluster import KMeans
6. from sklearn.metrics import silhouette_score
7. import pandas as pd
8. from sklearn.decomposition import PCA
9. from sklearn.metrics import accuracy_score
10. from sklearn.metrics import recall_score
11. from sklearn.metrics import roc_auc_score
12. from sklearn.metrics import precision_score
13. # Import data
14. data = pd.read_csv('C:/Users/lenovo/Desktop/cluster.csv')
15. X = data.iloc[:,:-1].values
16. y = data.iloc[:,-1].values
17. # train the clustering model
18.
    km=KMeans(init='k-means++',n_clusters=4,max_iter=300,n_ini
   t=10,random_state=0)
19. y_means = km.fit_predict(X) # training
20. y_means
21. # centroids = kmeans.cluster_centers_
22. # Calculate clustering scores
23. from sklearn import metrics
24. print(metrics.rand_score(y, y_means))
25. print(metrics.adjusted_rand_score(y, y_means))
```

```
26. print(metrics.adjusted_mutual_info_score(y, y_means))
27. print(accuracy_score(y, y_means))
28. plt.figure()
29. plt.scatter(X[:,0], X[:,1],marker='o', facecolors='none',
    edgecolors='k', s=30)
30. x_min, x_max = min(X[:,0]) - 1, max(X[:,0]) + 1
31. y_min, y_max = min(X[:,1]) - 1, max(X[:,1]) + 1
32. plt.title('Input data')
33. plt.xlim(x_min, x_max)
34. plt.ylim(y_min, y_max)
35. plt.xticks(())
36. plt.yticks(())
37. plt.show()
38. plt.scatter(X[y_means == 0, 0], X[y_means == 0, 1], s = 30,
    c = 'yellow', label = 'SSA') # Label 1
39. plt.scatter(X[y_means == 1, 0], X[y_means == 1, 1], s = 30,
    c = 'cyan', label = 'CFA') # Label 2
40. plt.scatter(X[y_means == 2, 0], X[y_means == 2, 1], s = 30,
    c = 'magenta', label = 'MSWBA') # Label 3
41. plt.scatter(X[y_means == 3, 0], X[y_means == 3, 1], s = 30,
    c = 'orange', label = 'MSWFA') # Label 4
```

```
# Modelling methodology
```

```
1. # Determine the optimal training-testing partition ratio
2. # Tunning range 30% - 90%
3. import time
4. startTime = time.time()
5. import numpy as np
6. import warnings
7. warnings.filterwarnings("ignore", category=Warning)
8. from sklearn.ensemble import RandomForestRegressor
9. from sklearn import preprocessing
10. import scipy.stats
11. from sklearn.model_selection import train_test_split
12. from sklearn.model_selection import cross_val_predict
13. # import data
14. data                                                          =
   np.genfromtxt('C:/Users/lenovo/Desktop/Data1.csv'  ,encodin
   g='utf-8',delimiter=',')
15. # # normalize features
16. # data = preprocessing.scale(data_origin)
17. Trainresult = []
18. Testresult = []
19. for p in range(10,50,5):
20.  single_Trainresult = []
21.  single_Testresult = []
22.  for j in range(0,50):
23.  np.random.shuffle(data)
24.  x_point = np.zeros((193,2))
25.  for i in range(2):
```

```
26. x_point[:,i] = data[:,i]
27. y_point = data[:,2].reshape(193,1).ravel()
28.
  X_train,X_test,y_train,y_test=train_test_split(x_point,y_p
  oint,test_size=(p/100))
29. clf = RandomForestRegressor()
30. clf = clf.fit(X_train,y_train)
31. y_training_pred = clf.predict(X_train)
32.
  single_Trainresult.append(scipy.stats.pearsonr(y_train,y_t
  raining_pred)[0])
33.
  single_Testresult.append(scipy.stats.pearsonr(y_test,clf.p
  redict(X_test))[0])
34.
35. Trainresult.append(single_Trainresult)
36. Testresult.append(single_Testresult)
```

```
1. # Determine the number of repetitions for converged modelling
   performance
2. import tensorflow
3. import keras
4. import pandas as pd
5. import numpy as np
6. from                     keras.layers                     import
   Dense,Activation,Input,BatchNormalization,Dropout
7. from keras.models import Sequential,Model
8. from sklearn.metrics import mean_squared_error # MSE
9. from sklearn.metrics import r2_score # R square
10. import scipy.stats
11. from keras.optimizers import SGD,Adam
12. from keras.layers.advanced_activations import LeakyReLU
13. import time
14. startTime = time.time()
15. from sklearn import preprocessing
16. from sklearn.model_selection import cross_val_predict
17. import warnings
18. warnings.filterwarnings("ignore", category=Warning)
19.
20. # import data
21. data=
    np.genfromtxt('C:/Users/lenovo/Desktop/data.csv',delimiter
    =',')
22.
23. trainaverage = []
24. trainsd = []
```

```
25.
26. for p in range(2,52,2):
27.  single_trainaverage = []
28.  for j in range(0,p):
29.  np.random.shuffle(data)
30.  data = preprocessing.scale(data)
31.  X=data[:,:-1]
32.  y=data[:,-1]
33.
34.  X = X.astype('float32')
35.  y = y.astype('float32')
36.
37.  def DNN():
38.  model = Sequential()
39.
   model.add(Dense(150,input_shape=(X.shape[1],),activation='
   relu'))
40.  model.add(Dropout(0.25))
41.  model.add(BatchNormalization())
42.  model.add(Dense(200,activation='relu'))
43.  model.add(Dropout(0.35))
44.  model.add(BatchNormalization())
45.  model.add(Dense(100,activation='relu'))
46.  model.add(Dropout(0.25))
47.  model.add(Dense(50,activation='relu'))
48.  model.add(Dense(1))
49.  adam = Adam(lr = 0.0005)
50.  model.compile(optimizer= adam, loss='mse')
```

```
51. return model
52.
53. result_dir ='C:/Users/lenovo/Desktop/picuture'
54.
55. model = DNN()
56.
  model.fit(X,y,validation_split=0.1,batch_size=128,epochs=5
  00,shuffle=True)
57. D_pred1 = model.predict(X)
58. D_pred1 =np.squeeze(D_pred1)
59. y=np.squeeze(y)
60. print(scipy.stats.pearsonr(y,D_pred1)[0])
61.
  single_trainaverage.append(scipy.stats.pearsonr(y,D_pred1)
  [0])
62.
63. trainaverage.append(np.mean(single_trainaverage))
64. trainsd.append(np.std(single_trainaverage))
65.
66. print(trainaverage)
67. print(trainsd)
```

```
1. # Modelling performance evaluation with 50 repetitions (using
   DNN as an example)
2. import tensorflow
3. import keras
4. import pandas as pd
5. import numpy as np
6. from                    keras.layers                    import
   Dense,Activation,Input,BatchNormalization,Dropout
7. from keras.models import Sequential,Model
8. from sklearn.metrics import mean_squared_error # MSE
9. from sklearn.metrics import mean_absolute_error # MAE
10. from sklearn.metrics import r2_score # R square
11. import scipy.stats
12. from keras.optimizers import SGD,Adam
13. from keras.layers.advanced_activations import LeakyReLU
14. import time
15. startTime = time.time()
16. import numpy as np
17. # np.random.seed(10)
18. import warnings
19. warnings.filterwarnings("ignore", category=Warning)
20. import scipy.stats
21. # ==========Import data================#
22.
    df=np.genfromtxt('C:/Users/lenovo/Desktop/modelling_data/D
   ata1.csv' ,encoding='utf-8',delimiter=',')
23. R_Trainresult = []
24. R_Testresult = []
```

```
25. R2_Trainresult = []
26. R2_Testresult = []
27. MSE_Trainresult = []
28. MSE_Testresult = []
29. MAE_Trainresult = []
30. MAE_Testresult = []
31. for j in range(0,1):
32.  np.random.shuffle(df)
33. # Scale
34.  from sklearn import preprocessing
35.  df = preprocessing.scale(df)
36.  Xtrain=df[:156,:-1]
37.  Xtest=df[156:183,:-1]
38.  ytrain=df[:156,-1]
39.  ytest=df[156:183,-1]
40.  Xtrain = Xtrain.astype('float32')
41.  ytrain = ytrain.astype('float32')
42.  Xtest = Xtest.astype('float32')
43.  ytest = ytest.astype('float32')
44.
45. # ==========Build and train the model===============
46.  def DNN():
47.  model = Sequential()
48.  model.add(Dense(8,
   input_shape=(Xtrain.shape[1],),activation='relu'))
49.  model.add(Dense(32,activation='relu'))
50.  model.add(BatchNormalization())
51.  model.add(Dense(64,activation='relu'))
52.  model.add(Dropout(0.15))
```

```
53.  model.add(Dense(16,activation='relu'))
54.  model.add(BatchNormalization())
55.  model.add(Dense(8,activation='relu'))
56.  model.add(Dense(1))
57.  adam = Adam(lr = 0.0005)
58.  model.compile(optimizer= adam, loss='mse')
59.  return model
60.
61.  result_dir ='C:/Users/lenovo/Desktop/picuture'
62.  import matplotlib.pyplot as plt
63.  def plot_history(history, result_dir):
64.  plt.plot(history.history['loss'], marker='.')
65.  plt.plot(history.history['val_loss'], marker='.')
66.  plt.title('model loss')
67.  plt.xlabel('epoch')
68.  plt.ylabel('loss')
69.  plt.legend(['loss', 'val_loss'], loc='upper right')
70.  plt.show
71.  model = DNN()
72.  history=model.fit(Xtrain,ytrain,validation_split=0.1,
   batch_size=128,epochs=500,shuffle=True)
73.  plot_history(history, result_dir)
74.  D_pred1 = model.predict(Xtrain)
75.  R2_Trainresult.append(r2_score(ytrain, D_pred1))
76.  MSE_Trainresult.append(mean_squared_error(ytrain,
   D_pred1))
77.  MAE_Trainresult.append(mean_absolute_error(ytrain,
   D_pred1))
```

```
78. # Model prediction and evaluation
79.  D_pred = model.predict(Xtest)
80.  R2_Testresult.append(r2_score(ytest, D_pred))
81.  MSE_Testresult.append(mean_squared_error(ytest, D_pred))
82.  MAE_Testresult.append(mean_absolute_error(ytest, D_pred))
83.  ytrain=np.squeeze(ytrain)
84.  D_pred1 =np.squeeze(D_pred1)
85.  ytest= np.squeeze(ytest)
86.  D_pred =np.squeeze(D_pred)
87.  R_Trainresult.append(scipy.stats.pearsonr(ytrain,
   D_pred1)[0])
88.  R_Testresult.append(scipy.stats.pearsonr(ytest,
   D_pred)[0])
```

```
# Hyper-parameters tuning
```

```
1. # Optimising heper-parameters of GBRT using PSO
2. # learning_rate = (0.01, 1) float
3. # n_estimators = (50, 2000) int
4. # max_depth = (3, 15) int
5. # min_samples_split = (2, 15) int
6. # min_samples_leaf = (1, 15) int
7. # max_features = (0.4, 1) float
8. import time
9. startTime = time.time()
10. import numpy as np
11. # np.random.seed(10)
12. import warnings
13. warnings.filterwarnings("ignore", category=Warning)
14. from sklearn.ensemble import GradientBoostingRegressor
15. import scipy.stats
16. from sklearn.model_selection import train_test_split
17. from sklearn.model_selection import cross_val_predict
18. # import data
19. data                                                        =
    np.genfromtxt('C:/Users/lenovo/Desktop/data.csv',delimiter
    =',')
20. from multiprocessing import Pool,Manager
21. import os, time
22. def son(x,i):
23.  np.random.shuffle(data)
24.  x_point = np.zeros((700, 9))
25.  for i in range(9):
```

```
26.  x_point[:, i] = data[:, i]
27.  y_point = data[:, 9].reshape(700, 1).ravel()
28.  X_train,        X_test,        y_train,        y_test        =
   train_test_split(x_point,
29.  y_point,
30.  test_size=0.15)
31.  estimator                                                      =
   GradientBoostingRegressor(learning_rate=x[0]/100,
32.  n_estimators=int(x[1]),
33.  max_depth=int(x[2]),
34.  min_samples_split=int(x[3]),
35.  min_samples_leaf=int(x[4]),
36.  max_features=x[5] / 10)
37.  y_training_pred  =  cross_val_predict(estimator,  X_train,
   y_train, cv=5)
38.  need=-scipy.stats.pearsonr(y_train,y_training_pred)[0]
39.  return need
40.
41. class pm_gbrt:
42.  def fitness(self,x):
43.  p = Pool()
44.  rt=[]
45.  for i in range(20):
46.  r=p.apply_async(son, args=(x,i))
47.  rt.append(r)
48.  p.close()
49.  p.join()
50.  result=[]
51.  for r in rt:
```

```
52.  result.append(r.get())
53.  return [np.mean(result)]
54.
55.  def get_bounds(self):
56.  return ([1,50,3,2,1,4],[100,2000,15,15,15,10])
57.
58. import pygmo as pg
59. algo = pg.algorithm(pg.pso(gen = 50, eta1 = 1.49618, eta2 =
   1.49618))
60. algo.set_verbosity(1)
61. prob = pg.problem(pm_gbrt())
62. pop = pg.population(prob, 200)
63. pop = algo.evolve(pop)
64. uda = algo.extract(pg.pso)
65. np.savetxt('PCA.txt',uda.get_log(),delimiter=' ')
66. print(pop.champion_f)
67. print(pop.champion_x)
68. print ('The script took {0} second !'.format(time.time() -
   startTime))
```

```
# Model interpretation
```

```
1. # Feature importance
2. import numpy as np
3. import pandas as pd
4. from scipy.stats import spearmanr
5. from sklearn.ensemble import RandomForestRegressor
6. from sklearn.inspection import permutation_importance
7. from sklearn.model_selection import train_test_split
8. data =
   pd.read_csv('C:/Users/lenovo/Desktop.csv',delimiter=',')
9. X = data.iloc[:,:-1]
10. Y= data.iloc[:,-1]
11. # Dataset partition
12.
    rfr=RandomForestRegressor(n_estimators=66,max_depth=6,min_
   samples_split=2,min_samples_leaf=1,max_features=0.35842150
   )
13. importances_values=rfr.feature_importances_
14. importances = pd.DataFrame(importances_values,
   columns=["importance"])
15. feature_data = pd.DataFrame(X.columns, columns=["feature"])
16. importance = pd.concat([feature_data, importances], axis=1)
17. importance = importance.sort_values(["importance"],
   ascending=False)
18. importance.set_index('feature', inplace=True)
19.importance.plot.barh(facecolor ='#AAAAFF', alpha=0.7, rot=0,
   figsize=(6, 6))
20. plt.title('Feature Importance')
21. plt.show()
```

```
1. # Permutation importance using sklearn
2. import sklearn
3. import shap
4. from sklearn.model_selection import train_test_split
5. import warnings
6. import pandas as pd
7. import numpy as np
8. import matplotlib.pyplot as plt
9. from sklearn.inspection import permutation_importance
10.
    df=np.genfromtxt('C:/Users/lenovo/Desktop/Data1.csv',encod
   ing='utf-8',delimiter=',')
11. np.random.shuffle(df)
12. from sklearn import preprocessing
13. df= preprocessing.scale(df)
14. print(df.shape)
15. columns=['Installed                          capacity','Coal
   consumption','Genaration of coal fly ash ']
16. df1=pd.DataFrame(df,columns=columns)
17. print(df1)
18. # Dataset partition
19. X = df1.iloc[:,:-1]
20. Y= df1.iloc[:,-1]
21. from sklearn import svm
22. SVR = svm.SVR()
23. SVR.fit(X,Y)
24. result = permutation_importance(SVR, X, Y, n_repeats=10, )
25. print(result)
26. print(result.importances_mean)
27. perm_sorted_idx = result.importances_mean.argsort()
28. print(perm_sorted_idx)
```

```
1. # Permutation importance using eli5
2. def score(X, y):
3.  y_pred = model.predict(X)
4.  return mse(y,y_pred)
5. base_score,                  score_decreases                  =
   get_score_importances(score,X,y)
6. feature_importances = np.mean(score_decreases, axis=0)
7. print(feature_importances)
```

```
1. # SHAP analysis
2. import matplotlib.pyplot as plt
3. import seaborn as sns
4. sns.set_style("white")
5. explainer =shap.KernelExplainer(model=SVR.predict,data=X)
6. shap.initjs()
7. # Global interpretation
8. shap_values = explainer.shap_values(X)
9. shap.summary_plot(shap_values, X)
10. # Local interpretation
11. shap.initjs()
12. shap_values =explainer.shap_values(X.iloc[0,:])
13. shap.force_plot(explainer.expected_value,     shap_values,
   X.iloc[0,:])
```

Index

Note: Page numbers followed by "*f*" and "*t*" refer to figures and tables, respectively.

M

N

O

P

R

Printed in the United States
by Baker & Taylor Publisher Services